W0106811

Lecture Notes in Control and Information Sciences 194

Editor: M. Thoma

Xi-Ren Cao

Realization Probabilities

The Dynamics of Queuing Systems

Springer-Verlag London Ltd.

Series Advisory Board

Author

Xi-Ren Cao, PhD
Department of Electrical and Electronic Engineering,
The Hong Kong University of Sciences and Technology,
Clear Water Bay, Kowloon, Hong Kong

ISBN 978-3-540-19872-7 ISBN 978-3-540-39330-6 (eBook)
DOI 10.1007/978-3-540-39330-6

British Library Cataloguing in Publication Data
A catalogue record for this book is available from the British Library

Originally published by Springer-Verlag Berlin Heidelberg New York in 1994

Typesetting: Camera ready by author

69/3830-543210 Printed on acid-free paper

Preface

In recent years, many modern large-scale man-made systems (e.g., manufacturing systems, computer and communication networks, air and highway traffic networks, military C^3I/logistic systems, and multinational financial and insurance firms) have been emerging. These systems are called *discrete event systems* because of the discrete nature of the events. The theory of the operation of these systems largely belongs in the domain of *operations research*. On the other hand, recent research indicates that these man-made systems possess many properties that are similar to those of natural physical systems. In particular, the evolution of these man-made systems demonstrates some dynamic features; exploring these dynamic properties may lead to new perspectives concerning the behavior of discrete event systems. The increasing need for analyzing, controlling, and optimizing discrete event systems has initiated a new research area, the dynamics of discrete event systems.

The objective of this book is to present a new multidisciplinary approach to the study of queuing systems. The queuing system is one of the major models in operation research for discrete event systems. This book applies the dynamic point of view to study the sensitivity and optimization of queuing systems. The approach is similar to the linearization of a perturbed continuous variable dynamic system. *Realization probability* is the major concept throughout the book. Realization probability measures the final effect of a small perturbation on the system performance as the system evolves. It can be viewed as the "building block" of sensitivity analysis in the sense that any performance sensitivity can be decomposed into the sum of the final effects of small perturbations that can be measured by realization probabilities. Various sensitivity formulas are derived by using realization probability, and their application to optimization is discussed in this book. The book emphasizes steady-state performance sensitivities and quantitative analysis. The material covered in this book can serve as an example to show how a multidisciplinary view can provide new insights as well as new results.

The book is a summary of the results in a new research area. In a systematic way, it introduces a new multidisciplinary approach and presents recent results in this area. Most of the material covered in this book has not appeared in any other books. The position of the theory developed in this book, relative to the queuing system theory and the continuous dynamic system theory, is illustrated in Figure 1.1 of Chapter 1. To make the book self-contained, I have devoted a chapter to the discussion of the mathematical background required for understanding the topics.

The book can be viewed as a complement to the existing queuing theory textbooks, in the sense that this book provides a new approach and develops new formulas for the performance sensitivity for queuing systems. The algorithms for sensitivity formulas developed in this book are a complement to, for example, the books of Bruell and Balbo [11] and Conway and Georganas [43].

The primary readers of this book are first-year graduate students, undergraduate seniors, and researchers and practicing analysts in system theory, industrial engineering, computer and communication engineering, and operations research. A part of the book was used by me as course material for a first-year graduate class at Harvard University.

Several people have been greatly helped in my research and in the process of preparing this book. First of all, I would like to express my sincere thanks to Professor Y. C. Ho of Harvard University, his insights and inspiration have made a significant impact on my research. P. Glasserman extended the concept of realization probability to the case of general performance function in load-dependent networks; this extension and some further works form the basis of Chapter 4. Chapter 9 and a part of Chapter 4 are based on joint work with D. J. Ma. Chapter 8 includes joint work with Y. Dallery. Special thanks go to Professor P. Brémaud and W. B. Gong, whose comments and suggestions on an early draft of Chapter 2 helped to improve its presentation.

Finally, for its support, I thank Harvard University, where most concepts and methodologies in this book were developed, and I express my appreciation to Digital Equipment Corporation's support of my continuing research in this fascinating area.

Xi-Ren Cao
Acton, Massachusetts

Contents

Chapter 1

Introduction

1.1 An Overview

The conventional approach used in the queuing system theory to obtain the steady-state performance sensitivity is mainly based on the probability and the stochastic process theories. With this approach, the explicit form for steady-state probability is first derived; the formula for the steady-state performance is then obtained; and the performance sensitivity with respect to a system parameter is finally obtained by taking the derivative of the performance formula.

This book presents another approach to derive the steady-state performance sensitivity formulas. The approach is an analogue to the linearization of a perturbed nonlinear continuous variable system along its trajectories; it is based on the dynamic nature of queuing systems. The main idea of this approach is to study the final effect of a small perturbation in a sample path of a queuing system on a performance function. This final effect is measured by a quantity defined as the *realization probability* or its extension, the *realization factor*. (We shall, in a general sense, use the term "realization probability" to refer to both the realization probability and the realization factor.) A set of linear equations can be derived for realization probabilities, and performance sensitivity formulas can be obtained in terms of these realization probabilities. For some systems, explicit solutions have been found for the realization probability equations; in these cases, explicit forms for the performance sensitivities can be obtained, and computational algorithms have been developed.

The fundamental idea of this new approach is completely different from that of the classic queuing theory. The approach explores the dynamic feature of a queuing network and hence represents an interdisciplinary effort. Therefore, the results obtained by this approach provide a new perspective of queuing systems and can be applied to solve a set of problems that may be different from those in the classic queuing theory.

First, since the realization probability measures the final effect of a small perturbation on a performance function, it contains more information than the parametric sensitivity, which can be obtained via the standard results such as the product-form solution for some queuing systems. For example, realization probabilities can be used to determine the performance sensitivity in the case where each service time increases by an equal amount. In fact, the realization probability can be viwed as a "building block" of sensitivity analysis in the sense that any performance sensitivity can be decomposed into the sum of the final effects of all the small perturbations, and these final effects can be measured by realization probabilities (see Chapter 3 for a detailed discussion and Section 1.3 for a simple example).

Next, the new approach is based on studying the evolution of a sample path of a queuing system. The realization probability and other terms in the sensitivity formulas can be estimated by using a single sample path. More precisely, the performance sensitivity can be expressed in terms of the expected value of the realization probability, which in turn is the expected value of a random variable, called the *realization index*, defined on each sample path (see Section 3.2). This feature is very important because for most systems there is no closed form solution for the steady-state performance, and simulation is the only way to obtain the performance sensitivity. The sensitivity formulas based on the realization probability provide a new direction for simulating the performance sensitivity; different performance sensitivities can be estimated by using a single sample path of a queuing system. In this respect, the perturbation analysis algorithms for estimating performance sensitivities, (see Glasserman [57] and Ho and Cao [68]), can be viewed as efficient ways to carry out the estimation suggested by the sensitivity formulas in terms of realization probabilities.

Finally, based on this approach, explicit forms can be obtained for the sensitivity in some networks. It has been shown that these formulas have a factored form and thus are amenable to optimization study. As an application of the realization probability, we shall show that these sensitivity formulas lead to some important optimal control policies.

The book applies the dynamic point of view to the study of queuing networks and can be viewed as a complement to the existing queuing theory books. There are many books that cover different aspects of the queu-

Continuous Systems	Queuing Systems
Nonlinear Dynamic System Theory (e.g., Differential Equations)	Classic Queuing System Theory (e.g., Product-Form Solutions)
Linearization Equations Along Trajectories	formulas Based on REALIZATION PROBABILITIES

Figure 1.1: The Position of the Realization Theory

ing theory and its applications. General introductions to the theory can be found, for example, in Saaty [93], Cohen [42], Gross and Harris [65], Kleinrock [76], Courtois [47], Kelly [74], Cooper [46], Gelenbe and Pujolle [55], and Walrand [104]. Applications of the theory to different practical problems are discussed in Saaty [93], Prabhu [90], Newell [89], Kleinrock [77], Courtois [47], and Jain [72]. Bruell and Balbo [11] and Conway and Georganas [43] provide overviews of the exact computational algorithms for product-form queuing networks in equilibrium. Cassandras [36] contains an introduction to queuing theory and some other modeling concepts and methodologies for discrete-event dynamic systems.

Figure 1.1 illustrates the relation of the realization probability theory and the formulas derived from it to the theory of queuing systems that is covered by the existing books. More discussion on the analogy shown in this figure will be given in the following two sections.

The realization probability is an extension of perturbation propagation, in the study of steady-state performance senstivities. Perturbation propogation is a main concept of *perturbation analysis*, a performance sensitivity analysis technique developed by Y. C. Ho and his co-workers. The book by Ho and Cao [68] gives an overview of this area as of 1990; Glasserman's book [57] mainly deals with the unbiasedness and convergence of the gradient estimates. Both books contain a comprehensive list of related references. This book differs from these two books by its focus on steady-state performance sensitivities and the quantitative analysis based on realization probabilities. We derive formulas for the steady-state performance sensitivity. The perturbation analysis estimation algorithms can be viewed as means to emulate the system dynamics and to estimate the values of the formulas on a single sample path. The quantitative nature is the main theme that distinguishes the work presented in this book from other works in perturbation analysis. System dynamics is another theme throughout this book that links our results to continuous variable systems.

In the remainder of this chapter, we sketch out the fundamental ideas and describe the intuition that motivates the study. In Section 1.2, we first review the linearization theory of a perturbed nonlinear continuous system, and then we compare its dynamics with that of a queuing system. Topics special to queuing systems are discussed. In Section 1.3, we use a simple example, a two-server cyclic queuing network, to illustrate the main concepts, the methodology, and the main results of the book. The treatment is informal, with the intention to offer a clear picture of what we are going to achieve before engaging in mathematical derivations.

In Chapter 2, we give a brief review of the fundamental mathematics that are related to the topics presented in the book. This will help the readers to get familiar with the basic concepts and terminologies used throughout the book. The material reviewed covers the fundamental probability theory, stochastic processes, and queuing systems. The focus is on ideas and concepts. We intend to present these ideas and concepts in an intuitive yet rigorous manner. This chapter will also be benifit to readers who want to gain an overview of the topics discussed in the chapter per se.

Chapter 3 applies the dynamic approach to study closed Jackson networks. In Section 3.1, we briefly review some concepts, such as the irreducibility of a network and the ergodicity of the state process, that are related to the topics in this book. Section 3.2 gives an abstract definition of realization probability and derives a set of linear equations that determines the realization probabilities. In Section 3.3, the realization probability defined in Section 3.2 is related to the perturbations generated on account of the changes in mean service times, and the concepts of *sample performance function* and *sample sensitivity* are introduced. It is shown that the effect of a change in a mean service time can be decomposed into the sum of the effect of all the small perturbations, which can be measured by realization probabilities. Based on this idea, the sample sensitivity of the system throughput with respect to a mean service time can be expressed in terms of realization probabilities. In Section 3.4, we prove that the sample elasticity (the "normalized" derivative) of the throughput with respect to a mean service time in fact converges both in mean and with probability one to the elasticity of the steady-state throughput. From this, we conclude that the elasticity of the steady-state throughput simply equals the negative expected value of the realization probability. In Section 3.5, we show that realization probabilities can be used to derive formulas for other sensitivities. Three cases are discussed: the sensitivity to the service rate changes when the system states are in a particular set, the sensitivity to the number of customers, and the sensitivity to any arbitrary changes in service time distributions. These formulas can not be obtained by taking the derivative of the product-form solution. Furthermore, since these formulas are based on realization probabilities, their values can be estimated by using a single sample path of the network.

In Chapters 4 through 6, we extend the concepts and results in Chapter 3 to other queuing networks and to general performance functions. The basic principles of the approach are similar to those in Chapter 3; however, the concepts are more general, and the proofs are more complicated. We shall focus on the special aspects of the problems in these chapters that are different from those in Chapter 3.

In Chapter 4, we generalize the results in Chapter 3 in three aspects. First, we consider a wide class of general performance measures that covers throughput as a special case. Second, we study queuing networks in which the service requirements are exponentially distributed, but the service rates may depend on the system state (i.e., the numbers of customers in all the servers). We call such a network a *state-dependent* network. Finally, we study the sensitivity of performance in open queuing networks. In Section 4.1, we study the evolution of a perturbation on a sample path and find that in a state-dependent system, a perturbation will be eventually realized with a different size (i.e., all the servers in the network will obtain a perturbation with the same size that may be bigger or smaller than the original perturbation). We then generalize the concept of the realization probability to that of the realization factor. The realization factor measures the final effect of a perturbation on a given performance measure in a state-dependent network. A set of linear equations is derived for the realization factors. In Section 4.2, we derive formulas for the sensitivities of the general performance measures with respect to service parameters; we prove that the performance sensitivities equal the negative expected values of the realization factors. All the results in Chapter 3 become special cases. In Section 4.3, we further explore the analogy between the realization probability and the linearization theory. We show that for any single-class closed queuing network with state-dependent service rates, the sensitivities of any steady-state performance measure can be expressed in a simple algebraic form by using a sensitivity matrix, which is of a finite dimension. That is, all the performance sensitivities are linear combinations of the entries of the sensitivity matrix. In Section 4.4, we discuss open queuing networks, which consist of an infinite number of states. It is shown that with a mild condition on performance functions, called the *quasi-Lipschitz* condition, we can derive sensitivity formulas and other results that are similar to those for closed networks.

Chapter 5 deals with networks whose service or interarrival times have general distributions. A perturbation *generation function* is defined to characterize the perturbation generated in any infinitesimal period. The linear equations that specify the realization factors become linear differential equations. It is shown that in an integration form, the performance sensitivity with respect to a parameter of a service (or interarrival) time distribution can be expressed in terms of the expected value of the product of the realization factor and the perturbation generation function. The formulas in Chapter 3 become special cases.

In Chapter 6, we study multiclass networks and networks with blocking. The main issue for these two types of networks is that the sample sensitivity may not converge to the steady-state performance sensitivity. Thus, it is necessary to find the conditions under which the formulas and results similar to those in the previous chapters hold. In Section 6.1, we generalize the concept of irreducible networks to multiclass networks and derive a sufficient condition under which a perturbation will be realized with probability one. We also derive a necessary condition for the sensitivity formulas with realization probabilities. In Section 6.2, we show that if in a network no server can block more than one server simultaneously, then results similar to nonblocking networks hold.

In Chapter 7, we extend our results further to the more general setting of the generalized semi-Markov processes (GSMP). Queuing networks with general distributed service distributions, multiclass networks, and networks with blocking can be modeled as GSMP. Thus, this chapter helps in understanding the results in the previous chapters at a higher conceptual level.

Since the realization probability can be estimated with a single sample path, the approach based on realization probabilities naturally fits the framework of "operational analysis," an area introduced by Denning and Buzen [53] and Buzen and Denning [13]. In Chapter 8, we take an operational approach to derive the sensitivity formulas by using "operational variables" such as the *realization ratio*. Operational variables are variables defined as averages on a sample path. In operational analysis, no stochastic assumptions are used, and operational assumptions, which can be verified on a sample path, are made. This approach provides us an alternative perspective for the sensitivity analysis based on realization probabilities.

Chapter 9 deals with another important topic, the application of the

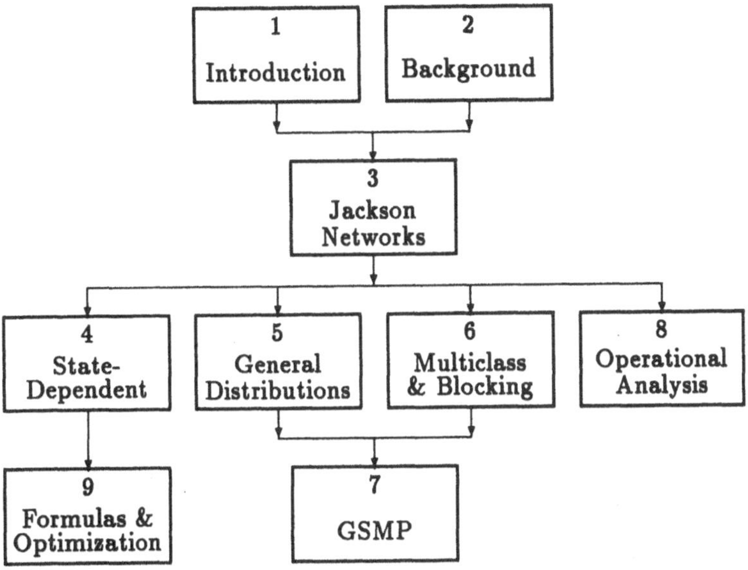

Figure 1.2: Dependency Among the Chapters

realization theory to the optimization of queuing networks. To this end, we first derive a closed-form solution to the realization factors for a two-server network. We then establish a property of equivalent for the performance sensitivity for Norton's aggregation (see Chandy et al. [37]) and use it to derive the explicit formulas for the sensitivities in a load-dependent closed Jackson network. Computational algorithms similar to those of Buzen's are also developed. It turns out that the explicit formulas possess a factored form, which is amenable to optimization analysis. Using this form, it is easily shown that the optimal policy for most performance measures in a load-dependent network is the bang-bang policy. This opens a new direction for the optimization and control of queuing networks.

Figure 1.2 illustrates the interdependency of the chapters. As the figure shows, there are several ways to use this book as a text. One may read only the introduction to gain a global sense of the new approach; one may study Chapters 1 and 3 to understand the fundamental results for Jackson networks; One may read Chapters 4-8 for further developments; or one may read Chapters 1, 3, 4, and 9 for applications to the optimization of queuing systems. Chapter 2 provides a review for the mathematical preliminaries.

1.2 Dynamics of Queuing Systems

The word "dynamics" is often used in the system theory to describe the behavior of system evolution. The theory of queuing systems belongs in the domain of *operations research.* Thus, the title of this section, "Dynamics of Queuing Systems," indicates the multidisciplinary nature of the topic. Indeed, the material presented in this book is the result of the effort in recent years by the author and his colleagues to apply the dynamical point of view to the study of the performance sensitivity of queuing systems.

When the dynamic property is studied, queuing systems are viewed as *discrete-event dynamic systems (DEDS).* Many practical systems, such as computer and communication networks, manufacturing systems, and traffic systems, can be modeled as DEDS (see e.g., Ho ed. [67]). The state of a DEDS changes at discrete and usually random instants of time that represent the occurrences of certain events such as the arrival or departure of a customer, the initiation or completion of a task, the start of transmission, and the end of receiving a message. In addition, the states are usually discrete; thus a change in state means a jump in the state trajectory. One can visualize a state trajectory of a DEDS by drawing a sample path of a queuing network.

Opposed to DEDS are the continuous variable dynamic systems (CVDS), where the system state changes continuously in time. Most physical systems, ranging from the movement of a satellite, which is governed by Newton's law, to the evolution of an electrical magnetic field, which obeys Maxwell's theory, are CVDS. The theory of CVDS has a long history and has ordinary and partial differential equations as its main mathematical tool. The success of the CVDS theory motivates the study of the dynamic nature of DEDS. The work presented in this book demonstrates that, by employing an approach that is different from the conventional one in queuing theory, one can obtain some fundamental results as well as new insights.

We shall focus on the performance sensitivity study of queuing networks, by using the dynamic point of view. First, we shall briefly review some related concepts in CVDS.

Dynamics of a Continuous Variable System

Consider a linear continuous dynamic system described by the following differential equation:

$$\dot{x}(t) = A(t)x(t) + B(t)u(t), \tag{1.1}$$

where $x(t)$ is an n-dimensional state variable, $u(t)$ is an m-dimensional control variable, and $A(t)$ and $B(t)$ are an $n \times n$ and an $n \times m$ matrix, respectively. Let x_0 be the initial state of the system at time $t = 0$; then

$$x(t) = \phi(t,0)x_0 + \int_0^t \phi(t,s)B(s)u(s)ds, \tag{1.2}$$

where $\phi(t,s)$ is the *state-transition matrix* of the system (see e.g., Section 2.4, or Szidarovszky and Bahill [102]).

The dynamic property of linear systems can be used to determine the effect of small perturbations in a system parameter. Consider a nonlinear system

$$\dot{x} = g(x, \theta(t), t), \tag{1.3}$$

where $\theta(t)$ is a time-varying parameter. Suppose that $\theta(t)$ changes to $\theta(t) + \Delta\theta(t)$, $|\Delta\theta(t)/\theta(t)| << 1$; $\Delta\theta(t)$ is called the *perturbation* of $\theta(t)$. Then

$$\Delta\dot{x}(t) = \frac{\partial}{\partial x}\{g(x,\theta(t),t)\}\Delta x + \frac{\partial}{\partial \theta}\{g(x,\theta(t),t)\}\Delta\theta(t). \tag{1.4}$$

The higher order terms of Δx and $\Delta\theta$ are omitted in the above equation. This equation is also known as the linearization of a nonlinear system along a trajectory.

Figure 1.3 illustrates the linearization of a nonlinear system. In the figure, the thick curve represents a trajectory of the original nonlinear system (1.3), and the thin curve represents a trajectory of the nonlinear system corresponding to a perturbation $\Delta\theta(t)$. The evolution of the Δx shown in the figure is then governed by Equation (1.4).

Assuming $\Delta x_0 = 0$ and applying Equation (1.2) to Equation (1.4), which is linear with respect to Δx, we have

$$\Delta x(t) = \int_0^t \phi(t,s)\{\frac{\partial}{\partial \theta} g(x,\theta(s),s)\}\Delta\theta(s)ds, \tag{1.5}$$

where $\phi(t,s)$ is the state transition matrix corresponding to $\frac{\partial}{\partial x} g(x,\theta(t),t)$.

Let

$$de(s) = [\frac{\partial}{\partial \theta} g(x,\theta(s),s)]\Delta\theta(s)ds. \tag{1.6}$$

Then Equation (1.5) can be rewritten in the following form:

$$\Delta x(t) = \int_0^t \phi(t,s)de(s). \tag{1.7}$$

This equation has the following intuitive explanation: $de(s)$ is the perturbation generated in $[s, s+ds]$ because of the change $\Delta\theta(s)$; $\phi(t,s)de(s)$ is

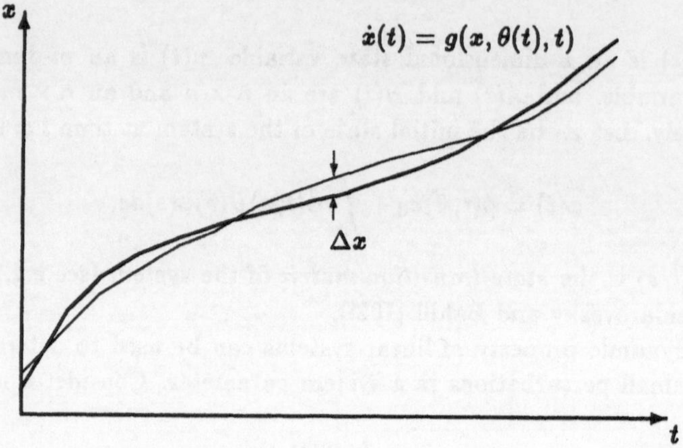

Figure 1.3: A Perturbed Trajectory of A Continuous Dynamic System

the effect of $de(s)$ to $\Delta x(t)$, or the perturbation propagated to $\Delta x(t)$ from $de(s)$; finally, $\Delta x(t)$ is the sum of all these effects in interval $[0, t]$.

Let the performance index for the dynamic system (1.3) be

$$J = \int_0^T f(x, t)dt.$$

The change in performance due to the change in $\theta(t)$, $\Delta\theta(t)$, is

$$\Delta J = \int_0^T \{\frac{\partial f}{\partial x}\Delta x(t)\}dt. \tag{1.8}$$

Therefore, the performance sensitivity $\Delta J/\Delta\theta$ can be obtained by using Equations (1.7) and (1.8). Note that in this method it is not required to solve the nonlinear differential equation (1.3) for $\theta + \Delta\theta$. It is an approach based on a single trajectory corresponding to θ. If $\partial f/\partial x$ is given on a trajectory, there is even no need to have an explicit solution for (1.3).

In the study of the system performance in an infinite time horizon, the performance index takes the form

$$J = \lim_{T\to\infty} \frac{1}{T}\int_0^T f(x, t)dt.$$

Assuming that the functions f and g are smooth enough so that one can change the order of operations freely, we get

$$\Delta J = \lim_{T\to\infty} \frac{1}{T}\int_0^T \{\frac{\partial f}{\partial x}\Delta x\}dt. \tag{1.9}$$

If we further assume that $\Delta\theta$ is a constant and that the state perturbation Δx is stable; i.e.,

$$\lim_{t\to\infty} \Delta x(t) = \Delta x, \tag{1.10}$$

where Δx is a constant, then

$$\frac{\Delta J}{\Delta\theta} = \lim_{T\to\infty}\{\frac{1}{T}\int_0^T \frac{\partial f}{\partial x}dt\}\frac{\Delta x}{\Delta\theta}. \tag{1.11}$$

There are not many CVDS for which (1.10) holds. As we shall see, however, a similar statement is true for most queuing systems. In fact, (1.10) is analogous to the concept of *realization probability* in queuing systems.

Issues for Queuing Systems

Viewed as DEDS, queuing systems possess dynamic properties similar to those described above for continuous systems (Cao [22]). Because of the specific nature of queuing systems, such as the randomness involved and the strong interconnection between customers and servers, there are some inherent properties associated with queuing systems. These properties are distinguishable from those of continuous variable systems.

Our goal is to develop a theory for queuing networks that is similar to the linearization theory described above. Let us first identify the main features of the approach and the special problems that we may encounter in the process of achieving our goal.

1. *A sample path approach:* As shown in the CVDS case, system dynamics are reflected by the evolution of a trajectory of the system. In this book, we shall use the term "sample path" to refer to a realization of the history of a queuing system. Thus, the approach is sample-path based. Because of the randomness involved in a queuing network, a sample path corresponds to a point in the underlying probability space, or a realization of a sequence of random vectors. We need to define precisely a sample path in a random environment.

2. *Perturbation generation, propagation, and realization:* These concepts correspond to (1.6), (1.7), and (1.10). They describe the behavior of each small perturbation on a sample path. We need to develop similar rules for queuing networks. We shall see, because of the strong interconnections between servers, a perturbed queuing network will eventually be stable; this leads to the concept of realization.

3. *Sensitivity formulas:* These are similar to Equation (1.11). We shall quantitatively determine the final effort of a perturbation and use it to derive the performance sensitivity formulas, which have forms different from those given by the conventional method and can be applied to more problems.

4. *Statistic properties:* Any data obtained from a single sample path is subject to statistic variations and can be used only as estmates of the steady-state values. We shall encounter issues, such as unbiasedness and consistency, that are common for the study of all stochastic systems. Especially, because the sample paths of a queuing system are usually discontinuous, the interchangeability assumed in (1.9) for CVDS may not hold for some queuing systems.

Some of the above issues, such as the interchangeability for a finite horizon problem in a general setting of generalized semi-Markov processes, have been well addressed in Glasserman [57] and Ho and Cao [68]. We shall keep the discussion of these issues to the minumum and shall limit our attention to the special part that is related to queuing systems. We shall devote most time to the derivation of our performance sensitivity formulas and their proofs. It should be noted that our approach for proving the convergence to the steady-state performance sensitivity is based on the concept of realization probability and is different from those based on regenerative theory.

1.3 A Two-Server Cyclic Queuing Network

Behind every mathematical or engineering concept, there are always some motivation and intuition. The notion of the realization probability is not an exception. The derivations of the results and the proofs of some theorems may look quite complicated, they just serve as a means to reach our ultimate goal set forward by the initial motivation or intuition. In this section, we shall use a simple example to show the intuitive ideas underlying the topics of this book.

For this purpose, we consider a cyclic, closed queuing network consisting of two servers with exponentially distributed service times. The routing probabilities of such a network are $q_{1,1} = q_{2,2} = 0$ and $q_{1,2} = q_{2,1} = 1$. To further simplify the discussion, we assume that the network consists of only two customers and that the mean service times of these two servers are equal, denoted as $\bar{s}_1 = \bar{s}_1 = \bar{s}$. This is equivalent to the servers having

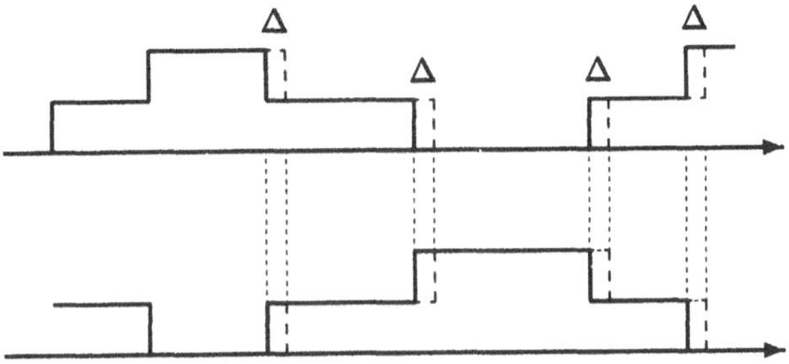

Figure 1.4: A Perturbation at $n = 2$ is Realized

a mean service rate $\lambda = 1/\bar{s}$ and the customers having service requests that are exponentially distributed with mean one. Using the notion of service rates sometimes helps to visualize the generation of perturbations.

Perturbation Realization

Figure 1.4 illustrates a part of a sample path of the network. A sample path of a queuing network is similar to a trajectory of a continuous system. Since the total number of customers in the two servers is fixed, the state of this system can be specified by the number of customers in the first server, denoted by n. In the figure, the top line represents the number of customers of the first server; the bottom line, of the second server. A sample path is completely determined by the initial state and the customer transitions, which occur at the service completion times.

A small *perturbation* of a sample path refers to a small change of a service completion time of a customer; this change is also called a perturbation of the customer or the server. For example, the dashed lines in Figure 1.4 show the perturbed completion times; the difference between a dashed line and its corresponding solid line is the size of the perturbation. In Figure 1.4, we assume that when $n = 2$, the service completion time of the first customer in server 1 is somehow delayed by a small amount Δ; i.e., server 1 obtains a perturbation at state $n = 2$. Figure 1.4 shows that, because at the same time server 2 is idle and is waiting to receive a customer from server 1 to start its service, server 2's service starting time has to be delayed by the same amount Δ. Thus, its service completion time is also delayed by Δ. In this case, we say that server 1's perturbation is *propagated* to server 2 through an idle period. An important observation is that after

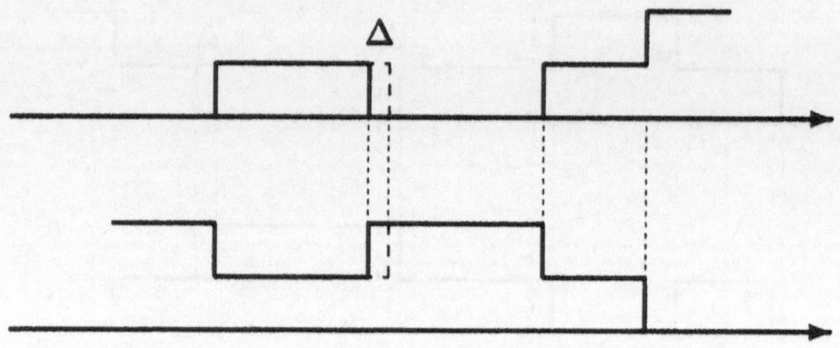

Figure 1.5: A Perturbation at $n = 1$ is Lost

server 2 obtains the perturbation, both servers in the network have the same perturbation. That is, all their service completion times thereafter will be delayed by the same amount, and the perturbed sample path looks just the same as the original one, except that the whole sample path is shifted to the right by Δ. In this case, we say that the perturbation is *realized* by the system. Figure 1.4 shows that a perturbation at $n = 2$ will always be realized.

Figure 1.5 shows a case where a customer in server 1 obtains a perturbation when $n = 1$ (this may happen, for example, when the service rate of server 1 is somehow slowed down a bit during the period of $n = 1$) and after its service completion the customer goes to server 2. The figure illustrates that although the arrival time of the customer to server 2 is delayed by Δ (which is assumed to be very small), the service completion times of the customers at server 2 are not affected at all. Only the waiting time of the customer at server 2 that was perturbed at server 1 decreases by Δ. After the perturbed customer enters server 2, the sample path looks exactly the same as the original one, as if nothing has happened. In this case, we say the perturbation is *lost.*

Figure 1.6 shows another case where a customer in server 1 obtains a perturbation when $n = 1$, but before the service completion of this customer, server 1 receives a customer from server 2. In this case, the service completion time of the customer at server 1, which takes place when $n = 2$, is delayed by Δ. In other words, the perturbation at $n = 1$ becomes a perturbation at $n = 2$. This was the case shown in Figure 1.4; Figure 1.6 also shows that the perturbation will be realized; i.e, the two servers will have the same perturbation Δ.

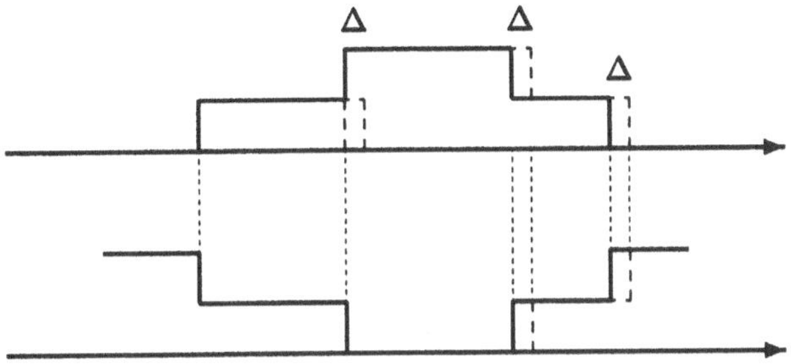

Figure 1.6: A Perturbation at $n = 1$ is Realized

Realization Probability

In summary, a perturbation of server 1 at $n = 2$ will always be realized. A perturbation of server 1 at $n = 1$ will be either realized or lost, depending on whether or not the perturbed customer leaves server 1 later than it receives a customer from server 2. Since the service times of the two servers have the same distribution and the same mean, the probability that a perturbation of server 1 at $n = 1$ is realized (or lost) is 1/2. The probability that a perturbation is realized is called the *realization probability* of the perturbation. The realization probability measures the average final effect of a perturbation on a sample path. It is the stochastic version of Equation (1.10) for queuing networks.

It is clear that the realization probability of a perturbation depends on the state of the network and the server of the perturbation. We use $c(n, i)$ to denote the realization probability of a perturbation at server i when the state is n. Thus, from the above discussion, we have the following equations for the two-server network:

$$c(2, i) = 1 \qquad i = 1, 2, \tag{1.12}$$

and

$$c(1, i) = \frac{1}{2} \qquad i = 1, 2. \tag{1.13}$$

Perturbation Generation

We have studied the evolution of a single perturbation and its final effect on the sample paths. To study the parametric sensitivity, we have to relate the perturbation of a server to the changes of a parameter. To illustrate this, we assume that the mean service time of server 1, $\bar{s}_1$, changes from $\bar{s}$ to $\bar{s} + \Delta\bar{s}$, or equivalently, the service rate changes from λ to $\lambda + \Delta\lambda$, $\Delta\lambda \approx -\lambda(\Delta\bar{s}/\bar{s})$. In this case, each customer's service time at server 1 will be prolonged by an amount that is proportional to its length. For example, if the service time of a customer is s, then after the service rate changes, the service time becomes

$$s + \Delta s = \frac{\lambda s}{\lambda + \Delta\lambda} \approx s - \frac{\Delta\lambda}{\lambda} s,$$

where λs can be viewed as the service request of the customer. Thus,

$$\Delta s \approx -(\frac{\Delta\lambda}{\lambda})s \approx (\frac{\Delta\bar{s}}{\bar{s}})s. \tag{1.14}$$

In general, the perturbation generated in any time period in which server 1 is busy because of the change $\Delta\bar{s}$ is proportional to the length of the period with a factor $\Delta\bar{s}/\bar{s}$. Equation (1.14) plays a similar role to that of (1.6) for CVDS.

Parametric Sensitivity Formulas

Now, consider a sample path of the network in $[0, T]$. For convenience, we consider T as a service completion time. Let L be the number of service completions of server 1 in $[0, T]$. The throughput of server 1 can be estimated by

$$\eta = \frac{L}{T}.$$

Suppose T is large enough so that the sample averages are close to the steady-state probabilities. Let $p(n)$ be the steady-state probability of n. The length of the time period in which the system is at state $n = 1$ is then approximately $Tp(1)$. According to (1.14), in this period there is a total of $Tp(1)(\Delta\bar{s}/\bar{s})$ perturbations generated because of $\Delta\bar{s}$. Among these perturbations, only a portion of them, determined by the realization probability $c(1,1)$, will have a final effect on the sample path; other perturbations are lost. This means that the perturbation generated when $n = 1$ shifts the sample path to the right by $Tp(1)(\Delta\bar{s}/\bar{s})c(1,1)$. Similarly, the perturbation generated when $n = 2$ shifts the sample path to the right by $Tp(2)(\Delta\bar{s}/\bar{s})c(2,1)$. Finally, we conclude that on the average, it will require the network $T + \Delta T$ to serve the same L customers, with

$$\Delta T = T\{p(1)c(1,1) + p(2)c(2,1)\}(\frac{\Delta\bar{s}}{\bar{s}}).$$

Thus, after the mean service time changes, the throughput of server 1 becomes

$$\begin{aligned}\eta + \Delta\eta &= \frac{L}{T+\Delta T} \approx \eta(1 - \frac{\Delta T}{T}) \\ &= \eta\{1 - [p(1)c(1,1) + p(2)c(2,1)](\frac{\Delta \bar{s}}{\bar{s}})\}.\end{aligned}$$

Therefore,

$$\frac{\bar{s}}{\eta}\frac{\Delta\eta}{\Delta\bar{s}} \approx -[p(1)c(1,1) + p(2)c(2,1)].$$

With the convention

$$c(0,1) = 0, \tag{1.15}$$

we can write

$$\frac{\bar{s}}{\eta}\frac{\Delta\eta}{\Delta\bar{s}} \approx -\sum_{n=0}^{2} p(n)c(n,1).$$

Letting $\Delta\bar{s} \to 0$, we obtain a formula for the *elasticity* of the throughput with respect to $\bar{s}_1$, expressed in realization probabilities:

$$\frac{\bar{s}_1}{\eta}\frac{\partial\eta}{\partial\bar{s}_1}\Big|_{\bar{s}_1=\bar{s}} = -\sum_{all\ n} p(n)c(n,1), \tag{1.16}$$

where $c(n,1)$, $n = 0,1,2$, are given in (1.12), (1.13), and (1.15). Equation (1.16) is a counterpart of (1.11) to CVDS. With $p(0) = p(1) = p(2) = 1/3$, (1.16) gives 1/2, and we can easily check that this is the same as the value given by taking directly the derivative of $\eta = p(1)/\bar{s}_1 + p(2)/\bar{s}_2$ with respect to $\bar{s}_1$.

Other Sensitivity Formulas

As mentioned in Section 1.1, the realization probability contains more information than the parametric sensitivity formula. In this subsection, we study two other examples that illustrate the application of the realization probability.

First, we consider the following question: What if server 1's service rate changes only when $n = 1$? Denote the service rate of server 1 when the state is n as $\lambda_{1,n}$, $n = 1,2$. We want to know the sensitivity of throughput η when $\lambda_{1,1}$ changes from λ to $\lambda + \Delta\lambda$, and $\lambda_{1,2}$ is unchanged. Following the same argument as for the sensitivity with respect to $\bar{s}_1$, we know that all the perturbations generated in $[0,T]$ at server 1 are $-Tp(1)(\Delta\lambda/\lambda)$, and among them only $-Tp(1)(\Delta\lambda/\lambda)c(1,1)$ are realized. Continuing the argument, we get the sensitivity in this case as follows.

$$\frac{\lambda_{1,1}}{\eta}\frac{\partial\eta}{\partial\lambda_{1,1}}|_{\lambda_{1,1}=\lambda} = p(1)c(1,1). \tag{1.17}$$

In the next example, we ask a different question: How about each customer's service time at server 1 increases by an equal amount of time? This may happen in practice, for example, when a fixture requires equal additional (or less) time to fix each part in a manufacturing system, or when a packet switch needs an equal amount of time to process each packet in a communication network. In this case, server 1 gets an equal amount of perturbation, denoted as Δ, at each service completion time.

Let $c_t(n,1)$ be the realization probability of a perturbation at the service completion time of server 1 when the state is n. Note that $c_t(n,1)$ is, in general, different from $c(n,1)$ because at a service completion time of a customer, this customer will definitely leave the server. It is, however, easy to check that $c_t(2,1) = c(2,1) = 1$ by the definition of realization. It is also easy to see that $c_t(1,1) = 0$, since the customer will join server 2 and the perturbation is then lost (see Figure 1.5).

Recall that there are L service completions at server 1 in $[0,T]$. Among these there are $L/2$ happened at $n=2$ and another $L/2$ at $n=1$. (This can be shown, for example, by the arrival theorem, which states that the distribution seen by a departure customer equals the stationary distribution of a system with one less customer. See Sevcik and Mitrani [95], or Melamed [86].) Thus, there are $(L/2)\Delta$ perturbations generated at $n=2$ and realized by the system; i.e.,

$$\Delta T = \frac{1}{2}L\Delta.$$

Continuing the same reasoning as before, we have

$$\frac{\bar{s}_1}{\eta}\frac{\Delta\eta}{\Delta} \approx -\frac{1}{2}\eta\bar{s}_1.$$

Since $\eta = \{p(1)+p(2)\}/\bar{s}_1 = \frac{2}{3}\frac{1}{\bar{s}_1}$, the above equation yields

$$\frac{\bar{s}_1}{\eta}\frac{\partial\eta}{\partial\bar{s}_1}|_{\Delta\bar{s}_1\equiv\Delta} = -\frac{1}{3}, \tag{1.18}$$

which is smaller than the sensitivity given by (1.16). It is worth noting that even for closed Jackson networks, (1.18) cannot be obtained from the existing product-form solution.

Remarks

Using a simple example, we have explained the intuitions behind the main concepts and derived (informally) the sensitivity formulas. The approach resembles the linearization of a CVDS. Ho and Cao [70] takes the same informal approach and obtains equations similar to (1.16) for closed Jackson networks. The remaining chapters of this book will be devoted to a rigorous study of similar topics for various types of queuing networks. Concepts will be precisely defined, formulas will be formally derived, theorems will be rigorously proved, and conditions needed for the formulas and theorems will be discussed.

Chapter 2

Mathematical Background

In this chapter, we shall review some mathematical concepts and results that are related to the material presented in the remainder of the book. The chapter is not meant to be exhausive and the readers may consult the references cited in each section for detailed expositions. We shall present the ideas and concepts in an intuitive yet relatively rigorous manner. Some of the results, e.g., the performance sensitivity formulas for queuing networks, are still new in the area.

2.1 Probability Theory

2.1.1 The Elementary Concepts

There are many good textbooks on probability; the material in this section is mainly based on Billingsley [5] and Chung [40].

σ-fields

Let Ω be an arbitrary space, or a set of points ω. For a set $A \subseteq \Omega$, the *complement* of A is denoted by $A^c = \Omega - A$. The *intersection* of two sets A and B is denoted by $A \cup B$; their *union* is denoted by $A \cap B$.

A class $\mathcal{F}$ of subsets of Ω is called a *field* if it contains Ω and is closed under the formation of complements and finite unions; i.e., if it satisfies the following conditions:

i. $\Omega \in \mathcal{F}$;

ii. If $A \in \mathcal{F}$, then $A^c \in \mathcal{F}$;

iii. If $A, B \in \mathcal{F}$, then $A \cup B \in \mathcal{F}$.

Since $A \cap B = (A^c \cup B^c)^c$, a field is also closed under the formation of finite intersections.

A field is called a σ-field if it is also closed under the formation of countable unions:

iv. If $A_1, A_2, \cdots \in \mathcal{F}$, then $A_1 \cup A_2 \cup \cdots \in \mathcal{F}$.

A set in $\mathcal{F}$ is called an $\mathcal{F}$-measurable set, or simply an $\mathcal{F}$ set. The pair $(\Omega, \mathcal{F})$ is called a *measurable space*.

Let $\mathcal{G}$ be an arbitrary class of subsets in Ω. The smallest σ-field containing $\mathcal{G}$ is called the σ-field generated by $\mathcal{G}$. It is the intersection of all the σ-fields containing $\mathcal{G}$. Intuitively, any set in the σ-field generated by $\mathcal{G}$ can be obtained by implementing finite or countable operations (i.e., complement, union, or intersection) of the sets in $\mathcal{G}$.

The σ-field generated by all the open rectangular parallelepiped in the n-dimensional real space R^n is called the *Borel σ-field* and is denoted as $\mathcal{B}^n$ (when $n = 1$ we simply write it as $\mathcal{B}$). A set in a Borel field is called a *Borel set.*

Probability Spaces

A *set function* on a class of subsets of Ω is a mapping that assigns a real number to each set in the class.

A set function $P(\cdot)$ on a σ-field $\mathcal{F}$ is called a *probability measure* if it satisfies the following conditions:

1. $P(A) \geq 0$ for all $A \in \mathcal{F}$;

2. $P(\Omega) = 1$ and $P(\emptyset) = 0$, where $\emptyset$ is the set that contains no points;

3. If $A_i \in \mathcal{F}$, $i = 1, 2, \cdots$, and $A_i \cup A_j = \emptyset$ for $i \neq j$, then

$$P(\cup_{k=1}^{\infty} A_k) = \sum_{k=1}^{\infty} P(A_k). \qquad (2.1)$$

The third condition is called *countable additivity*. Letting $A_{n+1} = A_{n+2} = \cdots = \emptyset$ in (2.1), we get

$$P(\cup_{k=1}^{n} A_k) = \sum_{k=1}^{n} P(A_k).$$

This is called *finite additivity*. It is easy to verify that $P(A) \leq 1$ for all $A \in \mathcal{F}$.

If $\mathcal{F}$ is a σ-field in Ω and P is a probability measure on $\mathcal{F}$, the triple $(\Omega, \mathcal{F}, P)$ is called a *probability space*. A *support* of P is any set $A \in \mathcal{F}$ for which $P(A) = 1$. Any set B (including a sigleton, which consists only one point) in $\mathcal{F}$ is called an *event*. $P(B)$ is then the probability of the event.

If the condition $P(\Omega) = 1$ is not required in the definition, then P is called a *measure* of the measurable space $(\Omega, \mathcal{F})$. Moreover, if $P(\Omega) < \infty$, a probability measure can be obtained by normalizing, i.e., dividing every $P(A)$, $A \in \mathcal{F}$, by $P(\Omega)$.

The Lebesgue measure on $R = (-\infty, \infty)$ is the unique measure defined on the Borel field on R that assigns the value $b - a$ for any open interval (a, b), $a, b \in R$. (In fact, only some Lebesgue sets with measure zero may be non-Borel sets.) The Lebesgue measure on the n-dimensional space R^n is the measure defined on the Borel field on R^n that assigns any open rectangular parallelepiped with a value equal to the product of the length of each side. The Lebesgue measure defined on an n-dimensional cube $[0, 1)^n$ is a probability measure.

Intuitively speaking, the purpose of definitions of the σ-field, the probability measure, and the probability space is to justify the countable additivity equation (2.1): The sets for which P does not have a value or (2.1) does not hold are simply called not measurable and are excluded from our study.

Random Variables

Let $(\Omega, \mathcal{F})$ and $(\Omega', \mathcal{F}')$ be two measurable spaces. Consider a mapping $\mathcal{T} : \Omega \to \Omega'$. The inverse image of a set $A' \subseteq \Omega$ is denoted as $\mathcal{T}^{-1}A' = \{\omega \in \Omega : \mathcal{T}\omega \in A'\}$. The mapping $\mathcal{T}$ is said to be measurable $\mathcal{F}/\mathcal{F}'$ if $\mathcal{T}^{-1}A'$ is $\mathcal{F}$-measurable for each $\mathcal{F}'$-measurable set A'.

A real measurable function is a measurable mapping from $(\Omega, \mathcal{F})$ to $(R, \mathcal{B})$. Such a real measurable function is called a *random variable* on Ω. Equivalently, a random variabel can be defined as a function $X : \Omega \to R$ such that for every real number $a \in R$ the set

$$\{\omega \in \Omega : X(\omega) \leq a\}$$

belongs to $\mathcal{F}$.

The σ-field generated by a random variable X, denoted as $\mathcal{F}(X)$, is the smallest σ-field with respect to which X is measurable. It is the σ-field generated by the class $X^{-1}(-\infty, a)$, $a \in R$.

Transformations of Measures

Let $(\Omega, \mathcal{F})$ and $(\Omega', \mathcal{F}')$ be two measurable spaces and the mapping $\mathcal{T} : \Omega \to \Omega'$ be measurable $\mathcal{F}/\mathcal{F}'$. Given a probability measure P on $\mathcal{F}$, we can define a set function on $\mathcal{F}'$ as follows

$$P'(A') := P(\mathcal{T}^{-1}A').$$

P' is well-defined: if $A' \in \mathcal{F}'$, then $\mathcal{T}^{-1}A' \in \mathcal{F}$ because $\mathcal{T}$ is measurable. Since $\mathcal{T}^{-1}\{\cup_{k=1}^{\infty} A'_k\} = \cup_{k=1}^{\infty} \mathcal{T}^{-1}A'_k$ and the sets $\mathcal{T}^{-1}A'_k$ are disjoint in Ω if the sets A'_k are disjoint in Ω', the countable additivity (2.1) for P' follows from that for P. Therefore, P' is a measure on $(\Omega', \mathcal{F}')$. Next, let $A \in \Omega$ be any set; then

$$P'(\Omega) \geq P'(\mathcal{T}A) + P'(\mathcal{T}A^c) = P(A) + P(A^c) = 1.$$

On the other hand, we have $P'(\Omega') = P(\mathcal{T}^{-1}(\Omega')) \leq 1$. Thus, $P'(\Omega') = 1$. That is, P' is a probability measure.

Distribution Functions

A random variable X is a measurable mapping from $(\Omega, \mathcal{F}, P)$ to $(R, \mathcal{B})$. This mapping transforms the probability measure P on Ω to a probability measure P' on R. P' is the *distribution* or the *law* of the random variable X. The *distribution function* of X, $F(x)$, is defined as follows

$$F(x) := P[\omega :\ X(\omega) \leq x].$$

Clearly, $F(\cdot)$ is a nondecreasing function, and the left-hand limit $F(x-) = \lim_{y \uparrow x} F(y)$ exists. Also,

$$\lim_{x \to -\infty} F(x) = 0,$$

and

$$\lim_{x \to \infty} F(x) = 1.$$

In this book, we shall mostly deal with nonnegative random variables; thus $F(0-) = 0$.

A random variable and its distribution have a *density function* f if f is a nongenative function on R and

$$F(x) = \int_{\infty}^{x} f(x)dx. \tag{2.2}$$

If $f(x)$ is differentiable at x, then $f(x) = \frac{d}{dx}F(x)$ is the derivative of $F(x)$ with respect to x. We have the *normalizing condition:*

$$\int_{-\infty}^{\infty} f(x)dx = F(\infty) = 1.$$

If X represents the lifetime of an event, [1] then $f(x)$ is the rate that the event will ends in $[x, x + \Delta x)$.

The *mean* (or the *expected value*) of a random variable X is

$$E[X] = \int_{-\infty}^{\infty} x dF(x) = \int_{-\infty}^{\infty} x f(x)dx.$$

We also write $\bar{x} = E[x]$. The *variance* of X is

$$Var[X] := E[X - E(X)]^2 = E[X^2] - [E(X)]^2.$$

Hazard Rates

The hazard rate function is defined as

$$g(x) = \frac{f(x)}{1 - F(x)}. \tag{2.3}$$

Assuming that $F(x)$ is absolutely continuous in $(-\infty, +\infty)$, then $f(x) = \frac{d}{dx}F(x)$ and we can obtain

$$f(x) = g(x)e^{-\int_{-\infty}^{x} g(y)dy}. \tag{2.4}$$

In most cases, the hazard functions are used for nonnegative random variables; thus, the integration may be taken form zero to x.

If X is the lifetime of an event, then $g(x)$ is the rate that the event will end in $[x, x + \Delta x)$ given that the event survives until x.

[1] Note that the word "event" is used in two senses: a set in a probability space and a particular activity that sustains for a period of time. Since both usages are standard in the probability theory and the discrete-event system theory, respectively, we adopt these two meanings. Readers are cautioned to distinguish the meanings from the content of the text.

Examples

Example 2.1

i. *Uniform distribution:*

$$f(x) = \begin{cases} \frac{1}{c} & \textit{if } m - \frac{c}{2} \le x \le m + \frac{c}{2}, \\ 0 & \textit{otherwise.} \end{cases}$$

We have $E[X] = m$ and $var[X] = \frac{c^2}{12}$.

ii. *Exponential distribution:*

$$F(x) = 1 - e^{-x/\bar{x}} \qquad x \ge 0,$$

or

$$f(x) = \begin{cases} \frac{1}{\bar{x}} e^{-x/\bar{x}} & \textit{if } x \ge 0, \\ 0 & \textit{if } x < 0. \end{cases}$$

We have $E[X] = \bar{x}$ and $Var[X] = (\bar{x})^2$. From (2.3) and (2.4), a distribution is exponential if and only if its hazard rate is a constant.

iii. *Gaussian (or normal) distribution:*

$$f(x) = \frac{1}{\sqrt{2\pi}\sigma} e^{-\frac{(x-\bar{x})^2}{2\sigma^2}}.$$

We have $E(X) = \bar{x}$ and $Var(X) = \sigma^2$. Such a normal distribution is sometimes denoted as $N(\bar{x}, \sigma^2)$.

Example 2.2 *Coxian distribution:* Cox [45] proved that any distribution function whose Laplace transform is a rational function can be constructed by a series of service stages, each of them having an exponential service time distribution. Figure 2.1 shows the structure of such a construction. In the figure, a customer arrives at a station for service. The customer first enters stage 1, which has a service time with an exponential distribution with mean $\bar{s}_1$. After the completion of its service at stage 1, the customer goes to stage 2 with probability a_1 and leaves the station with probability $b_1 = 1 - a_1$. The customer stays at stage 2 for an exponentially distributed service time with mean $\bar{s}_2$; then it either enters stage 3 with probability a_2 or leaves the station with probability b_2, and so on. The customers service time at the station has a Coxian distribution.

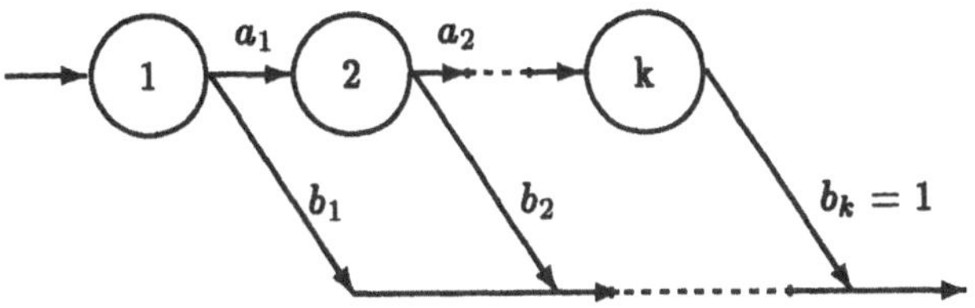

Figure 2.1: The Coxian Distribution

Example 2.3 *Phase type distribution:* Neuts [88] generalized the idea of the method of stages and introduced the phase-type distribution. Like a Coxian distribution, a phase-type distribution can be considered as the distribution of the service time of a customer at a service station consisting of k stages (or phases), each having an exponential distribution with mean $\bar{s}_i$, $i = 1, 2, \cdots, k$. The customer arriving at the station enters stage i with probability α_i, $\sum_{i=1}^{k} \alpha_i = 1$. After the completion of its service at stage i, the customer goes to stage j with probability $q_{i,j}$ and leaves the station with probability q_i, $\sum_{j=1}^{k} q_{i,j} + q_i = 1$. Let $Q = [q_{i,j}]$. We assume that $(I - Q)^{-1}$ exists so that the probability of a customer leaving the station is one.

A Coxian distribution is a phase-type distribution in which $\alpha_1 = 1$, $\alpha_i = 0$ for $i \neq 1$, $q_i = b_i$, $q_{i,i+1} = a_i$, and $q_{i,j} = 0$ for $k \neq i + 1$, $i = 1, \cdots, k - 1$, $q_k = b_k = 1$.

Using the phase-type model, we can express an exponential distribution with mean $\bar{s}$ as a service station with a server having an exponential distribution with mean $(1 - p)\bar{s}$ and a feedback rate $p < 1$, shown in Figure 2.2. In the model, after the service completion at the server, a customer has a probability p of being fed back to the same server and a probability of $1 - p$ of leaving the station. Hence $k = 1$, $\alpha_1 = 1$, $q_{1,1} = p$, and $q_1 = 1 - p$.

We prove that the distribution of the station shown in Figure 2.2 is exponential with mean $\bar{s}$. This can be easily seen by using the hazard rate. Note that no matter what the history is, a customer in the station has a rate of $\frac{1}{\bar{s}(1-p)}$ of ending its service at the server and has a rate of $(1 - p)\frac{1}{\bar{s}(1-p)} = \frac{1}{\bar{s}}$ of leaving the station. Thus, the hazard rate of the service time distribution at the station is a constant $\frac{1}{\bar{s}}$. This leads to our conclusion.

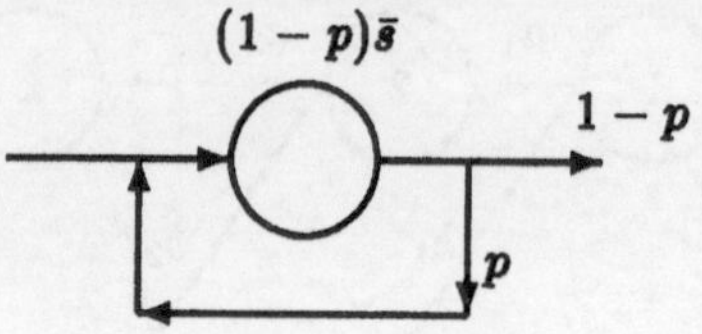

Figure 2.2: The Equivalent Exponential Distribution

Note that the Coxian distribution, which does not consist feedback loops, cannot be used to explain this equivalent model, which we shall use in the main text of this book.

Example 2.4 *Erlangian distribution and hyperexponential distribution:* An r-stage Erlangian distribution E_r is the distribution of a service station consisting of a series of r servers; each has an exponential distibuted service time with service rate $r\mu$. The probability density function is

$$f(x) = \frac{r\mu(r\mu x)^{r-1}e^{-r\mu x}}{(r-1)!} \qquad x \geq 0.$$

An R-stage hyperexponential distribution H_R is the distribution of a service station consisting of a parallel of R servers, each has an exponential distributed service time with service rate μ_i; and an external customer enters server i with probability α_i, $i = 1, 2, \cdots, R$. The probability density function is

$$f(x) = \sum_{i=1}^{R} \alpha_i \mu_i e^{\mu_i x} \qquad x \geq 0.$$

Random Vectors

An n-dimensional random vector $\mathbf{X} = (X_1, X_2, \cdots, X_n)$ on $(\Omega, \mathcal{F}, P)$ is an n-tuple of random variables $X_1, X_2, \cdots, X_n$, all of them are defined on $(\Omega, \mathcal{F}, P)$. Thus, an n-dimensional random vector is a measurable mapping from $(\Omega, \mathcal{F}, P)$ to $(R^n, \mathcal{B}^n)$. This mapping transforms the probability measure P to a probability measure P' on R^n. P' is called the distribution or the law of the random vector $\mathbf{X}$. The distribution function is

$$F(x_1, x_2, \cdots, x_n) = P(\omega \in \Omega : X_1(\omega) \leq x_1, X_2(\omega) \leq x_2, \cdots, X_n(\omega) \leq x_n).$$

The marginal distribution function of each random variable X_i is

$$F_i(x_i) = \lim_{x_k \to \infty,\ k \neq i} F(x_1, x_2, \cdots, x_n).$$

The probability density function is a function $f(x_1, \cdots, x_n)$ satisfying

$$F(x_1, x_2, \cdots, x_n) = \int_{-\infty}^{x_1} \int_{-\infty}^{x_2} \cdots \int_{-\infty}^{x_1} f(y_1, y_2, \cdots, y_n) dy_1 \cdots dy_n,$$

or

$$f(x_1, x_2, \cdots, x_n) = \frac{d}{dx_1} \cdots \frac{d}{dx_n} F(x_1, x_2, \cdots, x_n),$$

provided that the derivatives exist. The normalizing condition is

$$F(\infty, \infty, \cdots, \infty) = \int_{-\infty}^{\infty} \int_{-\infty}^{\infty} \cdots \int_{-\infty}^{\infty} f(y_1, y_2, \cdots, y_n) dy_1 \cdots dy_n = 1.$$

The random variables X_1, X_2, $\cdots$, X_n are said to be independent, if

$$F(x_1, x_2, \cdots, x_n) = \prod_{k=1}^{n} F_k(x_k),$$

or equivalently,

$$f(x_1, x_2, \cdots, x_n) = \prod_{k=1}^{n} f_k(x_k).$$

The mean of the random vector $\mathbf{X} = (X_1, X_2, \cdots, X_n)$ is the vector $E[\mathbf{X}] := (E[X_1], E[X_2], \cdots, E[X_n])$. The *covariance matrix* of the random vector $\mathbf{X} = (X_1, X_2, \cdots, X_n)$ is

$$Cov[\mathbf{X}] := \begin{bmatrix} E[(X_1 - \bar{x}_1)^2] & \cdots & E[(X_1 - \bar{X}_1)(X_n - \bar{x}_n)] \\ \cdots & \cdots & \cdots \\ E[(X_n - \bar{x}_n)(X_1 - \bar{x}_1)] & \cdots & E[(X_n - \bar{x}_n)^2] \end{bmatrix}$$

Two random vectors $\mathbf{X}$ and $\mathbf{Y}$ are said to be *uncorrelated*, if

$$E\{(X - E[X])(Y - E[Y])^T\} = 0,$$

where $(X - E[X])$ is a column vector and $(Y - E[Y])^T$ is a row vector, with "T" denoting the transpose.

Discrete Random Variables

We often encounter with random variables that take values in a countable set of numbers; such random variables are called *discrete random variables.* Let K be a discrete random variable and $\{1, 2, \cdots, k, \cdots\}$ be the set of possible values. The probability space on which K is defined can be partitioned as $\Omega = \{\omega_1, \omega_2, \cdots \omega_k, \cdots\}$, with $K(\omega_k) = k$. The probability measure can be specified by the probabilities of the elements ω_k, denoted as $p_1, p_2, \cdots, p_k, \cdots$, with $p_k := P(\omega_k)$.

The normalizing condition is

$$\sum_{k=1}^{\infty} p_k = 1.$$

If we view k as the value of a random variable K; i.e., $k := K(\omega_k)$, then the probability distribution function of K is a piecewise constant curve, which jumps at k with a height of p_k.

Example 2.5

i. *Bernoulli distribution:* $\Omega = \{0, 1\}$. $p_0 = P(0) = 1 - p$ and $p_1 = P(1) = p$, with $0 < p < 1$. The mean is p and the variance is $p(1-p)$.

ii. *Binomial distribution:* $\Omega = \{0, 1, 2, \cdots, n\}$.

$$p_k := P(k) = \binom{n}{k} p^k (1-p)^{n-k}, \qquad 0 < p < 1.$$

The mean is np and the variance is $np(1-p)$.

iii. *Geometric distribution:* $\Omega = \{1, 2, \cdots\}$. $p_k := P(k) = p(1-p)^{k-1}$, with $0 < p < 1$. The mean is $\frac{1}{p}$ and the variance is $\frac{1-p}{p^2}$.

Hybrid Random Variables

A *hybrid random variable* is a random variable whose elements can be specified by two part: a continuous part, denoted as X, and a discrete part, denoted as K. The distribution function of the random vector (K, X) is

$$F(k, x) = P[\omega : K(\omega) = k, \; X(\omega) \le x] \qquad k = 1, 2, \cdots.$$

The probability density function is $f(k, x) = \frac{d}{dx} F(k, x)$. The normalizing condition is

$$\sum_{k=1}^{\infty} \int_{-\infty}^{\infty} f(k, x) dx = \sum_{k=1}^{\infty} F(k, \infty) = 1.$$

We assume that the integration converges uniformly with respect to k, so that the order of $\sum_{k=1}^{\infty}$ and $\int$ in the above equation makes no difference.

2.1.2 Some Facts and Theorems

In this subsection, we review some results and theorems that are important to the understanding of the content of this book. First, we state one of the most commonly used random variable generation method, which is also the basis of the study in this book.

The Inverse Transformation Method

This method is used to generate a random variable with distribution function $F(x)$ from a random variable that is uniformly distributed on $[0,1)$. The method is based on the observation that $\xi = F(x)$ is uniformly distributed on $[0,1)$. Thus, to obtain random variable X with a given distribution $F(x)$, one first generates a random variable ξ, uniformly distributed on $[0,1)$, then sets

$$x = F^{-1}(\xi) = sup\{x : \ F(x) \le \xi\}.$$

This is illustrated in Figure 2.3.

Indeed, the random variable thus generated has the distribution $F(x)$:

$$P(X \le x) = P[F^{-1}(\xi) \le x] = P[\xi \le F(x)] = F(x).$$

The inverse transformation method will be used in this book to generate the service times and to determine the customer routings. When the same ξ is used to obtain two random variables X and Y with distribution functions $F(x)$ and $G(y)$, respectively, by setting $x = F^{-1}(\xi)$ and $y = G^{-1}(\xi)$, we say that a *common random variable* (ξ) is used.

For exponential distributions $F(x) = 1 - e^{-x/\bar{x}}$, we have

$$x = -\bar{x} ln(1-\xi). \tag{2.5}$$

For any discrete probability distributions with probabilities p_k, $k = 1,2,\cdots$, we partition the interval $[0,1)$ into small pieces: $[0,1) = [0,p_1) \cup [p_1, p_1+p_2) \cup \cdots \cup [\sum_{j=1}^{k-1} p_j, \sum_{j=1}^{k} p_j) \cdots$. If $\xi \in [\sum_{j=1}^{k-1} p_j, \sum_{j=1}^{k} p_j)$, with $\sum_{j=1}^{0} p_j = 0$, then set $K = k$.

In the inverse transformation method, we can think of $[0,1)$ as the underlying probability space with $\mathcal{B}$ as the σ-field and the Lebesgue measure as the probability measure. Thus, any random variable with a distribution function $F(x)$ can be viewed as a random variable defined on this probability space.

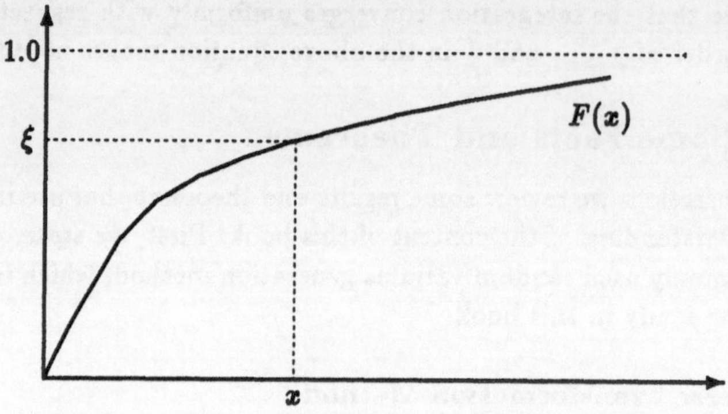

Figure 2.3: The Inverse Transformation Method

The Conditional Probability

Given two events A and B in Ω and $P(B) > 0$, the *conditional probability* of A given B is defined as

$$P(A|B) = \frac{P(A \cap B)}{P(B)}. \tag{2.6}$$

From (2.6), we have

$$P(A \cap B) = P(A|B)P(B) = P(B|A)P(A).$$

More general, we have the chain-rule formulas:

$$P(A \cap B \cap C) = P(C)P(B|C)P(A|B \cap C),$$

$$P(A \cap B \cap C \cap D) = P(D)P(C|D)P(B|C \cap D)P(A|B \cap C \cap D),$$

and so on.

Events A and B are *independent*, if $P(A \cap B) = P(A)P(B)$. For events with positive probabilities, this is equivalent to $P(A|B) = P(A)$ or $P(B|A) = P(B)$. More generally, a collection (infinite or finite) A_1, A_2, $\cdots$ of events is independent if

$$P(A_{k_1} \cap \cdots \cap A_{k_j}) = P(A_{k_1}) \cdots P(A_{k_j})$$

for any mutually exclusive subcollection A_{k_1}, $\cdots$, A_{k_j} of the events in the collection.

If B_1, B_2, $\cdots$ partition Ω (i.e., they are mutually exclusive and their union is Ω), then for any set A in Ω we have

$$P(A) = \sum_k P(A \cap B_k) = \sum_k P(B_k)P(A|B_k). \tag{2.7}$$

This is called the theorem of *total probability*. From (2.6) and (2.7), we obtain for every k

$$P(B_k|A) = \frac{P(B_k)P(A|B_k)}{\sum_k P(B_k)P(A|B_k)},$$

provided $P(A) > 0$ and $P(B_k) > 0$. This is called *Bayes' rule*.

Next, let Y be a random variable and $A \subset \Omega$ and B be a Borel set in R. The conditional probability $P(A|Y \in B)$ is well-defined by (2.6) only if $P(Y \in B) > 0$. Suppose we want to study the probability of A under the condition $Y = y$, and assume that $P(Y = y) = 0$. One way to define this conditional probability is by

$$P(A|Y = y) = \lim_{\Delta y \downarrow 0} \frac{P(A, Y \in (y - \Delta y, y + \Delta y))}{P(Y \in (y - \Delta y, y + \Delta y))}.$$

This definition requires that the limit exists. A better way is to define the conditional probability by using integration as follows (cf. the definition for probability density functions (2.2)).

The conditional probability of A given Y, denoted as $P(A|Y = y)$, is defined as any random variable on Ω, measurable $\mathcal{F}(Y)$, and satisfying

$$P(A, Y \in B) = \int_{y \in B} P(A|Y = y)dP \qquad \textit{for all } B \in \mathcal{B}. \tag{2.8}$$

There may be different versions of the conditional probability, but they differ only on a set with probability zero.

Conditional Probability Distributions

Given any set $B \in \Omega$ with $P(B) > 0$, (2.6) defines a probability measure, denoted as P_B, on Ω. Any random variable X defined on Ω transforms the probability measure P_B on Ω to a probability measure P'_B on R. P'_B is the *conditional distribution* of X given B. The *conditional distribution function* of X given B, $F(x|B)$, is then

$$F(x|B) = \frac{P(\{\omega : X(\omega) \le x\} \cap B)}{P(B)}.$$

Let X and Y be two random variables with a joint distribution function $F(x,y)$. The conditional distribution function of X given $Y \in [y, y+\Delta y)$ is

$$F[x|Y \in [y, y+\Delta y)] = \frac{P(\{X(\omega) \le x\} \cap \{Y \in [y, y+\Delta y)\})}{P(Y \in [y, y+\Delta + y))}.$$

Assuming the probability density function for X and Y exists, (when $F(x,y)$ is discontinuous, we may view the density function as having mass functions) we have for sufficient small Δy

$$F[x|Y \in [y, y+\Delta y)] = \frac{\{\frac{d}{dy}F(x,y)\}\Delta y}{\{\frac{d}{dy}F(\infty,y)\}\Delta y}.$$

Define the conditional distribution function of X given Y as

$$F(x|Y=y) = \frac{\frac{d}{dy}F(x,y)}{\frac{d}{dy}F(\infty,y)} = \frac{\frac{d}{dy}F(x,y)}{f(y)}.$$

We have

$$F(x|Y=y) = \lim_{\Delta y \to 0} F[x|Y \in [y, y+\Delta y)].$$

This is consistent with the definition of the conditional probability (2.8). Indeed, for any $B \in \mathcal{B}$, we have

$$\int_{y \in B} F(x|Y=y)F(dy) = \int_{y \in B} \frac{\frac{d}{dy}F(x,y)}{f(y)} F(dy) = F(x, Y \in B).$$

The conditional probability density function of X given Y is

$$f(x|Y=y) = F'_x(x|Y=y) = \frac{f(x,y)}{f(y)}.$$

The conditional mean (or expected value) of X given Y is

$$E[X|Y=y] = \int_{-\infty}^{\infty} x dF(x|Y=y) = \frac{1}{f(y)} \int_{-\infty}^{\infty} x f(x,y) dx.$$

This is a random variable on Ω. It is easy to verify that the mean of this random variable is $E[X]$:

$$E\{E[X|Y=y]\} = E[X].$$

The Memoryless Property of Exponential Distributions

For exponential distribution $F(x) = 1 - e^{-x/\bar{x}}$, the conditional distribution of x given $x \geq x_0$ is

$$F(x|x \geq x_0) = 1 - e^{-\frac{x-x_0}{\bar{x}}} = F(x - x_0). \tag{2.9}$$

Imagine X as the lifetime of an event. Equation (2.9) shows that if one knows that the event survives at $x_0 > 0$, then the residual lifetime at x_0, $x - x_0$, has the same distribution as the lifetime itself. This is called the *memoryless* property of the exponential distribution.

Convergences of Random Sequences

Let $X_1, X_2, \cdots, X_n, \cdots$, be a sequence of random variables defined on the same probability space $(\Omega, \mathcal{F}, P)$. There are four major concepts regarding the convergence of a random sequence. (Same definitions apply when the discrete index "n" changes to a continuous one, say "t".)

i. *Convergence in probability.* The random sequence $\{X_n\}$ converges in probability to a random variable X, if for any $\epsilon > 0$,

$$\lim_{n\to\infty} P[|X_n - X| \geq \epsilon] = 0.$$

ii. *Convergence with probability one (w.p.1).* $\{X_n\}$ converges with probability one to a random variable X, if

$$P(\omega : \lim_{n\to\infty} X_n = X) = 1,$$

or equivalently, for any $\epsilon > 0$,

$$\lim_{n\to\infty} P(|X_k - X| > \epsilon \ \textit{for some}\ k \geq n) = 0.$$

iii. *Convergence in mean or in mean square.* $\{X_n\}$ converges in mean or in mean square to a random variable X, if

$$\lim_{n\to\infty} E[|X_n - X|] = 0$$

or

$$\lim_{n\to\infty} E[|X_n - X|^2] = 0.$$

iv. *Convergence in distribution and weak convergence.* $\{X_n\}$ converges in distribution to a random variable X, if

$$\lim_{n\to\infty} F_n(x) = F(x),$$

where $F_n(x)$ and $F(x)$ are the distribution functions of X_n and X, respectively. The sequence of distribution functions $\{F_n(x)\}$ is said to converge weakly to $F(x)$.

Both convergence with probability one and convergence in mean (or in mean square) imply convergence in probability which, in turn, implies convergence in distribution (see, e.g., Billingsley [5]). Convergence with probability one and convergence in mean do not imply each other. However, if X_n are dominated by a random variable Y having a finite mean, then X_n converges to X w.p.1. implies that X_n converges to X in mean. This can be easily proved by using the Lebesgue donimated convergence theorem stated later.

Example 2.6 Consider a random sequence X_n with $P(X_n = 1) = \frac{1}{n}$ and $P(X_n = 0) = 1 - \frac{1}{n}$.

i. The sequence converges in probability to zero, since for any $1 > \epsilon > 0$,

$$\lim_{n\to\infty} P[|X_n| \geq \epsilon] = \lim_{n\to\infty} \frac{1}{n} = 0.$$

ii. The sequence does not converge w.p.1 to zero, since for any $1 > \epsilon > 0$ and any n, the probability of the event with $|X_k - 0| = 1 > \epsilon$ for some $k > n$ is $1 - \prod_{k=n}^{\infty}(1 - \frac{1}{k})$. Thus,

$$\lim_{n\to\infty} P(|X_k - X| > \epsilon \textit{ for some } k \geq n) = \lim_{n\to\infty} \{1 - \prod_{k=n}^{\infty}(1 - \frac{1}{k})\} = 1.$$

iii. The sequence does converge in mean and in mean square, because

$$\lim_{n\to\infty} E[X_n] = \lim_{n\to\infty} E[X_n^2] = 0.$$

Example 2.7 Consider a random sequence X_n with $P(X_n = 1) = \frac{1}{n^2}$ and $P(X_n = 0) = 1 - \frac{1}{n^2}$.

i. Clearly, the sequence converges in probability to zero.

ii. This sequence does converge w.p.1 to zero, since for any $1 > \epsilon > 0$ and any n, the probability of the event with $|X_k - 0| = 1 > \epsilon$ for some $k > n$ is $1 - \prod_{k=n}^{\infty}(1 - \frac{1}{k^2})$. Thus,

$$\lim_{n\to\infty} P(|X_k - X| > \epsilon \; for \; some \; k \geq n) = \lim_{n\to\infty}\{1 - \prod_{k=n}^{\infty}(1 - \frac{1}{k^2})\} = 0.$$

iii. The sequence converges in mean but not in mean square, because

$$\lim_{n\to\infty} E[X_n] = 0, \qquad \lim_{n\to\infty} E[X_n^2] = 1.$$

The law of large numbers

Suppose that $\{X_n\}$ is an independent sequence with $E[|X_n|] = 0$. If $\frac{Var[X_n]}{n^2} < \infty$, then

$$\lim_{n\to\infty} \frac{1}{n}\sum_{k=1}^{N} X_k = 0 \qquad w.p.1.$$

The following corollary is more often used in practice: Suppose that X_1, X_2, $\cdots$ are independent and identically distributed random variables with $E[X_n] = E[X]$, then

$$\lim_{n\to\infty} \frac{1}{n}\sum_{k=1}^{N} X_k = E[X] \qquad w.p.1.$$

Integrable Functions

Let f be a measurable function (random variable) $f : \Omega \to R$. The integration of f on $(\Omega, \mathcal{F}, P)$ is defined as

$$\int f dP := sup \sum_i \{[inf_{\omega\in A_i} \; f(\omega)]P(A_i)\},$$

where the supremum is over any finite decomposition $\{A_i\}$ of Ω into $\mathcal{F}$-sets.

Changing the variable from $\omega \in \Omega$ to $x \in R$, we get the Lebesgue integration over R:

$$\int f dP = \int_{-\infty}^{\infty} x dF(x).$$

For any measurable function f, define

$$f^+(\omega) = \begin{cases} f(\omega) & if \; 0 \leq f(\omega) \leq \infty, \\ 0 & if \; -\infty \leq f(\omega) \leq 0, \end{cases}$$

and $f^-(\omega) = f(\omega) - f^+(\omega)$. If $\int f^+ dP$ and $\int f^- dP$ are both finite, f is said to be *integrable*.

A sequence $\{f_n\}$ is said to be *uniformly integrable*, if

$$\lim_{\alpha\to\infty} sup_n \int_{\{|f_n|\geq\alpha\}} |f_n| dP = 0. \tag{2.10}$$

Theorem 2.1 *If $\{f_n\}$ is uniformly integrable and $|g_n| \leq |f_n|$ for all n, then $\{g_n\}$ is also uniformly integrable.*

Proof: From $|g_n| \leq |f_n|$, we get $\{|g_n| \geq \alpha\} \subseteq \{|f_n| \geq \alpha\}$ for any $\alpha > 0$. Thus for all n we have

$$0 \leq \int_{\{|g_n|\geq\alpha\}} |g_n| dP \leq \int_{\{|f_n|\geq\alpha\}} |g_n| dP \leq \int_{\{|f_n|\geq\alpha\}} |f_n| dP.$$

Therefore,

$$0 \leq sup_n \int_{\{|g_n|\geq\alpha\}} |g_n| dP \leq sup_n \int_{\{|f_n|\geq\alpha\}} |f_n| dP.$$

Taking the limit of $\alpha \to \infty$ in the above inequality and using (2.10), we obtain

$$\lim_{\alpha\to\infty} sup_n \int_{\{|g_n|\geq\alpha\}} |g_n| d\mu = 0.$$

Thus $\{g_n\}$ is uniformly integrable. □

Theorem 2.2 *Suppose f_n are integrable, $\lim_{n\to\infty} f_n = f$ in probability, then $\lim_{n\to\infty} \int f_n dP = \int f dP$ is equivalent to that $\{|f_n|\}$ is uniformly integrable.*

The prove of this theorem can be found in, e.g., Chung [40]. The theorem says that if f_n are integrable and $\{f_n\}$ converges to f in probability, then the uniform integrability is a necessary and sufficient condition for the interchange of limit and integration.

Convergence Theorems

Theorem 2.3 *(Lebesgue's dominated convergence theorem.) If $|f_n| \leq g$ with probability one, where g is integrable, and if $\{f_n\}$ converges to f with probability one, then f and f_n are integrable and*

$$\lim_{n\to\infty} \int f_n dP = \int f dP.$$

Theorem 2.4 *If $\{f_n\}$ is uniformly integrable and $\{f_n\}$ converges to f with probability one, then f is integrable and*

$$\lim_{n\to\infty} \int f_n dP = \int f dP.$$

The fact that a random variable X is integrable is equivalent to $E[|X|] < \infty$. Thus, the above four theorems can be written in a form with random variables. For example, in Theorems 2.3 and 2.4, we have

$$\lim_{n\to\infty} E[X_n] = E[\lim_{n\to\infty} X_n].$$

2.2 Stochastic Processes

Let $(\Omega, \mathcal{F}, P)$ be a probability space. A *stochastic process* defined on $(\Omega, \mathcal{F}, P)$ is a collection of random variables defined on Ω and indexed by a parameter regarded as representing time. We use n for the discrete index and t for the continuous one. Thus, a stochastic process is denoted as $\{X_n;\ n = 0, 1, \cdots\}$ or $\{X_t;\ t \in [a, b]\}$, with $[a, b] \subset (-\infty, \infty)$. In most cases, we consider $a = 0$ and $b = \infty$. Let Φ be the set from which the random variables take their values. Φ is called the *state space*. For any t (or n), X_t (or X_n) is a random variable (or a random vector) from Ω to Φ. For any $\omega \in \Omega$, $X_t(\omega)$ (or $X_n(\omega)$) is a collection (or a sequence) of points in Φ. The collection is called a *realization*, or a *sample path*, of the stochastic process.

The sequences studied in the last section for the convergence property are in fact stochastic processes.

The material of this section is based on Çinlar [41] and Breiman [9].

2.2.1 Markov Chains

If the future behavior of a stochastic process is conditionally independent of its past history provided that the present state is known, then the stochastic process is said to possess the *Markov property*.

In this subsection, we study the stochastic processes with a discrete index and a discrete (i.e., countable) state space. For convenience, we denote the states in Φ as integers.

Definition 2.1 *The stochastic process $\mathbf{X} = \{X_n : n \in (0, 1, \cdots)\}$ is called a Markov chain with state space Φ provided that*

$$P[X_{n+1} = j | X_0, \cdots, X_n] = P[X_{n+1} = j | X_n]$$

for all $j \in \Phi$ and $n \in (0, 1, \cdots)$.

State Transitions

The probabilities $Q_n(i,j) := P(X_{n+1} = j | X_n = i)$ are called the *transition probabilities* of the Markov chain. If $Q_n(i,j) = Q(i,j)$ for all n, the Markov chain is said to be *time-homogeneous.* We shall restrict ourselves to time-homogeneous Markov chains in this book.

The matrix $Q = [Q(i,j)]$ is called the *transition matrix.* It may be of a finite or an infinite dimension. Note that $Q(i,j) \geq 0$ and $\sum_j Q(i,j) = 1$ for all i. Such a matrix is called a *Markov matrix.*

From the definition of $Q(i,j)$, the Markov property, and the equations for conditional probabilities, we have

$$\begin{aligned} & P(X_0, X_1, X_2, \cdots, X_k) \\ = \; & P(X_0)Q(X_0, X_1)Q(X_1, X_2)\cdots Q(X_{k-1}, X_k). \end{aligned} \tag{2.11}$$

Let $Q^{(k)}(i,j)$ be the elements in Q^k. Then it is easy to prove that

$$P(X_{n+k} = j | X_n = i) = Q^{(k)}(i,j). \tag{2.12}$$

Thus, Q^k is called the *k-step transition matrix.*

Obviously, we have

$$Q^{n+m} = Q^n Q^m.$$

This is equivalent to

$$Q^{(n+m)}(i,j) = \sum_{k \in \Phi} Q^{(n)}(i,k) Q^{(m)}(k,j), \qquad i,j \in \Phi.$$

This is called th *Chapman-Kolmogorov equation.*

State Classification

A set of states is said to be *closed* if no state outside the set can reach from any state in the set; i.e., $\Phi_0 \subset \Phi$ is a closed set if $Q(i,j) = 0$ for all $i \in \Phi_0$ and $j \in \Phi - \Phi_0$.

A state $i \in \Phi$ with $Q(i,i) = 1$ is called an *absorbing state.*

A closed set is *irreducible* if no proper subset of it is closed.

A Markov chain is said to be *irreducible* if its only closed set is Φ. A Markov chain is irreducible if and only if all states can be reached from each other. That is, for any $i, j \in \Phi$, there exist two integers k_1 and k_2 such that $Q^{(k_1)}(i,j) > 0$ and $Q^{(k_2)}(j,i) > 0$. Such a Markov matrix is called an *indecomposable matrix*.

Suppose that the Markov chain starts with an initial state j. Let N_j be be the time of the first visit to state j.

State j is said to be *recurrent* if $P[N_j < \infty | X_0 = j] = 1$; otherwise, if $P[N_j = \infty | X_0 = j] > 0$, then j is said to be *transient*.

A recurrent state is said to be *null*, if $E[N_j | X_0 = j] = \infty$; otherwise, it is said to be *non-null*.

A recurrent state j is said to be *periodic* with *period* d, if $d \geq 2$ is the largest integer for which

$$P[N_j = kd \; for \; some \; k \geq 1] = 1;$$

otherwise, if there is no such $d \geq 1$, j is said to be *aperiodic*.

If j is a transient state, then for any $i \in \Phi$,

$$\lim_{n \to \infty} Q^{(n)}(i,j) = 0.$$

A Markov chain may have several irreducible closed sets. If a Markov chain starts with a state in an irreducible closed set, then the Markov chain will stay within the same chain. If a Markov chain starts with a transient state, the Markov chain will eventually (with probability one) enter an irreducible closed set and then will remain there.

The following two theorems are very useful in characterizing the states in a Markov chain with a countable state space.

Theorem 2.5 *In an irreducible Markov chain, either all states are transient, or all are recurrent null, or all are recurrent non-null. Either all states are aperiodic, or else if one state is periodic with period d, then all states are periodic with the same period d.*

Theorem 2.6 *If Φ is finite, then all states of an irreducible Markov chain are recurrent non-null.*

Stationarity

A Markov chain is said to be *stationary*, if for any $k \geq 0$

$$P(X_0, X_1, \cdots) = P(X_k, X_{k+1}, \cdots).$$

Taking the marginal distribution for X_k, we get $P(X_k) = P(X_0)$ for all $k > 0$. That is, the distribution functions of the states of a stationary Markov chain at any time are the same. This distribution is called the *stationary distribution* and is denoted as $\pi(j)$, $j \in \Phi$. Let $\pi = (\pi(1), \pi(2), \cdots)$ be the vector of the stationary probabilities. By definition,

$$\pi^T = \pi^T Q = \pi^T Q^2 = \cdots = \pi^T Q^k \qquad \textit{for all } k > 0.$$

Theorem 2.7 *Let $\{X_n\}$ be an irreducible Markov chain. Then all states are recurrent non-null if and only if the set of linear equations*

$$\pi^T = \pi^T Q \tag{2.13}$$

and

$$\sum_i \pi(i) = 1$$

has a unique solution. Furthermore, $\pi(j) > 0$ for all $j \in \Phi$.

If the initial state distribution is the stationary distribution, then from (2.11), the Markov process is stationary. From Theorems 2.6 and 2.7, for an irreducible Markov chain with a finite space, there exists a unique stationary distribution.

Note that the probability distribution $P(X_n)$ for any n is a distribution of a random vector, X_n, which is defined on $(\Omega, \mathcal{F}, P)$. While the stationary distribution is merely a probability measure on the state space Φ.

Theorem 2.8 *Let $\{X_n\}$ be stationary and $\psi(x_1, x_2, \cdots)$ be measurable on Φ^∞, then the process $\{Y_n\}$ defined by $Y_n = \psi(X_n, X_{n+1}, \cdots)$ is stationary.*

Asymptotical Stationarity

If for any initail state X_0, $\lim_{n\to\infty} P(X_n = j|X_0) = \pi(j)$, for all $j \in \Phi$, then the Markov process is said to be *asymptotically stationary.*

Theorem 2.9 *If a Markov chain is irreducible, aperiodic, and recurrent non-null, then*

$$\lim_{n\to\infty} Q^{(n)}(i,j) = \pi(j)$$

for all $i, j \in \Phi$.

The theorem implies that as $n \to \infty$, Q^n converges to a matrix whose rows are identical. From (2.12), such a Markov chain is aymptotically stationary.

Remark 2.1 A stationary process is not asymptotically stationary when it is periodic. For periodic processes, the stationary distribution (i.e., the solution to (2.13)) may exist; the process with a stationary initial distribution is stationary. However, because of the periodic behavior, the distributions of X_n, given any initial state, may not have a limit.

Ergodicity

The ergodicity of a stochastic process offers a very important tool in studying the property of a stochastic system by using one of its sample path. The concept looks like sophiscated, yet the intuition is rather simple. Roughly speaking, ergodicity implies that in a long run all the possible sample paths behave similarly, or in a somewhat formal language, the average of the values of a random variable defined on Ω, taken over any sample path, equals the ensemble average of the random variable.

We first consider a more general problem, which may shed light on the meaning of ergodicity. Let $\mathcal{T}$ be a transformation of $(\Omega, \mathcal{F}, P)$ onto itself. Let $\mathcal{T}^k$ be the transformation of applying $\mathcal{T}$ consecutively k times on Ω and X be a random variable defined on Ω. Starting from any point $\omega \in \Omega$, we apply the transformation $\mathcal{T}$ repeately. As the point ω moves following the path $\omega \rightarrow \mathcal{T}(\omega) \rightarrow \mathcal{T}^2(\omega) \rightarrow \cdots$, the value of X jumps along the path $X(\omega) \rightarrow X(\mathcal{T}\omega) \rightarrow X(\mathcal{T}^2\omega) \rightarrow \cdots$. Now, we ask the following question: under what conditions does the average

$$\frac{1}{n}\sum_{k=1}^{n} X(\mathcal{T}^{k-1}\omega) \tag{2.14}$$

converges with probability one to $E(X)$ as n goes to infinity? In general, this may require the sequence $\{\mathcal{T}^k\omega, k = 0, 1, \cdots\}$ to have a positive probability of going through every point in Ω (unless it so happens that the conditional mean of X on a subset of Ω is the same as $E(X)$). This may also require $E[X(\mathcal{T}^k\omega)]$ to be the same for all k; otherwise, the information about X (i.e., the initial distribution of the sequence $\{X(\omega), X(\mathcal{T}\omega), \cdots\}$) may be lost during the transformations. (In this case, using (2.14) to estimate $E(X)$ is just like "shooting a moving target.") $E[X(\mathcal{T}^k\omega)]$ will be the same if the transformation $\mathcal{T}$ preserves the measure. It turns out that these two conditions are sufficient for (2.14) to converge to $E(X)$. We formally state the problem below.

A measurable transformation $\mathcal{T}$ on $\Omega \rightarrow \Omega$ is said to be *measure-preserving*, if $P(\mathcal{T}^{-1}A) = P(A)$ for all $A \in \mathcal{F}$.

A set $A \in \mathcal{F}$ is *invariant* under the transformation $\mathcal{T}$, if $\mathcal{T}^{-1}A = A$.

Definition 2.2 *A measure-preserving transformation T on $(\Omega, \mathcal{F}, P)$ is said to be ergodic, if for every invariant set A, $P(A) = 0$ or 1.*

Suppose that A is an invariant set and $0 < P(A) < 1$. Then starting with $\omega \in A$, the sequence $T\omega, T^2\omega, \cdots$ can never reach A^c, and $P(A^c) > 0$. This explains the need for $P(A) = 1$ or 0. Next, let $\omega' = T^k(\omega)$. With a measure-preserving transformation T, we have $E[X(\omega')] = E[X]$ for any k. The above two points intuitively explain the following *ergodic theorem.*

Theorem 2.10 *If T is an ergodic transformation of $(\Omega, \mathcal{F}, P)$, then for any random variable X with $E|X| < \infty$,*

$$\lim_{n\to\infty} \frac{1}{n} \sum_{k=1}^{n-1} X(T^k\omega) = E(X) \qquad w.p.1.$$

Now, we return to stochastic processes defined on $(\Omega, \mathcal{F}, P)$ and with a state space Φ. Let $(x_1, x_2, \cdots) \in \Phi^\infty$. We define a *shift transformation* $\vartheta : \Phi^\infty \to \Phi^\infty$ by $\vartheta(x_1, x_2, \cdots) = (x_2, x_3, \cdots)$. The stochastic process transforms the probability measure P to a probability measure $\hat{P}$ on Φ^∞. It is easy to see that ϑ preserves $\hat{P}$ measure if the process $\{X_n\}$ is stationary.

The shift transformation ϑ is defined on Φ^∞. To define the invariant set, it is nice to work on Ω. An event (set) $A \in \mathcal{F}$ is invariant (under the shift transformation) if there exists a set B in Φ^∞ such that for every $n \geq 1$

$$A = \{(X_n, X_{n+1}, \cdots) \in B\}.$$

Definition 2.3 *A stationary process $\{X_n\}$ is ergodic if every invariant event has probability one or zero.*

Then, by Theorem 2.10, we have

Theorem 2.11 *Let $\{X_n\}$ be a stationary and ergodic process and $E|X_n| < \infty$. Then*

$$\lim_{n\to\infty} \frac{1}{n} \sum_{k=1}^{n-1} X_k = E(X) \qquad w.p.1. \tag{2.15}$$

The following theorem will be used in the text.

Theorem 2.12 *Let $\{X_n\}$ be a stationary and ergodic process, $\phi(x_1, x_2, \cdots)$ be a measurable function on Φ^∞, then the process $\{Y_n\}$ defined by $Y_n = \phi(X_n, X_{n+1}, \cdots)$ is ergodic.*

The following theorem specifies the conditions for a Markov chain to be ergodic.

Theorem 2.13 *An irreducible stationary Markov chain is ergodic.*

Remark 2.2 The stationarity is required in Theorem 2.11 to guarantee that $E(X_k)$ does not depend on k and thus $E(X)$ is well-defined. In fact, if we let $E(X)$ on the right-hand side of (2.15) be the mean of the stationary distribution (assume it exists), then (2.15) holds even if the initial distribution is not the stationary one (i.e., the process is not stationary). Based on this, the ergodic process can be defined in a slightly weak way: A stochastic process is ergodic if the stationary distribution exists and every invariant set has probability zero or one (cf. the point process theory reviewed in Section 2.2.5).

Example 2.8 Consider a simple Markov chain with two states 1, 2, and $Q(1,1) = Q(2,2) = 0$, $Q(1,2) = Q(2,1) = 1$. The Markov chain is periodic with period $d = 2$. It is irreducible and the stationary distribution exists, with $\pi(1) = \pi(2) = 0.5$. The Markov chain with this distribution as the initial distribution is stationary. However, $\lim_{k\to\infty} P(X_k|X_0)$ does not exist; thus, the Markov chain is not asymptotically stationary. In fact, if $X_0 = 1$, then for even k, $P(X_k = 1|X_0) = 1$ and for odd k, $P(X_k = 0|X_0) = 1$. The Markov chain is ergodic; i.e., (2.15) holds, where $E(X)$ is the mean of the stationary distribution.

2.2.2 Markov Processes

In this subsection, we study the stochastic processes $\{Y_t\}$ with a continuous index $t \in R_+$, $R_+ = [0,\infty)$, and a countable state space Φ. Any point $\omega \in (\Omega, \mathcal{F}, P)$ represents a sample path of the process. For any t, Y_t is a random variable on Ω.

Definition 2.4 *The stochastic process $\mathcal{Y} = \{Y_t : t \in R_+\}$ is called a Markov process with state space Φ provided that for any $t, s \geq 0$ and $j \in \Phi$,*

$$P[Y_{t+s} = j|Y_u; u \leq t] = P[Y_{t+s} = j|Y_t].$$

Transition Functions

When $P[Y_{t+s} = j|Y_t] = P_s(i,j)$ is independent of $t \geq 0$ for all $i, j \in \Phi$ and $s \geq 0$, the Markov process $\mathcal{Y}$ is said to be *time-homogeneous*. We shall discuss only time-homogeneous processes.

For any fixed pair $(i,j) \in \Phi$, the function $t \to P_t(i,j)$ is called a *transition function.*

We observe that the transition functions have the following properties:

$$P_t(i,j) \geq 0; \tag{2.16}$$

$$\sum_{k\in\Phi} P_t(i,k) = 1; \tag{2.17}$$

and for every $i, j \in \Phi$ and $t, s \geq 0$,

$$\sum_{k\in\Phi} P_t(i,k)P_s(k,j) = P_{t+s}(i,j). \tag{2.18}$$

This is the *Chapman-Kolmogorov equation* for the continuous time case.

The Structure of a Markov Process

Let R_t be the length of the time in which the process $\{Y_t\}$ remains in the state being occupied at the instant t; i.e.,

$$R_t(\omega) := inf\{s > 0 : \; Y_{t+s}(\omega) \neq Y_t(\omega)\}.$$

Since given Y_t, the future does not depend on the past, R_t possesses the "memoryless" property. Thus, R_t has an exponential distribution; i.e.,

$$P(R_t > u | Y_t = i) = e^{-\lambda(i)u}, \qquad u \geq 0, \tag{2.19}$$

for some number $\lambda(i) \in [0,\infty)$ (If $\lambda(i) = \infty$, then $P(R_t > u|Y_t = i) = 0$ for all $u > 0$). $\lambda(i)$ is called the *state transition rate.*

Equation (2.19) indicates that a Markov process stays in a state for an exponentially distributed time and then jumps to another state. Let $T_0 = 0$, T_1, T_2, $\cdots$ be the instants of transitions for the process $\{Y_t\}$ and X_0, X_1, X_2, $\cdots$ be the successive states visited by $\{Y_t\}$. $T_{n+1} - T_n$ is called the *sojourn time* in state $X_n = i$. We assume that the sample paths are right-continuous; i.e.,$X_n = Y_{T_n}$ is the state after the transition that occurs at T_n.

For any n, any $j \in \Phi$, and $u \geq 0$, if $X_n = i$ then we have

$$\begin{aligned} &P[X_{n+1} = j, T_{n+1} - T_n > u | X_0, \cdots X_n; T_0, \cdots, T_n] \\ &= Q(i,j)e^{-\lambda(i)u}, \end{aligned} \tag{2.20}$$

where $Q(i,j) \geq 0$, $Q(i,i) = 0$, and $\sum_{j\in\Phi} Q(i,j) = 1$.

From (2.20), the sequence $\{X_0, X_1, X_2, \cdots\}$ of the successive states visited by the Markov process $\{Y_t\}$ is a Markov chain with the transition matrix $Q = [Q(i,j)]$. $\{X_n\}$ is called the *embedded Markov chain* of $\{Y_t\}$.

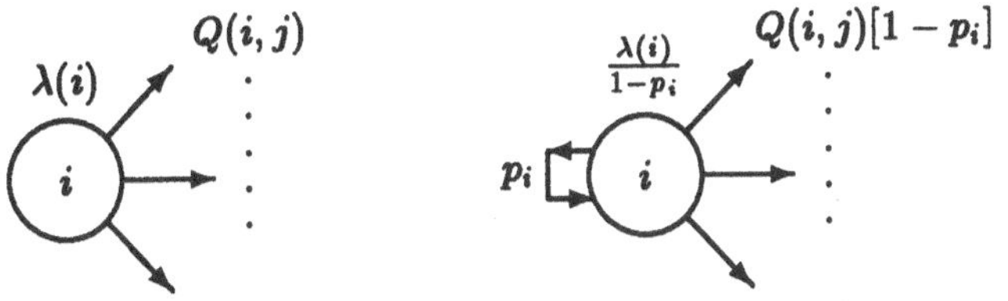

Figure 2.4: The Equivalent State by Adding a Feedback

Also from (2.20), the sojourn times $T_{n+1} - T_n$, $n = 0, 1, \cdots$, are exponentially distributed.

By definition, $Q(i,i) = 0$ for all i. Next, as shown in Figure 2.2, an exponential distribution with rate $\lambda(i)$ is equivalent to an exponential distribution with rate $\lambda(i)/(1-p_i)$ and a feedback rate p_i. Thus, we can artificially add transition points from state i to itself, corresponding to the feedback shown in Figure 2.2, on the Markov process $\{Y_t\}$. In this way, we can obtain an embedded Markov chain with $Q'(i,j) = Q(i,j)[1-p_i]$, $i \neq j$, $Q'(i,i) = p_i$ and $\lambda'(i) = \lambda(i)/(1-p_i)$. Figure 2.4 illustrates a state i and its equivalent state with a feedback loop added.

Suppose $sup\{\lambda(i),\ i \in \Phi\} \leq c < \infty$. If we choose

$$p_i = 1 - \frac{\lambda(i)}{c}, \tag{2.21}$$

then $\lambda'(i) = c$ for all $i \in \Phi$. That is, the state transition rates can be made equal by adding feedback jumps. This is called the *uniformization* of the Markov process. (see, e.g., Dijk [54]).

Another important application of the equivalent embedded Markov chain is that we can always construct an aperiodic embedded Markov chain, by adding feedbacks.

Some transition rates of a Markov process, $\lambda(i)$, may be infinity. In this case, a Markov process will immediately jump to other states when it reaches state i. In an even worse situation, a Markov process may jump infinite times in a finite period. To formally define this, let

$$\hat{T}(\omega) = sup\{T_n(\omega),\ all\ n\}.$$

If $\hat{T}(\omega) < \infty$, then there are infinitely many transitions on the sample path $\omega \in \Omega$ within any small neighborhood of $\hat{T}(\omega)$.

A Markov process is said to be *regular*, if $\hat{T}(\omega) = \infty$ with probability one. We only study regular Markov processes in this book. If $\lambda(i) \leq c < \infty$ for all $i \in \Phi$, then the Markov process is regular.

If $\lambda(i) = 0$, then i is called an *absorbing state*. For an absorbing state, we may redefine $Q(i,i) = 1$. In this way, an absorbing state of a Markov process corresponds to an absorbing state in its embedded Markov chain.

The Infinitesimal Generator

It is convenient to construct a Markov process by using its embedded Markov chain, with transition matrix Q, and the exponentially distributed sojourn times, with rates $\lambda(i)$. The transition functions $P_t(i,j)$ can be obtained by $Q(i,j)$ and $\lambda(i)$, $i,j \in \Phi$.

Let

$$\mathcal{A}(i,j) = \begin{cases} -\lambda(i)[1-Q(i,i)] & \text{if } i = j, \\ \lambda(i)Q(i,j) & \text{if } i \neq j. \end{cases} \tag{2.22}$$

This definition allows a state to jump to itself. We have

$$\sum_{j \in \Phi} \mathcal{A}(i,j) = 0, \qquad i \in \Phi.$$

$\mathcal{A}(i,j)$ are the same for all equivalent embedded Markov chains. In fact, $\lambda(i)Q(i,j) = \lambda'(i)Q'(i,j)$ and $\lambda(i)[1-Q(i,i)] = \lambda'(i)[1-Q'(i,i)]$ for all $i,j \in \Phi$.

We have the following *Kolmogorov's* equation:

$$\begin{aligned} \frac{d}{dt}P_t(i,j) &= \sum_{k \in \Phi} \mathcal{A}(i,k)P_t(k,j) \\ &= \sum_{k \in \Phi} P_t(i,k)\mathcal{A}(k,j). \end{aligned}$$

In the matrix form, the equation is

$$\frac{d}{dt}P_t = \mathcal{A}P_t = P_t\mathcal{A}, \tag{2.23}$$

where $P_t := [P_t(i,j)]$ and $\mathcal{A} := [\mathcal{A}(i,j)]$. We have $P_0(i,i) = 1$ and $P_0(i,j) = 0$ if $i \neq j$; i.e., P_0 is an identity matrix. The matrix $\mathcal{A}$ is called the *infinitesimal generator* of the Markov process $\mathcal{Y}$.

The solution to (2.23) is

$$P_t = e^{t\mathcal{A}} := \sum_{n=0}^{\infty} \frac{t^n}{n!}\mathcal{A}^n.$$

In the matrix form, (2.18) can be written as

$$P_{t+s} = P_t P_s.$$

Another equation is

$$\frac{d}{ds} P_{t+s} = P_t \mathcal{A} P_s.$$

Limit Theorems

The stationarity, asymptotically stationarity, and ergodicity for stochastic processes with continuous index t can be defined in a way similar to those for stochastic processes with discrete index n. For example, an invariant set $A \in \mathcal{F}$ is a set for which if ω belongs to A, then for all $s \geq 0$, the point ω' such that $Y_t(\omega') = Y_{t+s}(\omega)$ also belongs to A. A process is said to be ergodic if the stationary distribution exists and every invariant set has probability one or zero.

Let $p_t(i) = P(Y_t = i)$ and $p_t^T = (p_t(1), p_t(2), \cdots)$. Then by the theorem of total probability,

$$p_{t+s}^T = p_t^T P_s \tag{2.24}$$

Taking the derivative with respect to s, setting $s = 0$, and applying Kolmogorov's equation yield

$$\frac{d}{dt} p_t^T = p_t^T \mathcal{A}.$$

Thus, p^T is a stationary distribution of the Markov process if and only if

$$p^T \mathcal{A} = 0. \tag{2.25}$$

The solution to (2.25) is the same as the stationary distribution of the embedded Markov chain of the uniformized Markov process. Indeed, in the uniformized Markov process, $\lambda'(i) = c$ for all $i \in \Phi$. Thus, from (2.22), (2.25) is equivalent to

$$p^T Q' = p^T, \tag{2.26}$$

where Q' is the transition matrix for the embedded Markov chain of the uniformized Markov process. Therefore, p^T is the its stationary distribution.

Let $\pi^T = (\pi(1), \pi(2), \cdots)$ be the stationary distribution of the embedded Markov chain of the original Markov process $\mathcal{Y}$. In order to construct a uniformized Markov procees, we have to add a beedback transition with a rate $p_i = 1 - [\lambda(i)/c]$ to state i, $i \in \Phi$ (see (2.21)). This is equivalent to increasing the number of transitions from state i by, on the average,

$1/(1-p_i) = c/\lambda(i)$ times. Thus, the stationary distribution of the embedded Markov chain of the uniformized Markov process is

$$p(i) = c\frac{\pi(i)}{\lambda(i)} \qquad i \in \Phi. \tag{2.27}$$

Using the normalizing condition, we obtain

$$c = \sum_{i\in\Phi} \pi(i)\lambda(i) = \frac{1}{\sum_{i\in\Phi} \frac{\pi(i)}{\lambda(i)}} := \lambda. \tag{2.28}$$

λ is the average state transition rate.

From (2.26), we have the following theorem.

Theorem 2.14 *If the embedded Markov chain is irreducible and recurrent, then there is a unique stationary distribution satisfying (2.25) and $p(i) > 0$ for all $i \in \Phi$.*

The theorem follows directly from Theorem 2.7 and the fact that if the embedded Markov chain of the uniformized Markov process is irreducible and recurrent, so is that of the original process.

For the asymptotically stationarity, we have

Theorem 2.15 *If the embedded Markov chain is irreducible and recurrent, then*

$$\lim_{t\to\infty} P(Y_t = i) = p(i).$$

This simply follows from Theorem 2.9 and the fact that the embedded Markov chain can be chosen as aperiodic.

Finally, by definition, any invariant set of $\{Y_t\}$ must be invariant for its embedded Markov chain. Thus, if the embedded Markov chain is ergodic, then so is the Markov process. Thus, we have

Theorem 2.16 *If the embedded Markov chain is ergodic, then for any function f on Φ*

$$\lim_{T\to\infty} \frac{1}{T}\int_0^T f(Y_t)dt = E[f(Y)],$$

where E denotes the mean of the stationary distribution on Φ.

2.2.3 Semi-Markov Processes

Any stochastic process $\{Y_t\}$ with a discrete state space "jumps" from states to states. Suppose the process is right continuous. Let T_0, T_1, $\cdots$ be the sequence of the instants of transitions and $X_n = Y_{T_n}$, $n = 0, 1, \cdots$.

A process $\{Y_t\}$ is a *semi-Markov process*, if

$$P[X_{n+1} = j, T_{n+1} - T_n \le t | X_0, \cdots, X_n; T_0, \cdots, T_n]$$
$$= P[X_{n+1} = j, T_{n+1} - T_n \le t | X_n]$$

for all $n = 0, 1, \cdots$, $j \in \Phi$, and $t \ge 0$.

$Q(i, j, t) = P[X_{n+1} = j, T_{n+1} - T_n \le t | X_n = i]$ is called a *semi-Markov kernel* on Φ. A semi-Markov process enjoys some Markov property: at any transition instant T_n, the future behavior is conditionally independent of the past provided the present state X_n is known.

Let $Q(i, j) = \lim_{t \to \infty} Q(i, j, t)$. It is an easy exercise to prove that $\{X_n, n = 0, 1, \cdots\}$ is a Markov chain with transition Martix $Q = [Q(i, j)]$. $\{X_n\}$ is called the *embedded Markov chain* of $\{Y_t\}$.

If

$$Q(i, j, t) = Q(i, j)[1 - e^{-\lambda(i)t}],$$

then a semi-Markov process is a Markov process.

A common mistake is to view a semi-Markov process simply as a process that has an embedded Markov chain and has the intertransition times being independent and identically distributed with a general distribution function. This over-simplified understanding may lead to the following wrong conclusion: a semi-Markov process with exponential intertransition times is a Markov process. This is not true, because in a semi-Markov process the transition probability from X_n to X_{n+1} may depend on $T_{n+1} - T_n$. An example of such processes is

$$Q(i, j, t) = G(i, j, t)[1 - e^{-\lambda(i)t}],$$

with $\sum_{j \in \Phi} G(i, j, t) = 1$, $P[X_{n+1} = j | T_{n+1} - T_n \le t, X_n = i] = G(i, j, t)$ and $P[T_{n+1} - T_n \le t | X_n = i] = 1 - e^{-\lambda(i)t}$.

Finally, if the embedded Markov chain is ergodic, then Theorem 2.16 holds for a semi-Markov process. This can be proved by using the embedded Markov chain, or by using Theorem 2.17 stated in the next subsection for regenerative processes.

2.2.4 Regenerative Processes

Renewal Processes

A sequence of time instants $\{T_n, n = 0, 1, \cdots\}$ is called a *renewal process*, if $S_n = T_{n+1} - T_n$, $n = 0, 1, 2, \cdots$, are independent and identically distributed non-negative random variables. T_n, $n = 1, 2, \cdots$, are called the *renewal times*, or the *arrival points*.

A renewal process $\{T_n\}$ is said to be *recurrent*, if $S_n < \infty$ for every n with probability one; otherwise, it is called *transient*. $\{T_n\}$ is said to be *periodic*, if S_n are discrete random variables taking values 0, δ, 2δ, $\cdots$; otherwise, it is called *aperiodic*.

If S_n, $n = 0, 1, 2, \cdots$, are exponentially distributed with mean $1/\lambda$, then the renewal process is called a *Poisson process*. λ is the rate of the arrivals. Because of the memoryless property of the exponential distribution, the time until next arrival from any time t is also exponentially distributed with mean $1/\lambda$.

Regenerative Processes

Let $\mathcal{F}_t$ be the σ-field generated by $\{Y_s,\ s \le t\}$. A non-negative random variable τ is called a *stopping time* of the stochastic process $\{Y_t, t \ge 0\}$ if

$$\{\omega : \tau(\omega) \le t\} \in \mathcal{F}_t, \qquad t \ge 0.$$

In words, a random variale τ is a stopping time if for every $0 \le t < \infty$, the occurrence or non-occurrence of the event $\{\tau \le t\}$ can be determined from the history $\{Y_s,\ s \le t\}$.

Definition 2.5 *A process $\{Y_t;\ t \ge 0\}$ is said to be regenerative, if there exists a sequence $T_0 = 0$, T_1, $\cdots$ of stopping times such that*

i. *$\{T_n,\ n = 0, 1, \cdots\}$ is a renewal process, and*

ii. *for any positive integers n, m, t_1, $\cdots$, $t_n \in [0, \infty)$, and any bounded function f on Φ^n*

$$E[f(Y_{T_m+t_1}, \cdots, Y_{T_m+t_n})|Y_s, s \le T_m] = E[f(Y_{t_1}, \cdots, Y_{t_n})]. \qquad (2.29)$$

(2.29) is called the *regenerative property*, and T_0, T_1, $\cdots$, are called the *regenerative times*. By the definition, the behavior of a regenerative process starting from any regenerative time is statistically the same as that of the process starting from $T_0 = 0$. In particular, the behavior of a regenerative process in any period $T_{n+1} - T_n$ is statistically the same.

Theorem 2.17 *If $\{T_n\}$, $T_0 = 0$, is recurrent aperiodic and $E(T_1) < \infty$, then for any $i \in \Phi$*

$$\lim_{t \to \infty} P(Y_t = i) = \frac{E\{\int_0^{T_1} I(Y_t = i)dt\}}{E(T_1)}, \tag{2.30}$$

where $I(Y_t = i) = 1$ if $Y_t = i$, and $I(Y_t = i) = 0$ otherwise.

This theorem has an important application in simulation. It implies that the steady-state distribution of a regenerative process can be obtained by the ratio of the average of an integral over a regenerative period and the average length of a regenerative period. It can be easily seen that the right-hand side of (2.30) equals

$$\lim_{n \to \infty} \frac{\int_0^{T_n} I(Y_t = i)dt}{T_n}.$$

Thus, (2.30) yield the ergodicity equation.

2.2.5 Point Processes

The point process theory provides a powerful tool to prove many important properties of stochastic processes, ranging from the fundamental result of Little's law to the more elegant result of EAST (events see time average, see, e.g., Brémaud et al. [10]). The rigorness and the succinctness of the proofs make the theory very attractive. The most derivations of the results of this book do not require the point process theory, it, however, will provide a deeper understanding for the concepts and the main results, especially for those regarding the relationship between the sample-path average and the mean of the performance sensitivity.

In this subsection, we intend to give an intuitive conceptual review for some fundamentals in the point process theory. Readers are referred to Baccelli and Brémaud [1] and [2], Daley and Vere-Jones [49], and Karr [73] for more details.

Marked Point Processes

Let $(\Omega, \mathcal{F}, P)$ be a probability space. Defined on Ω, there is a sequence of random arrival times $0 \le T_1 \le T_2 \le \cdots$, where $T_n \in R_+$, $n = 1, 2, \cdots$. We assume that $\lim_{n \to \infty} T_n = \infty$ with probability one. Associated with each arrival time T_n, there is a *mark*, X_n, taking values in a space $\mathcal{M}$ (called the *mark space*). The random sequence $\mathcal{X} = \{(T_n, X_n) : n \geq 1\}$ is called a *marked point process.* $S_n := T_{n+1} - T_n$, $n \geq 1$, is the nth interarrival time.

In most applications, $\mathcal{M}$ is a discrete space (e.g., the state space of a discrete Markov process). In general, we assume that $\mathcal{M}$ contains a σ-field. Let $\mathcal{H}$ be a σ-field of the product space $R_+ \times \mathcal{M}$. For any fixed $\omega \in \Omega$, $\mathcal{X}(\omega)$ is a sequence of $\{(T_n(\omega), X_n(\omega)) : n \geq 1\}$, which is a sample path (or a realization) of the marked point process. Let $\mathcal{Z}$ be the collection of all such (deterministic) sequences. This is the *space of the sample paths.*

Let $M \subseteq \mathcal{M}$ be a subset of $\mathcal{M}$ and A be any set in R_+. For any integer $k \geq 0$, we define an elementary subset $\mathcal{E}(A, M, k)$ of $\mathcal{Z}$ as the collection of the sample paths that have k arrivals, carrying marks in set M, in set A. All these elementary subsets of $\mathcal{Z}$ generate a σ-field of $\mathcal{Z}$, denoted as $\mathcal{G}$. A marked point process $\mathcal{X}$ is then a measurable $\mathcal{F}/\mathcal{G}$ mapping $\mathcal{X} : \Omega \to \mathcal{Z}$. This mapping transforms the probability measure P on $(\Omega, \mathcal{F})$ to a probability measure P^0 on $(\mathcal{Z}, \mathcal{G})$; we call P^0 the *initial distribution* (or simply the distribution) of $\mathcal{X}$.

The Palm Version and the Stationary Version

For each $t > 0$, we define a shift mapping $\vartheta_t : \mathcal{Z} \to \mathcal{Z}$: for any $Z \in \mathcal{Z}$, $\vartheta_t Z$ is a sequence obtained from Z by relabeling t as the origin (and ignoring everything to the left of t). Define $\vartheta_{(n)} := \vartheta_{T_n}$ be the mapping obtained by shifting the origin to the arrival epoch T_n. We write $Z_t := \vartheta_t Z$ and $Z_{(n)} := \vartheta_{(n)} Z$. For any marked point process $\mathcal{X}$, we define $\mathcal{X}_t(\omega) := \vartheta_t \mathcal{X}(\omega)$ and $\mathcal{X}_{(n)}(\omega) := \vartheta_{(n)} \mathcal{X}(\omega)$. Note that the shifting t in $\mathcal{X}_t$ is fixed, while the shifting T_n in $\mathcal{X}_{(n)}$ is a random variable. The relationship of a sample path of $\mathcal{X}$ and that of $\mathcal{X}_t$ is shown in Figure 2.5.

Using the marked point process $\mathcal{X}$, we can define (assuming they exist) two other probability measures on $(\mathcal{Z}, \mathcal{G})$:

$$P^{(a)}(\mathcal{E}) := \lim_{n \to \infty} \frac{1}{n} \sum_{j=1}^{n} P^0(\mathcal{X}_{(n)} \in \mathcal{E}), \tag{2.31}$$

and

$$P^{(s)}(\mathcal{E}) := \lim_{t \to \infty} \frac{1}{t} \int_0^t P^0(\mathcal{X}_t \in \mathcal{E}) dt. \tag{2.32}$$

The limits in (2.31) and (2.32) exist if $\mathcal{X}$ is asymptotically stationary; i.e., $\lim_{n \to \infty} P^0(\mathcal{X}_{(n)} \in \mathcal{E})$ or $\lim_{t \to \infty} P^0(\mathcal{X}_t \in \mathcal{E})$ exist. (2.31) is called the *arrival stationary distribution*, or the *Palm distribution*, of $\mathcal{X}$; it is the distribution seen at the arrival instants after a long time period. (2.32) is

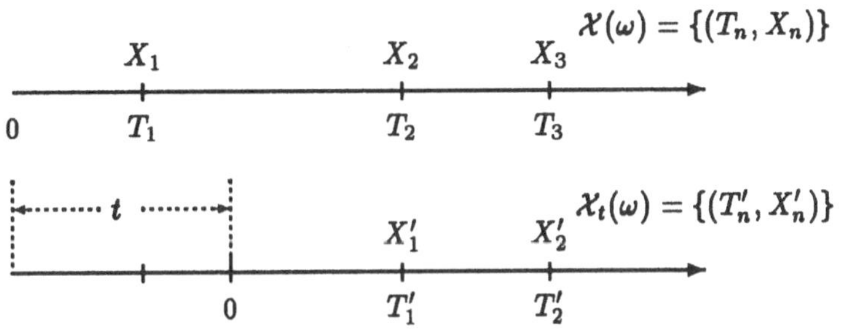

$$X'_1 = X_2,\ X'_2 = X_3,\ T'_1 = T_2 - t,\ T'_2 = T_3 - t, \cdots$$

Figure 2.5: A Sample Path of $\mathcal{X}$ and $\mathcal{X}_t$

called the *time stationary distribution*, or simply the stationary distribution, of $\mathcal{X}$; it is the distribution seen at any instant after a long time period.

When it is necessary, we denote the initial, the Palm, and the stationary distributions of $\mathcal{X}$ as $P^0_{\mathcal{X}}$, $P^{(a)}_{\mathcal{X}}$, and $P^{(s)}_{\mathcal{X}}$, respectively.

If a point process, denoted as $\mathcal{X}^{(a)}$, satisfies $P^0_{\mathcal{X}^{(a)}} = P^{(a)}_{\mathcal{X}}$, then the point process is called a *Palm version* of $\mathcal{X}$. If a point process, denoted as $\mathcal{X}^{(s)}$, satisfies $P^0_{\mathcal{X}^{(s)}} = P^{(s)}_{\mathcal{X}}$, then the point process is called a *stationary version* of $\mathcal{X}$. We shall use the superscripts (a) and (s) to denote the quantities associated with the Palm version and the stationary version, respectively. A point process is said to be *arrival (time) stationary* if it is a Palm (stationary) version of itself.

It is intuitively clear that for every n, $[\mathcal{X}^{(a)}]_{(n)}$ has the same distribution, whereas for evert $t > 0$, $[\mathcal{X}^{(a)}]_t$ has the same distribution.

For any set $A \subset R_+$, define $N(A)$ be the number of arrivals in A. Let $N(t) := N([0,t])$, and let $\mathcal{E}_0$ be the set in $\mathcal{Z}$ such that $N(0) = 1$. Then $P^{(a)}(\mathcal{E}_0) = 1$ and $P^{(s)}(\mathcal{E}_0) = 0$ for all $\mathcal{X}$. Thus, we can write $T^{(a)}_0 = 0$ for any Palm version.

Remark 2.3 If we view $\mathcal{X}_t$, i.e., the future sample path, as the "state" of the point process at time t, then we can make a simple analogy between the stationarity of a point process and that of a Markov process. That is, a point process $\mathcal{X}$ is a stationary version if and only if the (initial) distributions of $\mathcal{X}_t$ are the same for all $t \geq 0$.

Some Theorems

An arrival (time) stationary point process is said to be *ergodic*, if the invariant set under the shift mapping ϑ_{-1} (ϑ_{-t}) has probability either zero or one; i.e., if $\vartheta_{-1}(\mathcal{E}) = \mathcal{E}$, then $P^{(a)}(\mathcal{E}) = 0$ or 1 (if $\vartheta_{-t}(\mathcal{E}) = \mathcal{E}$, then $P^{(s)}(\mathcal{E}) = 0$ or 1).

A point process $\mathcal{X}$ is said to be *ergodic*, if its Palm and stationary versions $\mathcal{X}^{(a)}$ and $\mathcal{X}^{(s)}$ are ergodic.

Let $\lambda = 1/[E(S_0^{(a)})]$. Note that $S_0^{(a)} := T_1^{(a)} - T_0^{(a)}$ is a random variable defined on $(\mathcal{Z}, \mathcal{G}, P^{(a)})$. Thus, the expectation of $S_0^{(a)}$ is associated with the probability measure $P^{(a)}$. Similarly, the expectation in $E[N^{(s)}(1)]$, where $N^{(s)}(1)$ is the number of arrivals of $\mathcal{X}^{(s)}$ in $[0,1]$, is associated with the probability measure $P^{(s)}$. Here, and in some other places in the text, we use the same notation "E" to denote the expectation, with the understanding that it is taken in the probability space on which the random variable is defined.

The following fact is known: $\mathcal{X}^{(a)}$ exists and is ergodic with $0 < \lambda < \infty$ if and only if $\mathcal{X}^{(s)}$ exists and is ergodic with $E[N^{(s)}(1)] < \infty$; in this case, $E[N^{(s)}(t)] = \lambda t$.

Having review the basic concepts, we are ready to state the main theorems.

Theorem 2.18 *If $\mathcal{X}$ is ergodic, then for any measurable function $f : \mathcal{Z} \to R_+$,*

$$\begin{aligned} \lim_{t\to\infty} \frac{1}{t} \int_0^t f(\mathcal{X}_s) ds &= E[f(\mathcal{X}^{(s)})] \qquad w.p.1 \\ &= \frac{E\{\int_0^{T_1^{(a)}} f[(\mathcal{X}^{(a)})_s] ds\}}{E(T_1^{(a)})}, \end{aligned} \tag{2.33}$$

and

$$\begin{aligned} \lim_{n\to\infty} \frac{1}{n} \sum_0^n f(\mathcal{X}_{(n)}) &= E[f(\mathcal{X}^{(a)})] \qquad w.p.1 \\ &= \frac{E\{\sum_{j=1}^{N^{(s)}(1)} f[(\mathcal{X}^{(s)})_{(j)}]\}}{E[N^{(s)}(1)]}, \end{aligned} \tag{2.34}$$

where $E(T_1^{(a)}) = \frac{1}{\lambda}$.

The first equation of (2.33) is the ergodic equation of the point process; if we choose $\mathcal{X}_t = X_{[t]}$, where $[t]$ denotes the integer part of t, then the equation is the same as the ergodic equation for Markov processes. The second equation of (2.33) resembles (2.30) for regenerative processes. This equation indicates that by working on the Palm version, the regenerative structure can be avoid to achieve an equation similar to (2.30). The first equation of (2.34) shows that the mean with respect to the Palm distribution can be obtained by averaging the values at the arrival instants. The second part of (2.34) is similar to the regenerative property of the stationary version.

Let f be the indicator function, then we get the equations specifying the relationship between the Palm and the stationary distributions:

$$P^{(s)}(\mathcal{E}) = \frac{E\{\int_0^{T_1^{(a)}} I[(\mathcal{X}^{(a)})_s \in \mathcal{E}]ds\}}{E(T_1^{(a)})}, \tag{2.35}$$

and

$$P^{(a)}(\mathcal{E}) = \frac{E\{\sum_{j=1}^{N^{(s)}(1)} I[(\mathcal{X}^{(s)})_{(j)} \in \mathcal{E}]\}}{E[N^{(s)}(1)]}, \tag{2.36}$$

Example 2.9 Let $R(t) := T_{N(t)+1} - t$ be the residual arrival time, i.e., the time until the next arrival after t. Setting $f = R(0)$ in (2.33), we have $f[(\mathcal{X}^{(a)})_s] = T_1^{(a)} - s$, and

$$\int_0^{T_1^{(a)}} f[(\mathcal{X}^{(a)})_s]ds = \int_0^{T_1^{(a)}} [T_1^{(a)} - s]ds = \frac{1}{2}[T_1^{(a)}]^2.$$

Therefore, from (2.33) we get

$$E[T_1^{(s)}] = \frac{E[T_1^{(a)}]^2}{2E[T_1^{(a)}]}.$$

This result is well known for renewal processes.

It is clear that the Palm version and the stationary version of a point process $\mathcal{X}$ can be constructed on R instead of R_+. In such Palm (or stationary) versions, the distributions of the process starting from any T_n, $-\infty < n < \infty$, (or any t, $-\infty < t < \infty$) are the Palm (or the stationary) distributions of $\mathcal{X}$.

Theorem 2.19 *Campbell's Theorem: If $\mathcal{X}$ is ergodic with Palm and stationary versions $\mathcal{X}^{(a)}$ and $\mathcal{X}^{(s)}$ on R, then for any measurable function $f : R \times \mathcal{M} \to R_+$ we have*

$$E\{\sum_{n=-\infty}^{\infty} f(T_n^{(s)}, X_n^{(s)})\} = \lambda \int_{-\infty}^{\infty} \int_{\mathcal{M}} f(t, x) P^{(a)}(dx)dt. \tag{2.37}$$

The theorem can be explained as follows. Since $\mathcal{X}^{(s)}$ is time stationary, we can replace the interval $[0,1]$ in (2.36) by $[t, t+dt)$ and obtain

$$P^{(a)}(\mathcal{E}) = \frac{E\{\sum_{j=N^{(s)}(t)+1}^{N^{(s)}(t+dt)} I[(\mathcal{X}^{(s)})_{(j)}]\}}{E[N^{(s)}\{[t, t+dt)\}]}.$$

Setting $\mathcal{E}$ be the set with $x_0 \in (x, x+dx)$, we have

$$\lambda P^{(a)}[x_0 \in (x, x+dx)] = E\{\sum_{n=-\infty}^{\infty} I[T_n^{(s)} \in (t, t+dt), X_n \in (x, x+dx)]\},$$

where I is the indicator function. Multiplying this equation by $f(t,x)$ and integrating over $R \times \mathcal{M}$ yield (2.37).

Example 2.10 A Markov process $\mathcal{Y} = \{Y_t,\ t \geq 0\}$ can be viewed as a point process $\{(T_n, X_n)\}$, with T_n being the transition epoches and X_n being the state at T_n. A Markov process starting from an arbitrary initial distribution is not stationary. However, the time average and the arrival average (provided they exist) give the stationary distribution of the Markov process, $p(i)$, and that of the embedded Markov chain, $\pi(i)$. Any Markov process starting with the $\pi(i)$ is a Palm version of the original Markov process and is not time stationary. Any Markov process starting with the $p(i)$ is a stationary version of the original Markov process, and its embedded Markov chain is not stationary.

Equations (2.27) and (2.28) can be obtained by using either (2.35) or (2.36). Let $\mathcal{E}$ be the set with $x_0 = i$. Then in (2.35), we have

$$\begin{aligned} E\{\int_0^{T_1^{(a)}} I[(\mathcal{Y}^{(a)})_s \in \mathcal{E}]ds\} &= E\{\int_0^{\infty} I[X_0 = i] \times I[s \leq T_1^{(a)}]ds\} \\ &= \pi(i)E[T_1^{(a)}|X_0 = i] = \frac{\pi(i)}{\lambda(i)}. \end{aligned}$$

In (2.36), $E\{\sum_{j=1}^{N^{(s)}(1)} I[(\mathcal{Y}^{(s)})_{(j)} \in \mathcal{E}]\}$ is the average number of arrivals in $[0,1]$ that carry a mark i. Thus,

$$E\{\sum_{j=1}^{N^{(s)}(1)} I[(\mathcal{Y}^{(s)})_{(j)} \in \mathcal{E}]\} = p(i)\lambda(i).$$

Furthermore, λ in (2.28) is the average arrival rate. Finally, (2.27) follows immediately.

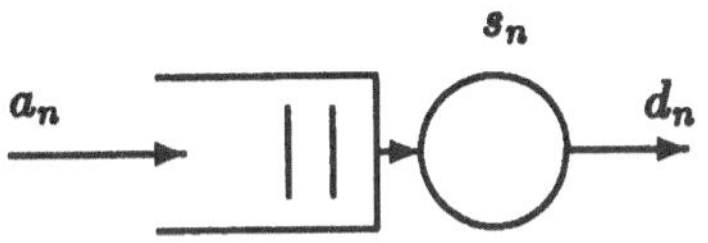

Figure 2.6: A Single-Server Queue

2.3 Queuing Systems

This section reviews some material in queuing theory that is relevant to the topics in this book. The section also contains some recent results about the performance sensitivity formulas and their computational algorithms (see, e.g., Liu and Nain [80] and Cao and Ma [35]). The readers are refered to the books listed in Chapter 1 for more details.

2.3.1 The Single-Station Queue

The model

A single-station queue is the basic component of queuing systems. Figure 2.6 illustrates a model of a single-station queue, where the station contains only one server. In the figure, the circle represents the server; the open box represents a buffer. Customers arrive at a sequence of random times $0 \le a_1 \le a_2 \le \cdots$. The nth customer, arriving at the server at a_n and departing from the server at d_n, requires a certain amount of service that will take the server s_n time units to process. $\tau_n := a_{n+1} - a_n$ is called the *interarrival time*, and $d_{n+1} - d_n$, the *interdeparture time*.

A multiserver station contains multiple servers sharing a common queue.

The queue may have an infinite capacity or a finite capacity with a size K. A customer arriving at a full buffer will be lost (e.g., in the case of a single-server queue) or will have to wait elsewhere (e.g., in some other servers in a network of queues).

It is conventional to adopt the four-part description $A/B/m/K$ to specify a single-station queue. In the notation, m denotes the number of servers in the station, K the buffer size. A and B specify the distributions of interarrival times τ_n and the service times s_n, respectively. The most common single-server queue is the $M/M/1$ queue, where M/M indicates that both the interarrival time and the service time are independent and identically distributed with exponential distributions. (Hence the arrival process is a Poisson process.) The last letter $K = \infty$ is omitted for simplicity. Examples of other queues are the $M/GI/1$ queue, where GI indicates that the service times are independent and identically distributed with nonexponential distribution, and the $GI/M/1$ queue, where GI indicates that the the arrival times form a renewal process. Other commonly used symbols for A and B are: D for deterministic distributions, E_r for r-stage Erlangian distributions, H_R for R-stage hyperexponential distributions, and PH for phase-type distributions.

Service Disciplines

The service discipline determines which customer is served at any given time. We first assume that the *work-conservative* law holds; i.e., a server will provide service at its whole capacity as long as the queue is not empty. Thus, if any server in the service station is idle, then an arriving customer gets served immediately.

Some commonly used service disciplines for a server are as follows.

1. *First come first served (FCFS):* The customer who arrives at the queue first is served first by the server.

2. *The priority scheme:* Customers are assigned with different priorities and the customers with the highest priority in the queue gets served first. A priority scheme may be either *preemptive* or *nonpreemptive.* In a preemptive scheme, a customer in service is liable to be ejected from service when a customer with a higher priority enters the queue. In a nonpreemtive scheme, the arriving customer has to wait in the queue until the server completes its service to its current customer, even if the arriving customer has a higher priority. A preemptive scheme may be either *resume* or *nonresume.* In a preemptive resume scheme, the service that a preempted customer received is not lost when the service is resumed. While in a preemptive nonresume scheme, the preempted customer has to start the whole service again when it is resumed.

3. *Last come first served (LCFS):* The customer who arrives at the queue last receives the service first. This scheme has different versions: preemptive resume, preemptive nonresume, and nonpreemptive.

4. *Processor sharing (PS):* The service power of the server is shared equally by all the customers in the queue. Thus, if the service rate of a server is μ and there are n customers in the queue, then the service rate for each customer is μ/n. In modeling computer systems,

a *round-robin* scheme with a small quantum size can be considered as a PS scheme.

The M/M/1 Queue

Because of the memoryless property of the exponential distributions, the state of an $M/M/1$ queue can be simply chosen as the number of the customers in the queue, denoted as n.

Let λ and μ be the interarrival and the service rates, respectively. The stationary distribution of n, $p(n)$, can be easily obtained by solving the flow-balance equations; we have

$$p(n) = \rho^n(1-\rho) \qquad \rho = \frac{\lambda}{\mu}, \quad n = 0, 1, \cdots.$$

Let $\bar{n} := \sum_{n=0}^{\infty} np(n)$ be the average number of customers in the queue (including the customer in service), $\bar{n}_b$ be the average number of customers in the buffer (excluding the customer in service), T be the average time that the customers spend in the system (including the service time), and W be the average waiting time (excluding the service time) of the customers. Then

$$\bar{n} = \frac{\rho}{1-\rho},$$

$$\bar{n}_b = \frac{\rho^2}{1-\rho},$$

$$T = \frac{1}{\mu-\lambda},$$

and

$$W = T - \frac{1}{\mu} = \frac{\rho}{\mu-\lambda}.$$

Readers are referred to standard textbooks for formulas for other single-station queues.

Little's Law

We observe that for an M/M/1 queue,

$$\bar{n} = \lambda T \tag{2.38}$$

and

$$\bar{n}_b = \lambda W. \tag{2.39}$$

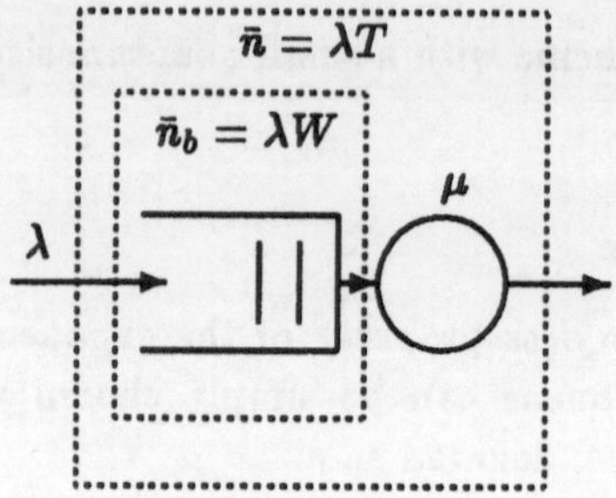

Figure 2.7: Little's Law for an M/M/1 Queue

These two equations are called *Little's Law*. Little's law applies to any networks or subnetworks. It says that the average number of the customers in any system or subsystem equals the product of the average time that a customer stays in the system or the subsystem and the arrival rate of the customers. The customer arrival process may be any point process. The customers may belong to different classes.

Applying Little's law to the subsystem consisting of the buffer, we obtain (2.39). For the system consisting of both the buffer and the server, we have (2.38). This is shown in Figure 2.7.

Sub-busy Periods

The idea of *sub-busy periods* can be helpful in some cases. In an M/M/1 queue, the period from the time that a customer arrives at an idle server to the time that the server becomes idle again is called a *busy period.* Because of the memoryless property of both the interarrival and the service time distributions, we can make the following observation: the period from the time that the system enters state n, $n > 0$, to the first time that the system enters state $n-1$ behaves statistically similar to a busy period. Such a period is called a *sub-busy period.* If $n = 1$, then the sub-busy period is a busy period.

Let T_B be the length of a busy period and $\bar{T}_B$ be its mean. As shown in Figure 2.8, a busy period of an M/M/1 queue consisting of more than one service time has two sub-busy periods. The first one starts from the second arrival time (if it occurs before the first customer leaves the server) and ends at the time when the state of the system enters 1 for the first time; the second one consists of the rest of the busy period. The mean of the length of a sub-busy period equals that of a busy period. Thus, it is

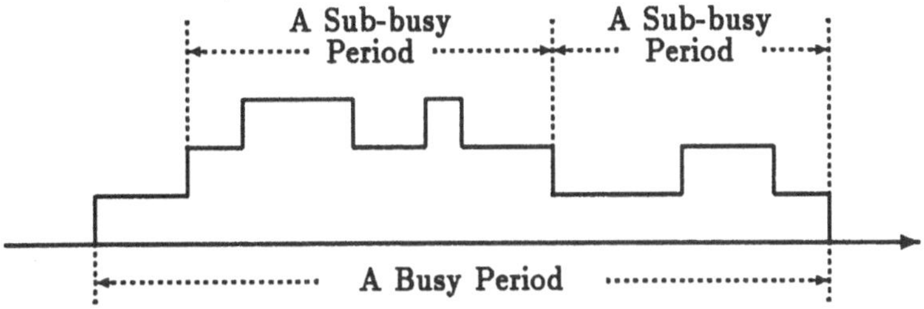

Figure 2.8: Sub-busy Periods in an M/M/1 Queue

easy to see that

$$\bar{T}_B = \frac{1}{\lambda+\mu} + \frac{\lambda}{\lambda+\mu}(\bar{T}_B + \bar{T}_B).$$

The first term on the right-hand side is the expected value of the minimum of the length of an interarrival time and that of a service time. This is the expected value of the length of the time period between the first arrival to the next arrival, (in the case where the next arrival happens before the service completion of the first custmer), or the time period between the first arrival to the first departure (in the case where there is only one customer in the busy period). This time period is followed with probability $\frac{\lambda}{\lambda+\mu}$ by two sub-busy periods shown in Figure 2.8. From the above equation, we have

$$\bar{T}_B = \frac{1}{\mu-\lambda}.$$

Let N_B be the number of customers served in a busy period. Then, by the same reasoning, we have

$$\bar{N}_B = \frac{\mu}{\lambda+\mu} \times 1 + \frac{\lambda}{\lambda+\mu}[1 + \bar{N}_B + (\bar{N}_B - 1)].$$

In the equation, $\frac{\mu}{\lambda+\mu}$ is the probability that there is only one customer in the busy period; the first number one in the square brackets represents the first arrival in the busy period; the "-1" in the brackets is due to the fact that the starting point of the second busy period should not be counted as an arrival. Thus,

$$\bar{N}_B = \frac{\mu}{\mu-\lambda}.$$

2.3.2 Queuing Networks

A queuing network is a system consisting of a number of service stations. Customers in a network may belong to different classes, meaning that they may have different routing mechanisms and different service time distributions. A queuing network may belong to one of the three types: open, closed, or mixed. In an open network, customers arrive at the network from outside and eventally leave the network; in a closed network, customers circulate among stations and no customer arrives or leaves the network; A mixed network is open for some classes of customers and is closed for others.

Jackson Networks

We consider an open network consisting of M single-server stations and N single-class customers. Each server has an buffer with an infinite capacity and the service discipline is FCFS. Customers arrive at server i in a Poisson process with a rate $\lambda_{0,i}$, $i = 1, 2, \cdots, M$. After receiving the service at server i, a customer enters server j with probability $q_{i,j}$ and leaves the network with probability $q_{i,0}$. We have $\sum_{j=0}^{M} q_{i,j} = 1$, $i = 1, 2, , \cdots, M$. The service time of server i is exponentially distributed with mean $\bar{s}_i = 1/\mu_i$, $i = 1, 2, \cdots, M$. Such a network is called an (open) *Jackson network* (Jackson [71]).

The system state is $\mathbf{n} = (n_1, n_2, \cdots, n_M)$, where n_i is the number of customers in server i. Let λ_i be the arrival rate of the customers to server i. Then

$$\lambda_i = \lambda_{0,i} + \sum_{j=1}^{M} \lambda_j q_{j,i} \qquad i = 1, 2, \cdots, M.$$

It is known that in an acyclic open Jackson network, the arrival process to each server is a Poisson process. However, if there are feedback loops in the network, the arrival process to a server is generally not a Poisson process. Thus, each server behaves differently from an M/M/1 queue. Nevertheless, the steady-state distribution, $p(\mathbf{n})$, of the network looks the same as if each server is an M/M/1 queue. In fact, we have

$$p(\mathbf{n}) = p(n_1, n_2, \cdots, n_M) = \prod_{k=1}^{M} p(n_k) \tag{2.40}$$

with

$$p(n_k) = (1 - \rho_k)\rho_k^{n_k} \qquad \rho_k = \frac{\lambda_k}{\mu_k} \qquad k = 1, 2, \cdots, M.$$

This shows that in an open Jackson network the steady-state distributions of the numbers of the customers in the servers are independent of each other.

Closed Jackson (Gordon-Newell) Networks

A load-independent closed Jackson network is similar to the open Jackson network described above, except that there are N customers circulating among servers according to the routing probabilities $q_{i,j}$, $\sum_{k=1}^{M} q_{i,k} = 1$, $i = 1, 2, \cdots, M$. We have $\sum_{k=1}^{M} n_k = N$. In a load-dependent closed Jackson network, a customer requires a certain amount of service from a server. We assume that the service requirement is exponentially distributed with a mean equal to one. The service rate of a server may depend on the load of the server, i.e., the number of customers in the server. Let μ_{i,n_i} be the service rate of server i when there are n_i customers in the server, $0 \leq \mu_{i,n_i} < \infty$, $n_i = 1, 2, \cdots, N$, $i = 1, 2, \cdots, M$. Note that the assumption of the mean service requirement being one does not lose generality, since a service requirement with mean $\bar{r} > 0$ and service rate μ_{i,n_i} is equivalent to a service requirement with mean one and a service rate $\mu_{i,n_i}/\bar{r}$.

In a load-independent network, $\mu_{i,n_i} \equiv \mu_i$ for n_i, $i = 1, 2, \cdots, M$. In this case, the service time at server i is exponentially distributed with mean $\bar{s}_i = 1/\mu_i$, $i = 1, 2, \cdots, M$.

The Product-Form Solution

The state of such a system is $\mathbf{n} = (n_1, n_2, \cdots, n_M)$. Let Φ denote the state space; it contains $\binom{N + M - 1}{M - 1}$ states. Let $p(\mathbf{n})$ be the steady-state probability of state $\mathbf{n}$. We use $\mathbf{n}_{i,j} = (n_1, \cdots, n_i - 1, \cdots, n_j + 1, \cdots, n_M)$, $n_i > 0$, to denote a neighboring state of $\mathbf{n}$. Let

$$\epsilon(n_k) = \begin{cases} 1 & if\ n_k > 0, \\ 0 & if\ n_k = 0, \end{cases}$$

and let

$$\mu(\mathbf{n}) = \sum_{k=1}^{M} \epsilon(n_k)\mu_{k,n_k}. \tag{2.41}$$

Then the flow balance equation for $p(\mathbf{n})$ is

$$\mu(\mathbf{n})p(\mathbf{n}) = \sum_{i=1}^{M}\sum_{j=1}^{M} \epsilon(n_j)\mu_{i,n_i+1}q_{i,j}p(\mathbf{n}_{j,i}). \tag{2.42}$$

Gordon and Newell [64] derived a solution to the above equations. Let $y_i > 0$, $i = 1, 2, \cdots, M$, be the *visit ratio* to server i, i.e., a solution (within a multiplicative constant) to the equation

$$y_i = \sum_{j=1}^{M} q_{j,i} y_j, \qquad j = 1, 2, \cdots, M. \tag{2.43}$$

The solution to the above equations is not unique. In fact, if $(y_1, y_2, \cdots, y_M)$ is a set of visit ratios, then for any $\alpha > 0$, $(\alpha y_1, \alpha y_2, \cdots, \alpha y_M)$ is also a set of visit ratios. Let $A_i(0) = 1$, $i = 1, 2, \cdots, M$, and

$$A_i(k) = \prod_{j=1}^{k} \mu_{i,j}, \qquad i = 1, 2, \cdots, M,$$

and for every $n = 1, 2, \cdots, N$ and $m = 1, 2, \cdots, M$, let

$$G_m(n) = \sum_{n_1 + \cdots + n_m = n} \prod_{i=1}^{m} \frac{y_i^{n_i}}{A_i(n_i)}.$$

Then we have

$$p(\mathbf{n}) = \frac{1}{G_M(N)} \prod_{i=1}^{M} \frac{y_i^{n_i}}{A_i(n_i)}. \tag{2.44}$$

Equation (2.44) is often called a *product-form* solution.

For load-independent networks, $\mu_{i,n_i} \equiv \mu_i$, $i = 1, 2, \cdots, M$. The product-form solution becomes

$$G_m(n) = \sum_{n_1 + \cdots + n_m = n} \prod_{i=1}^{m} x_i^{n_i},$$

and

$$p(\mathbf{n}) = \frac{1}{G_M(N)} \prod_{i=1}^{M} x_i^{n_i}, \tag{2.45}$$

where $x_i = y_i / \mu_i = y_i \bar{s}_i$, $i = 1, 2, \cdots, M$.

Buzen's Algorithm

Buzen [14] developed an efficient algorithm for calculating $G_M(N)$. The algorithm is stated below and will be referred to as Buzen's algorithm.

$$G_m(n) = \sum_{k=0}^{n} \frac{y_m^k}{A_m(k)} G_{m-1}(n-k), \tag{2.46}$$

with

$$G_m(0) = 1, \qquad m = 1, 2, \cdots, M,$$

and

$$G_1(n) = (y_1)^n / A_1(n), \qquad n = 0, 1, \cdots, N.$$

The marginal distribution of the number of customers at server M is

$$p(n_M = k) = \frac{y_M^k}{A_M(k)} \frac{G_{M-1}(N-k)}{G_M(N)}. \tag{2.47}$$

For load-independent networks, Buzen's algorithm is

$$G_m(n) = G_{m-1}(n) + x_m G_m(n-1), \tag{2.48}$$

with

$$G_m(0) = 1, \qquad m = 1, 2, \cdots, M,$$

and

$$G_1(n) = (x_1)^n, \qquad n = 0, 1, \cdots, N.$$

Equation (2.47) becomes

$$p(n_i = k) = \frac{x_i^k}{G_M(N)} \{G_M(N-k) - x_i G_M(N-k-1)\}. \tag{2.49}$$

This equation holds for any $i = 1, 2, \cdots, M$. From this, we can obtain the mean queuing length of server i in an N-customer network, $\bar{n}_i(N)$:

$$\bar{n}_i(N) := E(n_i) = \sum_{k=1}^{N} x_i^k \frac{G_M(N-k)}{G_M(N)}, \tag{2.50}$$

where E denotes the expectation in steady state.

Besides the computational algorithms, some closed-form expressions have been found for the normalizing constant $G_M(N)$. For example, Gordon [63] derived a formula for $G_M(N)$ for load-independent closed Jackson networks.

Sensitivity Formulas

Let f be a performance function $f : \Phi \to R$. The steady-state mean of f is

$$E(f) = \sum_{\mathbf{n}} f(\mathbf{n}) p(\mathbf{n})$$

The elasticities of $E(f)$ with respect to $\mu_{i,k}$, $i=1,2,...,M$, $k=1,2,...,N$, can be obtained by taking the derivatives of the product-form formula. We have (see, e.g., Liu and Nain [80] and Cao and Ma [35])

$$\frac{\mu_{i,k}}{E(f)}\frac{\partial E(f)}{\partial \mu_{i,k}} = p_N(n_i \geq k) - \frac{E[f\chi(n_i \geq k)]}{E(f)} \tag{2.51}$$

where $\chi(n_i \geq k) = 1$ if $n_i \geq k$, and 0 otherwise; $p_N(n_i \geq k)$ is the steady-state probability of $n_i \geq k$ in a network with N customers.

The system throughput is defined as

$$\eta := \sum_{k=1}^{M} \mu(\mathbf{n})p(\mathbf{n}). \tag{2.52}$$

We have

$$\eta = \frac{G_M(N-1)}{G_M(N)}(\sum_{i=1}^{M} y_i).$$

Taking the derivatives, we have

$$\frac{\mu_{i,k}}{\eta}\frac{\partial \eta}{\partial \mu_{i,k}} = p_N(n_i \geq k) - p_{N-1}(n_i \geq k),$$

For load-independent networks, we have

$$\frac{\mu_i}{\eta}\frac{\partial \eta}{\partial \mu_i} = \sum_{k=1}^{N} \frac{\mu_{i,k}}{\eta}\frac{\partial \eta}{\partial \mu_{i,k}} = \bar{n}_i(N) - \bar{n}_i(N-1), \tag{2.53}$$

where $\bar{n}_i(N)$ is the steady-state mean of the number of customers in server i.

Other sensitivity formulas obtained from the product-form solution can be found in, for example, Williams and Bhandiwad [105], Suri [100], Tay and Suri [103], Liu and Nain [80], and Cao and Ma [35].

Another topic that is closely related to sensitivity analysis is the approximation of a performance curve. A notable result in this area is the Pade approximation method by Gong et al. [62].

Computational Algorithms for $E(f)$

In general, to calculate $E(f)$ one has to calculate $p(\mathbf{n})$ for every state $\mathbf{n}$. This is not efficient for large networks. However, for a large class of performance functions, algorithms based on Buzen's algorithm can be developed, and the computational effort can be significantly reduced (Cao and Ma [35]).

Suppose the performance function takes the following form:

$$f(\mathbf{n}) = f(n_1, ..., n_M) = \sum_{k=1}^{K} f_k(n_1, ..., n_M), \qquad K \geq 1,$$

where

$$f_k(n_1, ..., n_M) = \prod_{i=1}^{M} g_{k,i}(n_i) \qquad k = 1, 2, ..., K.$$

For each $i = 1, 2, ..., M$ and $k = 1, 2, ..., K$, $g_{k,i}$ is a function on $\{0, 1, ..., N\}$. We have

$$\begin{aligned} E(f) &= \frac{1}{G_M(N)} \sum_{k=1}^{K} \sum_{n_1+\cdot+n_M=N} \{f_k(n_1, ..., n_M) \prod_{i=1}^{M} \frac{y_i^{n_i}}{A_i(n_i)}\} \\ &= \frac{1}{G_M(N)} \sum_{k=1}^{K} \sum_{n_1+\cdot+n_M=N} \prod_{i=1}^{M} \frac{g_{k,i}(n_i) y_i^{n_i}}{A_i(n_i)}. \end{aligned} \tag{2.54}$$

For each $k = 1, 2, ..., K$, $m = 1, 2, ..., M$, and $n = 1, 2, ..., N$, define

$$G_{k,m}(n) := \sum_{n_1+...+n_m=n} \prod_{i=1}^{m} \frac{g_{k,i}(n_i) X_i^{n_i}}{A_i(n_i)}. \tag{2.55}$$

For each k, the $G_{k,m}(n)$'s can be recursively computed via the recursion

$$G_{k,m}(n) = \sum_{j=0}^{n} \frac{g_{k,m}(j) y_m^j}{A_m(j)} G_{k,m-1}(n-j), \tag{2.56}$$

with

$$G_{k,1}(n) = \frac{g_{k,1}(n) y_1^n}{A_1(n)}, \qquad n = 1, 2, \cdots, N,$$

and

$$G_{k,m}(0) = \prod_{i=1}^{m} g_{k,i}(0), \qquad m = 1, 2, \cdots, M.$$

This is exactly the same as Buzen's algorithm for $G_m(n)$, except with the additional weighting factor $g_{k,m}(j)$. Finally, we have

$$E(f) = \frac{1}{G_M(N)} \sum_{k=1}^{K} G_{k,M}(N). \tag{2.57}$$

Using this algorithm, the computational effort for calculating $E(f)$ reduces to that of applying Buzen's algorithm $K+1$ times (for $G_M(N)$ and $G_{k,M}(N)$, $k = 1, 2, \cdots, M$, respectively).

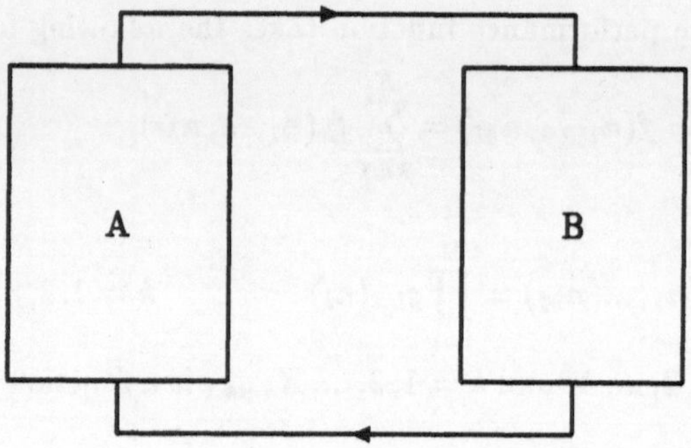

Figure 2.9: A Closed Jackson Network with Two Subsets

BCMP Networks

Another important class of networks whose steady-state distribution possesses a product-form is the *BCMP* network. In a BCMP network, there are M service stations and K classes of customers. While moving from stations to stations, customers may change their classes. The customers arrive to the network in Poisson processes. The service discipline may be FCFS, PS, LCFS preemptive-resume, and IS (a service station that has an infinite number of servers). The service time distribution for any PS, IS, and LCFS station may be any Coxian distributions, and that for a FCFS station must be exponential with the same mean for all classes of customers. It is shown in (Basket et. al [4]) that the stationary distribution of a BCMP network has a product form.

Aggregation: Norton's Theorem

For networks with product-form stationary distributions, there is a result similar to Norton's theorem for electrical circuits (Chandy et al. [37] and Balsamo and Iazeolla [3]). The result states that under some conditions, a subset of a network can be replaced by a sinlge server without affecting the stationary marginal distribution of the rest part of the network; the service rate of the single server can be determined by the thoughput of a network obtained by "shorting" (i.e., setting the service time equal zero) the complement of the subset in the original network.

We explain the theorem by using Figures 2.9-2.11. The network shown

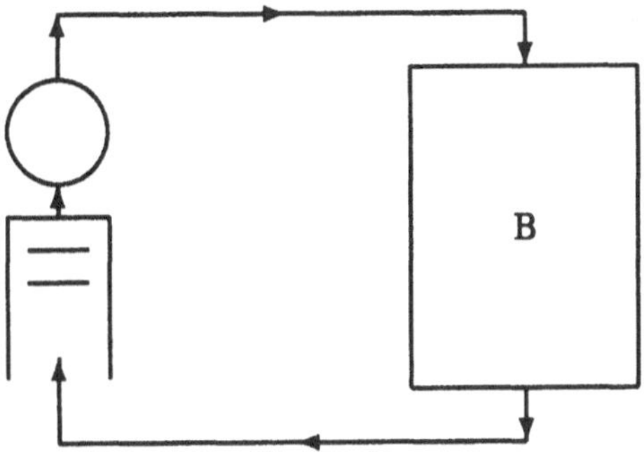

Figure 2.10: The Equivalent Network

in the figure consists of two subnetworks, A and B. The customer leaving subnetwork A goes to B, and vise versa. Using Norton's theorem, we can replace subnetwork A by a load-dependent server, and in the resulted network, shown in Figure 2.10, the stationary marginal distribution for subnetwork B is the same as that in the original network. The service rate μ_k of the equivalent server with k customers in it equals the throughput on the shorted path in the network that consists of k customers, shown in Figure 2.11.

2.4 Linear Differential Equations

The state of a continuous linear system is obtained by solving the governing linear different equations. Thus, the theory for linear differential equations forms the fundation for the linear systems theory (see, e.g, Szidarovszky and Bahill [102]).

Consider the linear ordinary differential equation

$$\dot{x} = A(t)x + f(t), \tag{2.58}$$

where $A(t)$ is an $n \times n$ matriz, $f(t)$ is an n-dimensional vector. We assume that all the elements in $A(t)$ and $f(t)$ are continuous functions of t on a closed interval $\mathcal{I} \in R$.

Since all the elements in $A(t)$ and $f(t)$ are continuous, they are bounded

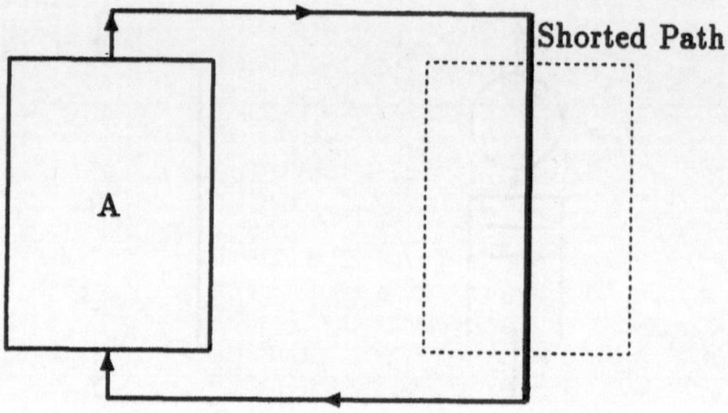

Figure 2.11: The Network with One Subset Shorted

in the closed interval. Thus, the *Lipschitz condition* holds:

$$||[A(t)x_2 + f(t)] - [A(t)x_1(t) + f(t)]|| \leq K||x_2 - x_1|| \qquad K < \infty$$

for all $t \in \mathcal{I}$ and $x_1, x_2 \in R^n$, where $||\cdot||$ denotes the norm in R^n. Therefore, Equation (2.58) with intitial condition $x(t_0) = x_0$, $t_0 \in \mathcal{I}$ and $x_0 \in R$ has a unique solution.

To solve (2.58), we first consider the corresponding homogeneous equation:

$$\dot{x} = A(t)x. \tag{2.59}$$

Let $x_k(t)$ be the solution of (2.59) with initial state $x(t_0) = e_k$, respectively, with e_k being the kth basis vector whose kth element equals one and all other elements equal zero. Let $\phi(t, t_0)$ be the matrix whose kth column is $x_k(t)$; i.e.,

$$\phi(t, t_0) := [x_1(t), x_2(t), \cdots, x_n(t)].$$

$\phi(t, t_0)$ is called the *state-transition matrix* of Equation (2.58) or (2.59).

If $A(t)$ is a constant matrix; i.e., $A(t) \equiv A$, then

$$\phi(t, t_0) = e^{A(t-t_0)}.$$

The general solution for (2.59) is

$$x(t) = \phi(t, t_0)d, \tag{2.60}$$

where $d = (d_1, d_2, \cdots, d_n)^T$ is a constant vector. This can be simply verified by substituting (2.60) into (2.59).

Using (2.59), we can prove that the transition matrix satisfies the following properties:

$$\phi(t_0, t_0) = I, \tag{2.61}$$

where I is the identity matrix;

$$\phi(t, t_1)\phi(t_1, t_0) = \phi(t, t_0); \tag{2.62}$$

$$\phi(t_1, t_0)^{-1} = \phi(t_0, t_1); \tag{2.63}$$

$$\frac{\partial}{\partial t}\phi(t, t_0) = A(t)\phi(t, t_0); \tag{2.64}$$

and

$$\frac{\partial}{\partial t}\phi(t_0, t) = -\phi(t_0, t)A(t); \tag{2.65}$$

To solve the inhomogeneous equation (2.58), we guess a sulotion of the form

$$x(t) = \phi(t, t_0)y(t). \tag{2.66}$$

Substituting this into (2.58), we get

$$[\frac{\partial}{\partial t}\phi(t, t_0)]y(t) + \phi(t, t_0)\dot{y}(t) = A(t)\phi(t, t_0)y(t) + f(t).$$

Using (2.63) and (2.64), we have

$$\dot{y}(t) = \phi(t_0, t)f(t).$$

Solving this equation yields

$$y(t) = \int_{t_0}^{t} \phi(t_0, s)f(s)ds + d,$$

with d being a constant vector. Finally, from (2.66), the solution for (2.58) with initial condition $x(t_0) = x_0$ is

$$x(t) = \phi(t, t_0)x_0 + \int_{t_0}^{t} \phi(t, s)f(s)ds. \tag{2.67}$$

Using (2.80), we can [illegible] the transition matrix satisfies the following equation:

[illegible]

where I is the identity matrix.

[illegible]

[illegible]

[illegible]

and

[illegible]

To solve the nonhomogeneous equation (2.84), we assume a solution of the form

[illegible]

Substituting this into (2.84), we get

[illegible]

Using (2.83) and (2.84), we have

[illegible]

[illegible]

[illegible] which is a constant vector. From (2.80), the solution of (2.84) with initial condition [illegible] is

[illegible]

Chapter 3

Realization Probabilities in Closed Jackson Networks

In this chapter, we shall apply our approach to study the closed Jackson networks with load-independent service rates. We shall formally define realization probabilities and use this concept to derive formulas for the sensitivities of throughput in several different cases. Section 3.1 contains the preliminary material for the study. In Section 3.2, we define realization probabilities and derive a set of linear equations that determine the realization probabilities. In Section 3.3, we define sample sensitivities and show that the sample elasticity of the system throughput with respect to a mean service time can be expressed in terms of realization probabilities. In Section 3.4, we study the convergence property of the sample sensitivities and prove that the elasticity of the steady-state throughput simply equals the negative expected value of the realization probability. In Section 3.5, we show that realization probability can be used to derive formulas for other sensitivities. The main references of this chapter are Cao [20], [23], [28], [29], and [31].

3.1 The Markov Process Model

The closed Jackson network and its product-form solution were reviewed in Section 2.3. In this section, we study the state processes of these networks.

The results of this section hold for closed networks with exponentially distributed service requirements and state dependent service rates; i.e., the service rate of a server may depend on the numbers of customers in other servers in the network. We define

$$\mu(\mathbf{n}) = \sum_{k=1}^{M} \epsilon(n_k)\mu_{k,\mathbf{n}}, \tag{3.1}$$

where $\mu_{i,\mathbf{n}}$ is the service rate of server i when the system state is $\mathbf{n}$. Equation (3.1) is similar to (2.41).

The evolution of the state of such a network can be described by a right-continuous pure-jump Markov process, denoted as $\mathcal{N}(t)$. Let $\{T_l\}_{l=0}^{\infty}$ be the transition times of the network. Then T_l, $l = 0, 1, \cdots$, are stopping times of $\mathcal{N}(t)$. (When $q_{i,i} \neq 0$ for some i, we consider that the Markov process $\mathcal{N}(t)$ can jump from a state to itself, corresponding to a customer feedback at server i.) Define $X_l := \mathcal{N}(T_l)$. Then $\mathbf{X} := \{X_l\}_{l=0}^{\infty}$ is a Markov chain embedded in $\mathcal{N}(t)$.

To study the property of the Markov process $\mathcal{N}(t)$ and its embedded chain $\mathbf{X}$, we need to make some mild restrictions on the system topology. First, we assume that all the networks discussed in this book are irreducible networks, which are defined below. Let $Q = [q_{i,j}]$ be the *routing probability matrix*. Q is a Markov matrix. Q is said to be *decomposable* (see Pullman [91]), if by reordering the indices of servers, Q can be rewritten in the following form:

$$Q = \begin{bmatrix} Q_{1,1} & O \\ Q_{2,1} & Q_{2,2} \end{bmatrix}, \tag{3.2}$$

where $Q_{1,1}$ is an $m \times m$ matrix, $0 < m < M$, and O is a matrix whose elements are all zeros (we denote this fact as $O = 0$). Otherwise, Q is said to be *indecomposable*. A closed queuing network is called an *irreducible network* if its routing probability matrix is indecomposable.

Let $\Gamma = \{1, 2, \cdots, M\}$ and Γ_1 and Γ_2 be two subsets of Γ corresponding to the matrices $Q_{1,1}$ and $Q_{2,2}$ in a decomposable queuing network. If $Q_{2,1} = 0$, then the customers in servers in Γ_1 (or Γ_2) can never visit the servers in Γ_2 (or Γ_1). If $Q_{2,1} \neq 0$, then the customers in Γ_2 may visit the servers in Γ_1 with a positive probability, but they can never get back to Γ_2 after entering Γ_1. Thus, in equilibrium the servers in Γ_2 will contain no customers. Therefore, the steady-state property of a reducible network is equivalent to the steady-state property of two (or more) irreducible subnetworks (if $Q_{2,1} = 0$) or one irreducible subnetwork (if $Q_{2,1} \neq 0$). On the other hand, a customer in any server of an irreducible network has a positive probability of entering any other server in the network, either directly or through some other servers.

Lemma 3.1 *The state process $\mathcal{N}(t)$ of an irreducible queuing network is an irreducible Markov process.*

Proof: The probability that server i completes the service to its current customer earlier than all the other servers is $\{\epsilon(n_i)\mu_{i,\mathbf{n}}\}/\mu(\mathbf{n})$. From this, we have

$$P(X_{l+1} = \mathbf{n}_{i,j} | X_l = \mathbf{n}) = \frac{\epsilon(n_i)\mu_{i,\mathbf{n}}}{\mu(\mathbf{n})} q_{i,j} \qquad \textit{for all } l.$$

Thus, if $q_{i,j} > 0$, then the probability that the state process goes from $\mathbf{n}$ with $n_i > 0$ to $\mathbf{n}_{i,j}$ is positive. Next, suppose $q_{i,j} = 0$. Because of the irreducibility of Q, there is at least one path, denoted as $i \rightarrow j_1 \rightarrow j_2 \rightarrow \cdots \rightarrow j_m \rightarrow j$ such that $q_{i,j_1} q_{j_1,j_2} \cdots q_{j_m,j} > 0$. Thus, the probability that the process goes from $\mathbf{n}$ to $\mathbf{n}_{i,j}$ through customer transitions following $i \rightarrow j_1 \rightarrow \cdots j_m \rightarrow j$ is positive. That is, one can always move one customer from a server to any other server with a positive probability. This indicates that all the states of this process can be reached from each other with a positive probability. □

Because the number of states is finite, all the states of $\mathcal{N}(t)$ are recurrent nonnull (see Section 2.2 or Çinlar [41]).

Next, we study the ergodicity of the Markov chain $\mathbf{X}$. The ergodicity will be used in proving the convergence properties related to realization probabilities. We first note that for some special networks, $\mathbf{X}$'s are periodic. A simple example of such networks is a cyclic network consisting of two tandem nodes and one customer. (If $X_0 = (0,1)$, then $X_l = (0,1)$ for an even integer l, and $X_l = (1,0)$ for an odd integer l.) However, if $q_{i,i} \neq 0$ for some i, then the network is aperiodic. This can be seen as follows: Any state $\mathbf{n}$ with $n_i > 0$ is certainly aperiodic since the probability that the system jumps from state $\mathbf{n}$ to $\mathbf{n}$ is positive. Next, all states of $\mathbf{X}$ are aperiodic since $\mathbf{X}$ is irreducible (see Section 2.2 or Çinlar [41]).

Now suppose that $q_{i,i} = 0$ for all i and the Markov chain $\mathbf{X}$ is periodic. We construct an auxiliary network by replacing any server, say server k, by an equivalent server. (This is in the same spirit as adding a feedback to a state in a Markov chain to form an equivalent state, see Figure 2.4.) The equivalent server has mean service rates $\mu'_{k,\mathbf{n}} = 2\mu_{k,\mathbf{n}}$ for all $\mathbf{n}$, and the routing probabilities are $q'_{k,k} = 1/2$, and $q'_{k,j} = q_{k,j}/2$, $j \neq k$. After the completion of the service in the equivalent server, a customer has a probability of 1/2 of being fed back to the same server. On the average, a customer in the equivalent server will visit the server $1\times(1/2)+2\times(1/2)^2+\cdots=2$ times before it leaves the server. According to the FCFS discipline, the fed-back customer would be put at the end of the queue in the server. However, since all customers are identical, we can exchange the position of the fed-back customer with that of the first customer in the queue; thus the fed-back customer can be considered as receiving successive services from the server. The average number of the services a customer receives from this equivalent server is 2. It is easy to prove that the total time that a customer stays in the equivalent server when the state is $\mathbf{n}$ is exponentially distributed with mean $2/\mu'_{k,\mathbf{n}} = 1/\mu_{k,\mathbf{n}}$. On the other hand, the conditional routing probability given that the customer leaves the equivalent server is $q'_{k,j}/(1/2) = q_{k,j}$, $j \neq k$. Thus, a server with mean service rates $2\mu_{k,\mathbf{n}}$, $\mathbf{n} \in \Phi$, and routing probabilities $q'_{k,j}$ for $j = 1, 2, \cdots, M$ is equivalent to a server with mean rates $\mu_{k,\mathbf{n}}$, $\mathbf{n} \in \Phi$, and routing probabilities $q_{k,j}$, $j = 1, 2, \cdots, M$. The embedded Markov chain of the auxiliary network is aperiodic. The state processes of the original and the auxiliary networks are the same (if one ignores the feedback transitions that do not change the system state). Therefore, by studying the auxiliary network, if necessary, we can assume that $\mathbf{X}$ is aperiodic. Thus, the embedded Markov chain $\mathbf{X}$ is asymptotically stationary and ergodic (see Section 2.2 or Çinlar [41]).

Let $\pi(\mathbf{n})$ be the steady-state probability of the embedded Markov chain $\mathbf{X}$. We have the following Lemma.

Lemma 3.2

$$p(\mathbf{n}) = \eta \frac{\pi(\mathbf{n})}{\mu(\mathbf{n})}, \tag{3.3}$$

where

$$\eta = \sum_{all\ \mathbf{n}} \mu(\mathbf{n}) p(\mathbf{n}) = \frac{1}{\sum_{all\ \mathbf{n}} \frac{\pi(\mathbf{n})}{\mu(\mathbf{n})}} \tag{3.4}$$

is the system throughput.

Proof: The theorem is a direct consequence of (2.27) and (2.28). Here, we offer an alternative proof.

The balance equations for $p(\mathbf{n})$ are

$$\mu(\mathbf{n})p(\mathbf{n}) = \sum_{i=1}^{M}\sum_{j=1}^{M} \epsilon(n_j)\mu_{i,\mathbf{n}_{j,i}} q_{i,j} p(\mathbf{n}_{j,i}). \tag{3.5}$$

On the other hand, $\pi(\mathbf{n})$ satisfies

$$\pi(\mathbf{n}) = \sum_{i=1}^{M}\sum_{j=1}^{M} \frac{\epsilon(n_j)\mu_{i,\mathbf{n}_{j,i}}q_{i,j}}{\mu(\mathbf{n}_{j,i})}\pi(\mathbf{n}_{j,i}). \tag{3.6}$$

Equation (3.5) can be rewritten as

$$\mu(\mathbf{n})p(\mathbf{n}) = \sum_{i=1}^{M}\sum_{j=1}^{M} \frac{\epsilon(n_j)\mu_{i,\mathbf{n}_{j,i}}q_{i,j}}{\mu(\mathbf{n}_{j,i})}\{\mu(\mathbf{n}_{j,i})p(\mathbf{n}_{j,i})\}. \tag{3.7}$$

Equation (3.7) is in the same form for $\mu(\mathbf{n})p(\mathbf{n})$ as (3.6) for $\pi(\mathbf{n})$. Since all the states of the Markov chain $\mathbf{X}$ are recurrent nonnull and aperiodic, we have (Corollary 2.11 of Çinlar [41]):

$$\eta\pi(\mathbf{n}) = p(\mathbf{n})\mu(\mathbf{n}),$$

where η is a normalizing constant which can be determined by $\sum_{all\ \mathbf{n}} \pi(\mathbf{n}) = 1$ and $\sum_{all\ \mathbf{n}} p(\mathbf{n}) = 1$. Summing up the above equation over all $\mathbf{n}$ and using the normalizing equations for both $\pi(\mathbf{n})$ and $p(\mathbf{n})$, we get

$$\eta = \sum_{all\ \mathbf{n}} \mu(\mathbf{n})p(\mathbf{n}) = \frac{1}{\sum_{all\ \mathbf{n}} \frac{\pi(\mathbf{n})}{\mu(\mathbf{n})}}.$$

This conclusion concludes the proof. □

3.2 Realization Probabilities

3.2.1 The Evolution of a Perturbation on a Sample Path

In this section, we investigate the evolution of a perturbation on a sample path of the network. First, let us define a sample path.

A Description of a Sample Path

Let $(\Omega, \mathcal{F}, P)$ be the probability space underlying the state process $\mathcal{N}(t)$. Any point $\omega \in \Omega$ corresponds to a sample path, or a realization, of $\mathcal{N}(t)$. Thus, a sample path of $\mathcal{N}(t)$ can be denoted as $\mathcal{N}(t, \omega)$. For any fixed t, $\mathcal{N}(t, \omega)$ is a random vector defined on Ω. For any ω, $\mathcal{N}(t, \omega)$ is a function of t, representing a sample path of the state process.

Using ω to represent a sample path does not offer enough intuitive explanation. It is always convenient to work on something that has some concrete meanings. To this end, we observe that the state process of a queuing network can be constructed by using a sequence of random numbers. Let us first consider customers' service times. Let $s_{i,k}$ be the service time of the kth customers that server i serves since the start of the sample path. According to the inverse transformation method for generating random variables (see Section 2.1 or Rubinstein [92] and Mitrani [87]), $s_{i,k}$ can be obtained by using the following transformation:

$$s_{i,k} = -\bar{s}_i ln(1 - \xi_{i,k}), \quad i = 1, 2, \cdots, M, \ k = 1, 2, \cdots, \tag{3.8}$$

where $\xi_{i,k}$ are independent random variables uniformly distributed over $[0, 1)$. Let $\xi_i = (\xi_{i,1}, \xi_{i,2}, \cdots)$, $i = 1, 2, \cdots, M$; then the random vector $(\xi_1, \xi_2, \cdots, \xi_M)$ determines, via Equation (3.8), all the service times of all the servers in the network.

Let $t_{i,k}$ be the service completion time of the kth customer of server i, $i = 1, 2, \cdots, M$ and $k = 1, 2, \cdots$. A sample path can be completely described by these $(t_{i,k})'s$ and the states between these transition times. These $(t_{i,k})'s$ can be determined by the following recursive formula.

$$t_{i,k} = \begin{cases} t_{i,k-1} + s_{i,k} & if \ n_i(t_{i,k-1}+) \neq 0, \\ t_{j,h} + s_{i,k} & if \ n_i(t_{i,k-1}+) = 0. \end{cases} \tag{3.9}$$

In the second part of the equation, we assume that an idle period of server i is terminated by the hth customer of server j (i.e., after the idle period, server i receives the hth customer from server j). Equation (3.9) simply says that if a customer starts a new busy period, then its service starting time is decided by its arrival time to the server $(t_{j,h})$; otherwise, it is decided by the service completion time of the previous customer. This equation shows that idle periods play a dominant role in determining the server-to-server interaction.

The customer routes can be determined by another random vector $(\zeta_1, \zeta_2, \cdots, \zeta_M)$, with $\zeta_i = (\zeta_{i,1}, \zeta_{i,2}, \cdots)$, $i = 1, 2, \cdots, M$, where $\zeta_{i,k}$ are independent random variables uniformly distributed on [0,1). The destination of the kth customer of server i is determined as follows: If $\sum_{l=1}^{j-1} q_{i,l} \leq \zeta_{i,k} < \sum_{l=1}^{j} q_{i,l}$ (with the convention $\sum_{l=1}^{0} q_{i,l} = 0$), then the kth customer of server i will go to server j after it leaves server i.

Let $\xi = (\xi_1, \xi_2, \cdots, \xi_M; \zeta_1, \zeta_2, \cdots, \zeta_M; \mathbf{n}_0)$, with $\mathbf{n}_0$ being the initial state of the system. A sample path of the system can be completely determined by a realization of ξ via Equations (3.8) and (3.9) and the routing rule described above. That is, ξ represents all the randomness of the network. Let Ξ be the space of ξ; i.e., $\Xi = [0,1)^\infty \times \Phi$, where Φ is the state space of the process. Then ξ is a measurable mapping from Ω to Ξ. Therefore, instead of using $\omega \in \Omega$, we can use $\xi \in \Xi$ to represent a sample path. Thus, a sample path can be written as $\mathcal{N}(t,\xi)$ (or simply denoted as ξ), with $\bar{s}_i$ and $q_{i,j}$, $i,j = 1,2,\cdots,M$, being the system parameters.

Recall that T_l, $l = 1,2,\cdots$, are the jump epochs of the Markov process $\mathcal{N}(t)$. Obviously, $\{T_l\}_{l=0}^{\infty} = \cup_{i=1}^{M} \cup_{k=1}^{\infty} \{t_{i,k}\}$. Let $S_l = T_{l+1} - T_l$ be the time period that the system remains in its lth state. The minimum value of all these periods in $[0, T_L)$ is

$$S_{L,min}(\xi) = min\{S_l : 0 \le l \le L\}.$$

For any finite L, we have

$$S_{L,min} > 0 \qquad w.p.1. \tag{3.10}$$

Perturbation Propagation

Now we introduce the notion of perturbation and perturbation propagation. We first define these concepts in an abstract manner; their practical implications will be discussed in the next sections. Let $\Gamma = \{1,2,\cdots,M\}$ and $\emptyset$ be the empty set. At any given time t, we partition the set Γ into two complementary subsets V and $\bar{V}$. If $i \in V$, then server i is said to have a *(real) perturbation*. Otherwise, it is said to have a *null* perturbation. We assume that sets V and $\bar{V}$ evolve along the process $\mathcal{N}(t)$ according to the following rules:

$$V(t) = \begin{cases} V(t-) + \{j\} & if\ \mathcal{N}(t-) = \mathbf{n},\ \mathcal{N}(t) = \mathbf{n}_{i,j},\ n_j = 0, \\ & i \in V(t-),\ j \in \bar{V}(t-), \\ V(t-) - \{j\} & if\ \mathcal{N}(t-) = \mathbf{n},\ \mathcal{N}(t) = \mathbf{n}_{i,j},\ n_j = 0, \\ & i \in \bar{V}(t-),\ j \in V(t-), \\ V(t-) & otherwise. \end{cases} \tag{3.11}$$

The set V is called a *perturbation set.* For any t, $V(t)$ is a mapping $t \to 2^\Gamma$, where 2^Γ is the power set of Γ. The above rules can be described verbally as the following propagation rules.

Perturbation propagation rules:

1. A server will keep its perturbation (real or null) until it meets an idle period (see the last equation in (3.11)).

2. If a customer from server i terminates an idle period of server j, then after this idle period server j will have the same perturbation (real or null) as server i (see the first two equations in (3.11)). In this case, we say that the perturbation of server i is propagated to server j.

Note that sets V and $\bar{V}$ change only at the transition instants T_l, $l = 1, 2, \cdots$.

Perturbation Realization

For convenience, in this section we consider a perturbation at time $t_0 = 0$. The perturbation set $V(t)$ depends on a particular sample path of process $\mathcal{N}(t, \xi)$; hence it is a random function of ξ. We denote it as $V(t, \xi)$ to indicate this explicitly.

Definition 3.1 *For any sample path ξ, the perturbation in $V(0, \xi)$ is said to be realized on the sample path, if $\lim_{l\to\infty} V(T_l, \xi) = \Gamma$; it is said to be lost, if $\lim_{l\to\infty} V(T_l, \xi) = \emptyset$.*

Let

$$\Xi_0 = \{\xi \in \Xi : \lim_{l\to\infty} V(T_l, \xi) = \emptyset\},$$

and

$$\Xi_1 = \{\xi \in \Xi : \lim_{l\to\infty} V(T_l, \xi) = \Gamma\}.$$

Note that sets $\emptyset$ and Γ are two "absorbing" sets with respect to the propagation rules; i.e., if $V(t_1, \xi) = \emptyset$ (or Γ) at some time t_1, then $V(t, \xi) = \emptyset$ (or Γ) for any $t > t_1$. This is because the propagation rule described in the third equation of (3.11) always applies after t_1. Thus, the limit in Definition 3.1 implies that if l is sufficiently large, then $V(T_l, \xi) = \Gamma$ or $\emptyset$.

There is something special to an idle server. The perturbation status of an idle server (i.e., whether it is in set V or set $\bar{V}$) after the idle period is completely determined by that of the server that terminates the idle period. Thus, an idle server can be considered as either in set V or in set $\bar{V}$; it will not affect the perturbation status of the network after the idle period.

In the sequel, we shall study the probabilities of $\xi \in \Xi_0$ and $\xi \in \Xi_1$, denoted as $P(\Xi_0)$ and $P(\Xi_1)$. First, let $P(\Xi_0|\mathbf{n}, V_0)$ and $P(\Xi_1|\mathbf{n}, V_0)$ be the conditional probabilities given the initial state $\mathcal{N}(0, \xi) = \mathbf{n}$ and the initial perturbation set $V(0, \xi) = V_0$. The following theorem follows directly from the ergodicity of the system.

Theorem 3.1 *In an irreducible closed queuing network,*

$$P(\Xi_0 \cup \Xi_1|\mathbf{n}, V_0) = P(\Xi_0|\mathbf{n}, V_0) + P(\Xi_1|\mathbf{n}, V_0) = 1. \qquad (3.12)$$

In words, a perturbation in an irreducible closed queuing network will, with probability one, be either realized or lost.

Proof: Since the state process $\mathcal{N}(t)$ is ergodic, starting from any state $\mathbf{n}$, the system will reach any other state with probability one. In particular, it will reach state $(N, 0, \cdots, 0)$ with probability one. If at this time, denoted as t, server 1 belongs to $V(t, \xi)$, then because all the other servers are idle, the perturbation is realized (from this time on, all the servers will be in set $V(t, \xi)$). On the other hand, if at this time server 1 is in $\bar{V}(t, \xi)$, then the perturbation is lost. In sum, a perturbation will be either realized or lost with probability one. □

Since (3.12) holds for all $\mathbf{n}$ and V_0, we have

$$P(\Xi_0 \cup \Xi_1) = P(\Xi_0) + P(\Xi_1) = 1. \qquad (3.13)$$

Equation (3.12) can also be writen is the following form:

$$\lim_{t \to \infty} V(t, \xi) = \{\emptyset \text{ or } \Gamma\} \qquad w.p.1,$$

or

$$\lim_{l \to \infty} V(T_l, \xi) = \{\emptyset \text{ or } \Gamma\} \qquad w.p.1.$$

Remark 3.1 The Markov chain $\mathbf{X} = \{X_l\}_{l=0}^{\infty}$ with a perturbation set $V(t)$ can be regarded as a new Markov chain with states $(\mathbf{n}, V)$. Theorem 3.1 implies that in this new chain with the augmented states there are two closed sets $\{(\mathbf{n}, \Gamma) : \textit{all } \mathbf{n}\}$ and $\{(\mathbf{n}, \emptyset) : \textit{all } \mathbf{n}\}$. All the states that do not belong to these two sets are transient.

3.2.2 Realization Probabilities

Realization Probabilities

Suppose that at time t_0 the system state is $\mathbf{n}$ and the perturbation set is V (the subscript 0 is dropped for simplicity). Theorem 3.1 enables us to make the following definition.

Definition 3.2 $c(\mathbf{n}, V) := P[\Xi_1|\mathcal{N}(t_0) = \mathbf{n}, V(t_0) = V]$ *is called the realization probability of the perturbation in set* V *at state* $\mathbf{n}$. *If* $V = \{i\}$, *it is simply denoted as* $c(\mathbf{n}, i)$.

The realization probability satisfies the properties stated in the following lemma.

Lemma 3.3

$$\textit{If } n_i = 0, \textit{ then } c(\mathbf{n}, i) = 0; \tag{3.14}$$

$$c(\mathbf{n}, \Gamma) = 1; \tag{3.15}$$

If $V_1 \cap V_2 = \emptyset$ *and* $V_1 \cup V_2 = V_3$, *then*

$$c(\mathbf{n}, V_1) + c(\mathbf{n}, V_2) = c(\mathbf{n}, V_3); \tag{3.16}$$

and

$$\sum_{k=1}^{M} c(\mathbf{n}, k) = 1. \tag{3.17}$$

Equation (3.14) involves a convention; i.e., we may consider an idle server as having a (real) perturbation. This perturbation will be lost immediately after the server receives a customer from any other server. Equation (3.15) follows directly from the "absorbing" property of set Γ. Equation (3.16) represents superposition. Equation (3.17) is a consequence of (3.15) and (3.16). Equation (3.16) is a direct consequence of Lemma 3.4 proved in the next subsection.

One direct corollary of Equation (3.16) is

$$c(\mathbf{n}, V) = \sum_{k \in V} c(\mathbf{n}, k). \tag{3.18}$$

The realization probabilities in a closed Jackson network satisfy a set of linear equations, which is stated in the following theorem.

Theorem 3.2 *For a closed Jackson network, the following equations hold:*

$$\{\sum_{i=1}^{M} \epsilon(n_i)\mu_i\} c(\mathbf{n}, k) = \sum_{i=1}^{M}\sum_{j=1}^{M} \epsilon(n_i)\mu_i q_{i,j} c(\mathbf{n}_{i,j}, k)$$

$$+ \sum_{j=1}^{M} \{1 - \epsilon(n_j)\} \mu_k q_{k,j} c(\mathbf{n}_{k,j}, j) \qquad n_k > 0,\ k \in \Gamma. \tag{3.19}$$

These equations are similar to the flow-balance equations. To prove this theorem, recall that a state $\mathbf{n}$ has the probability $[\epsilon(n_i)\mu_i / \sum_{j=1}^{M} \epsilon(n_j)\mu_j] q_{i,j}$ to transfer to state $\mathbf{n}_{i,j}$. The terms in the last summation in (3.19) are equal to zero except when $n_j = 0$. This term represents the perturbation propagation effect. The superposition property $c(\mathbf{n},\{k,j\}) = c(\mathbf{n},k) + c(\mathbf{n},j)$ is employed in the derivation of (3.19). Solving (3.19) together with (3.14) to (3.17), we can obtain the values of all $c(\mathbf{n},V)$.

Based on (3.18), the realization probabilities $c(\mathbf{n},V)$ for all $\mathbf{n}$ and V can be determined by $c(\mathbf{n},k)$ for all $\mathbf{n}$ and k. Thus, the total number of realization probabilties that we have to determine is $M \times \begin{pmatrix} N+M-1 \\ M-1 \end{pmatrix}$. In Chapter 4, we shall prove, in a more general setting, that Equations (3.14)-(3.17) and (3.19) provide a unique solution.

Example 3.1 Consider a closed Jackson network in which $\mu_i q_{i,j} = \mu_j q_{j,i}$. We can prove that the solution to Equations (3.14)-(3.17) and (3.19) is

$$c(\mathbf{n},V) = \frac{\sum_{k\in V} n_k}{N}, \qquad V \subseteq \Gamma. \tag{3.20}$$

Especially,

$$c(\mathbf{n},k) = \frac{n_k}{N} \qquad k \in \Gamma. \tag{3.21}$$

Obviously, (3.20) satisfies (3.14)-(3.17). To check that (3.20) also satisfies (3.19), we assume that in state $\mathbf{n}$, there are $0 \le M_0 < M$ idle servers. Without loss of generality, we name these servers as servers $1, 2, \cdots, M_0$ and consider $k > M_0$. Thus, $\epsilon(n_i) = 0$, if $i \le M_0$; $\epsilon(n_i) = 1$ if $i > M_0$. Now, the right-hand side of (3.19) is

$$\sum_{i=M_0+1,\ i\neq k}^{M} \sum_{j\neq k}^{M} \mu_i q_{i,j} c(\mathbf{n}_{i,j},k) + \mu_k q_{k,k} c(\mathbf{n},k) + \sum_{i=M_0+1,\ i\neq k}^{M} \mu_i q_{i,k} c(\mathbf{n}_{i,k},k)$$

$$+ \sum_{j\neq k} \mu_k q_{k,j} c(\mathbf{n}_{k,j},k) \ + \sum_{j=1}^{M_0} \mu_k q_{k,j} c(\mathbf{n}_{k,j},k)$$

$$\begin{aligned} = \quad & \sum_{i=M_0+1,\ i\neq k}^{M} \sum_{j\neq k}^{M} \mu_i q_{i,j} \frac{n_k}{N} + \mu_k q_{k,k} \frac{n_k}{N} + \sum_{i=M_0+1,\ i\neq k}^{M} \mu_i q_{i,k} \frac{n_k+1}{N} \\ & + \sum_{j\neq k} \mu_k q_{k,j} \frac{n_k-1}{N} + \sum_{j=1}^{M_0} \mu_k q_{k,j} \frac{1}{N} \\ = \quad & \sum_{i=M_0+1} \sum_{j=1}^{M} \mu_i q_{i,j} \frac{n_k}{N} + \{ \sum_{i=M_0+1,\ i\neq k}^{M} \mu_i q_{i,k} \frac{1}{N} \end{aligned}$$

$$-\sum_{j\neq k}^{M}\mu_k q_{k,j}\frac{1}{N}+\sum_{j=1}^{M_0}\mu_k q_{k,j}\frac{1}{N}\}$$

$$=\frac{n_k}{N}\{\sum_{i=M_0+1}^{M}\mu_i\}, \tag{3.22}$$

which equals the left-hand side of (3.19). By $\mu_k q_{k,j}=\mu_j q_{j,k}$, it is easy to check that the terms in the braces sum up to zero.

Note that in Equation (3.20), $c(\mathbf{n},k)$ is proportional to n_k and is independent of n_i, $i\neq k$.

Example 3.2 In this example, we show some numerical values of the realization probabilities. Consider a system with $M=3$, $N=5$, $\bar{s}_1=10$, $\bar{s}_2=8$, $\bar{s}_3=5$, and routing matrix

$$Q=\begin{bmatrix}0 & 0.5 & 0.5\\ 0.8 & 0 & 0.2\\ 0.3 & 0.7 & 0\end{bmatrix}.$$

The realization probabilities are obtained by solving Equations (3.14)-(3.19). The results are listed in Table 3.1.

Realization Indices

As shown in the previous section, the perturbation realization depends on a sample path (a realization of the Markov process $\mathcal{N}(t)$) of the network. The realization index is the sample-path version of the realization probability. Suppose that at the lth state of a sample path the perturbation set is V. On the sample path, we define:

Definition 3.3 *Let V be the perturbation set at the lth state of $\mathcal{N}(t,\xi)$ and $X_l=\mathbf{n}$. The realization index of this perturbation on the sample path $\mathcal{N}(t,\xi)$ is*

$$RI(\mathbf{n},V,l)=\begin{cases}1 & \textit{if the perturbation is realized along } \mathcal{N}(t,\xi),\\ 0 & \textit{otherwise.}\end{cases}$$

By the Markov property, given $X_l=\mathbf{n}$, the conditional expected value of $RI(V,\mathbf{n},l)$ does not depend on l. By definition,

$$E\{RI(\mathbf{n},V,l)|\mathbf{n}\}=c(\mathbf{n},V) \qquad \textit{for all } l. \tag{3.23}$$

Four properties similar to (3.14)-(3.17) hold for realization indices. In particular, the following property holds.

n	p(n)	c(n, 1)	c(n, 2)	c(n, 3)
(5,0,0)	0.19047	1.00000	0.00000	0.00000
(4,0,1)	0.06644	0.90584	0.00000	0.09416
(4,1,0)	0.15061	0.89385	0.10615	0.00000
(3,0,2)	0.02318	0.78826	0.00000	0.21174
(3,1,1)	0.05254	0.77060	0.17336	0.05604
(3,2,0)	0.11908	0.74279	0.25721	0.00000
(2,0,3)	0.00809	0.62556	0.00000	0.37444
(2,1,2)	0.01833	0.60901	0.25029	0.14070
(2,2,1)	0.04154	0.58286	0.37574	0.04141
(2,3,0)	0.09416	0.54528	0.45472	0.00000
(1,0,4)	0.00282	0.38089	0.00000	0.61911
(1,1,3)	0.00639	0.37327	0.34810	0.27863
(1,2,2)	0.01449	0.35728	0.51926	0.12346
(1,3,1)	0.03285	0.33315	0.62079	0.04606
(1,4,0)	0.07445	0.29754	0.70246	0.00000
(0,0,5)	0.00098	0.00000	0.00000	1.00000
(0,1,4)	0.00223	0.00000	0.48951	0.51049
(0,2,3)	0.00505	0.00000	0.71510	0.28490
(0,3,2)	0.01146	0.00000	0.83485	0.16515
(0,4,1)	0.02597	0.00000	0.91819	0.08181
(0,5,1)	0.05887	0.00000	1.00000	0.00000

Table 3.1: A Numerical Example of Realization Probabilities

Lemma 3.4 *If $V_1 \cap V_2 = \emptyset$ and $V_1 \cup V_2 = V_3$, then*

$$RI(\mathbf{n}, V_1, l) + RI(\mathbf{n}, V_2, l) = RI(\mathbf{n}, V_3, l) \qquad \text{for all } l. \tag{3.24}$$

Proof: To prove this equation, it is convenient to use an M-dimensional 0-1 vector to denote set V. The ith component of this vector is 1 if $i \in V$ and 0 otherwise. As an illustrative example, assume that at the lth state V_1, V_2, and V_3 correspond to the following three vectors, respectively,

$$g_1 = (0, 1, 0, 0, 1, 1, 0, 0, 0);$$

$$g_2 = (0, 0, 1, 0, 0, 0, 0, 1, 1);$$

$$g_3 = (0, 1, 1, 0, 1, 1, 0, 1, 1).$$

Note that $V_1 \cap V_2 = \emptyset$ and $V_1 \cup V_2 = V_3$ is equivalent to

$$g_{3,k} = g_{1,k} + g_{2,k} \qquad \text{for } 1 \leq k \leq M, \tag{3.25}$$

where $g_{i,k}$ is the kth component of g_i, $i = 1, 2, 3$. By propagation rule (3.11), if no server meets an idle period, then g_1, g_2, and g_3 remain the same. Also, if server j receives a customer from server i after an idle period, server j will possess the same perturbation as server i after this customer transition. This is simply copying the jth column of the above array to its ith column. Thus, Equation (3.25) always holds at any time after the lth state. In particular, as l goes to infinity, (3.25) still holds. However, by Theorem 3.1, g_1, g_2 and g_3 will converge to a vector whose components are either all zeros or all ones. From (3.25), only three cases may happen: (i) all g_1, g_2 and g_3 are vectors with all zeros; or (ii) g_1 and g_3 are vectors with all ones, and g_2 is a vector with all zeros; or (iii) g_2 and g_3 are vectors with all ones, and g_1 is a vector with all zeros. For all these cases, (3.24) holds.

Taking the conditional expectation on $\mathbf{n}$ of the both sides of (3.24) and using (3.23), we obtain (3.16). □

3.3 Sample Sensitivities

3.3.1 Sample Performance Functions and Their Derivatives

In Section 3.2.1, we denote a sample path of $\mathcal{N}(t)$ as $\mathcal{N}(t, \xi)$, where ξ is a random vector defined on the underlying probability space Ω. The system parameters are $\bar{s}_i$ and $q_{i,j}$, $i, j = 1, 2, \cdots, M$. In sensitivity analysis, we are interested in the derivative of a performance function with respect to a parameter or the gradient with respect to some parameters. Denote these parameters as θ (it may be a vector, but in most cases, we consider it a scalar). In this respect, a sample path can be represented by (θ, ξ). We use Ω_θ to denote the dependency of Ω on θ. The performance measured on a sample path (θ, ξ) is called a *sample performance*. For example, the time required for all the servers in the network to serve L customers can be denoted as $T_L(\theta, \xi)$, and the system throughput measured in $[0, T_L)$ on a sample path is $\eta_L(\theta, \xi) = L/T_L(\theta, \xi)$. For any fixed θ, $T_L(\theta, \xi)$ and $\eta_L(\theta, \xi)$ are random variables on Ω_θ; for any fixed ξ, $T_L(\theta, \xi)$ and $\eta_L(\theta, \xi)$ are functions of θ. Such functions with a fixed ξ are called *sample performance functions* (or simply called *sample functions*, if there is no confusion); their derivatives with respect to θ, the *sample derivatives*. Precisely, for $\eta_L(\theta, \xi)$, the sample derivative of $\eta_L(\theta, \xi)$ with respect to θ is defined as follows:

$$\frac{\partial \eta_L(\theta, \xi)}{\partial \theta} = \lim_{\Delta\theta \to 0} \frac{\eta_L(\theta + \Delta\theta, \xi) - \eta_L(\theta, \xi)}{\Delta\theta}. \tag{3.26}$$

Note that the same random vector ξ is used for both θ and $\theta + \Delta\theta$.

When θ changes, the underlying probability space changes. Let Ω_θ and $\Omega_{\theta'}$ be the two probability spaces corresponding to the two networks with parameters θ and θ', respectively. The two random sequences, ξ_θ and $\xi_{\theta'}$, that represent the randomness of the two networks take the same values in space Ξ. The relationship among Ω_θ, $\Omega_{\theta'}$, Ξ, and the sample performance functions $\eta_L(\theta, \xi)$, $\eta_L(\theta', \xi)$ is shown in Figure 3.1. Suppose that $\theta = \bar{s}_i$ is the parameter of interest. Then $\omega_\theta \in \Omega_\theta$ represents a sample path of a network with a mean service time θ for server i. In this case, the mapping ξ_θ consists of the following submappings:

$$(\xi_\theta)_{i,k} = 1 - e^{-s_{i,k}/\theta}.$$

It is worthwhile to note that the space Ξ is not unique. One may choose different random sequences to represent the randomness of a network, resulting in different Ξ's and different sample performance functions. Each $\xi_\theta : \Omega_\theta \to \Xi$ is called a *representation* of a network. Interested readers are referred to Glasserman [58] and Cao [26] for further discussions and examples.

3.3.2 Perturbations Due to Parameter Changes

So far, the discussion has been focused on the realization probabilities and the realization indices. The perturbation is abstractly introduced as a per-

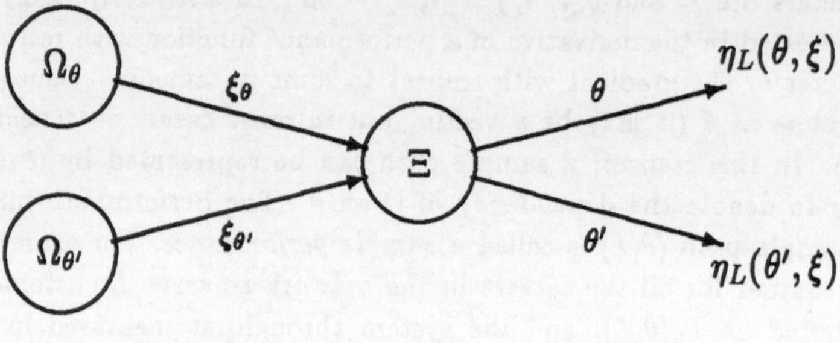

Figure 3.1: The Concept of a Sample Performance Function

turbation set V. It does not have any physical meaning yet. In this subsection and the next, we shall study the sample derivative of the system throughput with respect to a mean service time in a closed Jackson network.

Assume that the mean service time of server v changes from $\bar{s}_v$ to $\bar{s}'_v = \bar{s}_v + \Delta\bar{s}_v$, with $\Delta\bar{s}_v/\bar{s}_v << 1$. In this case, the parameter of interest is $\theta = \bar{s}_v$. The change in $\bar{s}_v$ will induce a series of perturbations at the service completion times of the customers in all the servers. In other words, T_l, $l = 1, 2, \cdots$, will change by a small amount ΔT_l because of this $\Delta\bar{s}_v$. The change ΔT_l is called a *perturbation* at time T_l. The system with mean service times $\bar{s}_v$ is called the *nominal system*; that with $\bar{s}_v + \Delta\bar{s}_v$, the *perturbed system*. Their corresponding sample paths are called the *nominal path* and the *perturbed path*, respectively. We use superscript "$'$" to denote the quantities associated with the perturbed path.

The service completion times $(t_{i,k})$'s can be determined by Equation (3.9). Let $t'_{i,k}$, $i = 1, 2, \cdots, M$, $k = 1, 2, \cdots$, be the service completion times of the perturbed system. We have

$$t'_{i,k} = \begin{cases} t'_{i,k-1} + s'_{i,k} & if\ n_i(t'_{i,k-1}+) \neq 0, \\ t'_{j,h} + s'_{i,k} & if\ n_i(t'_{i,k-1}+) = 0, \end{cases} \tag{3.27}$$

where

$$s'_{i,k} = \begin{cases} s_{i,k} & if\ i \neq v, \\ -(\bar{s}_v + \Delta\bar{s}_v)\{ln(1-\xi_{v.k})\} & if\ i = v. \end{cases} \tag{3.28}$$

Equation (3.27) is similar to Equation (3.9). Note that the same random vector ξ is used for both the nominal and the perturbed paths in Equations (3.8) and (3.28).

Taking the differences between both sides of Equations (3.9) and (3.27), we get

$$\Delta_{i,k} = \begin{cases} \Delta_{i,k-1} + \Delta s_{i,k} & if\ n_i(t_{i,k-1}+) \neq 0, \\ \Delta_{j,h} + \Delta s_{i,k} & if\ n_i(t_{i,k-1}+) = 0, \end{cases} \tag{3.29}$$

where $\Delta s_{i,k} = 0$ for all $i \neq v$ and

$$\begin{aligned} \Delta s_{v,k} &= s'_{v,k} - s_{v,k} = -\Delta \bar{s}_v \{ln(1 - \xi_{v.k})\} \\ &= \frac{\Delta \bar{s}_v}{\bar{s}_v} s_{v,k} = \gamma s_{v,k}, \end{aligned} \tag{3.30}$$

with $\gamma = \Delta \bar{s}_v / \bar{s}_v$. $\Delta s_{v,k}$ is the perturbation *generated* in the service time period of the kth customer of server v, and $\Delta_{i,k} = t'_{i,k} - t_{i,k}$ is the perturbation of the service completion time $t_{i,k}$, or that of server i at time $t_{i,k}$.

It is important to note that Equation (3.29) holds only if the change $\Delta \bar{s}_v$ does not cause any changes in the order of the occurrences of the service completion times; i.e., the sequence of the state $\{\mathcal{N}(T_0, \xi), \mathcal{N}(T_1, \xi), \cdots, \mathcal{N}(T_L, \xi)\}$ is the same for both the nominal and the perturbed paths. Otherwise, because of the changes in $\bar{s}_v$, it is possible that for some (i, k), the first equation of (3.9) holds for $t_{i,k}$, but the second equation (instead of the first one) of (3.27) holds for $t'_{i,k}$. In this case, taking the difference of (3.9) and (3.27) does not result in (3.29).

To precisely specify the situation discussed above, we introduce the following definition.

Definition 3.4 *Two sample paths in $[0, T_L)$ and $[0, T'_L)$ are said to be similar if the two sequences $\{X_l\}_{l=0}^{L}$ and $\{X'_l\}_{l=0}^{L}$ are the same.*

Thus, if the nominal and the perturbed paths are similar, then (3.29) holds. Lemma 3.5 below ensures us that we can apply (3.29) with a small $\Delta \bar{s}_v$.

Lemma 3.5 *For any sample path ξ of a closed Jackson network and any $L < \infty$, one can always choose a $\Delta \bar{s}_v$ (the size of which depends on ξ) which is small enough so that the nominal and the perturbed paths are similar.*

Proof: Assume $\Delta \bar{s}_v > 0$ (the case $\Delta \bar{s}_v < 0$ is similar). From (3.29) and (3.30), we can prove that if $t_{i,k} \leq T_L$, then

$$\Delta_{i,k} \leq \frac{\Delta \bar{s}_v}{\bar{s}_v} T_L. \tag{3.31}$$

Thus, if we choose

$$\Delta \bar{s}_v < \frac{S_{L,min} \bar{s}_v}{T_L},$$

then $\Delta_{i,k} < S_{L,min}$ for all $t_{i,k} \le T_L$. For this $\Delta\bar{s}_v$, $t'_{i,k} = t_{i,k} + \Delta_{i,k}$ lies in between $t_{i,k}$ and the next transition time of $\mathcal{N}(t,\xi)$. That is, the nominal and the perturbed paths are similar. From (3.10), $S_{L,min} > 0$ with probability one. Therefore, there is, with probability one, a $\Delta\bar{s}_v > 0$ (the size of which depends on ξ) such that the two paths are similar. □

Equation (3.30) describes the perturbation generation rule, which specifies the perturbation generated in each service period of server v because of the change $\Delta\bar{s}_v$. The perturbations obtained by one server will be propagated to other servers through idle periods. To study the propagation, we set $\Delta s_{i,k} = 0$ in (3.29) (i.e., we assume that there is no perturbation generated in the kth customer's service time at server i). Then we have

$$\Delta_{i,k} = \begin{cases} \Delta_{i,k-1} & \text{if } n_i(t_{i,k-1}+) \neq 0, \\ \Delta_{j,h} & \text{if } n_i(t_{i,k-1}+) = 0. \end{cases} \tag{3.32}$$

This equation describes the evolution of a single perturbation at the service completion times. The verbal explanation of this equation is exactly the same as the propagation rules stated on page 81. Therefore, all the results developed in Section 3.2 apply to the perturbation defined in this section.

Finally, Equation (3.29) indicates that the perturbation propagation possesses the superposition property. That is, the perturbation propagated to a server along a sample path because of two perturbations generated at two different instants equals the sum of the two perturbations propagated to the server; each of them is due to one of the perturbations propagated separately along the same sample path. This can be precisely stated as follows. Let $T_{l_1} < T_{l_2} < T_{l_3}$ be three transition times. Suppose that T_{l_1} is perturbed by a small amount Δ_1. This perturbation will be propagated along a given sample path to T_{l_3}, and T_{l_3} will obtain a perturbation $\Delta_{3,1}$. Also, suppose that T_{l_2} is perturbed by a small amount Δ_2, and T_{l_3} obtains a perturbation $\Delta_{3,2}$ through the propagation of Δ_2 along the same sample path. Then, if we propagate both the perturbations Δ_1 of T_{l_1} and Δ_2 of T_{l_2} simultaneously on the same sample path, T_{l_3} will obtain a perturbation $\Delta_3 = \Delta_{3,1} + \Delta_{3,2}$. Of course, the superposition is contingent on all the perturbed paths being similar to the nominal one. This can be achieved by choosing $\Delta\bar{s}_v$ small enough. By the superposition rule, the effect of many perturbations on a sample path can be decomposed into the sum of the effects of each perturbation on the same sample path alone.

3.3.3 Calculation of Sample Derivatives of Throughputs

Sample Derivative Formulas

In this subsection, we shall derive formulas for the sample derivatives of the system throughput. The sample function is defined as

$$\eta_L(\bar{s}_v, \xi) = \frac{L}{T_L(\bar{s}_v, \xi)}. \tag{3.33}$$

The sample *elasticity* of $\eta_L(\bar{s}_v, \xi)$ with respect to $\bar{s}_v$ is defined as

$$\frac{\bar{s}_v}{\eta_L(\bar{s}_v, \xi)} \frac{\partial \eta_L(\bar{s}_v, \xi)}{\partial \bar{s}_v} = \lim_{\Delta \bar{s}_v \to 0} \frac{\bar{s}_v}{\eta_L(\bar{s}_v, \xi)} \frac{\Delta \eta_L(\bar{s}_v, \xi)}{\Delta \bar{s}_v}. \tag{3.34}$$

Because of the change $\Delta \bar{s}_v$ in $\bar{s}_v$, the perturbed system will need $T_L + \Delta T_L$ to serve the same number of customers (a total of L service completions). ΔT_L is the perturbation of T_L. Therefore, the throughput of the system with $\bar{s}_v + \Delta \bar{s}_v$ for the same ξ is

$$\eta_L(\bar{s}_v, \xi) + \Delta \eta_L(\bar{s}_v, \xi) = \frac{L}{T_L + \Delta T_L}.$$

From this and (3.33), we have

$$\begin{aligned} \Delta \eta_L(\bar{s}_v, \xi) &= \frac{L}{T_L + \Delta T_L} - \frac{L}{T_L} \\ &= \eta_L(\bar{s}_v, \xi)\{\frac{1}{1 + \frac{\Delta T_L}{T_L}} - 1\}. \end{aligned}$$

Thus,

$$\lim_{\Delta \bar{s}_v \to 0} \frac{1}{\eta_L(\bar{s}_v, \xi)} \frac{\Delta \eta_L(\bar{s}_v, \xi)}{\Delta \bar{s}_v} = -\lim_{\Delta \bar{s}_v \to 0} \frac{1}{T_L} \frac{\Delta T_L}{\Delta \bar{s}_v}.$$

That is,

$$\frac{\bar{s}_v}{\eta_L(\bar{s}_v, \xi)} \frac{\partial \eta_L(\bar{s}_v, \xi)}{\partial \bar{s}_v} = -\lim_{\Delta \bar{s}_v \to 0} \frac{\bar{s}_v}{T_L} \frac{\Delta T_L}{\Delta \bar{s}_v} = -\frac{\bar{s}_v}{T_L} \frac{\partial T_L}{\partial \bar{s}_v}. \tag{3.35}$$

Equation (3.35) indicates that the elasticity

$$\frac{\bar{s}_v}{\eta_L(\bar{s}_v, \xi)} \frac{\partial \eta_L(\bar{s}_v, \xi)}{\partial \bar{s}_v}$$

does not depend explicitly on the number L. Thus if the numerator of (3.33) is redefined as the number of service completions of any particular server,

Equation (3.35) still holds. To be more precise, let L_i be the number of the service completions of server i, $i = 1, 2, \cdots, M$, in $[0, T_L)$ and $\eta_{i,L}(\bar{s}_v, \xi) = L_i/T_L(\bar{s}_v, \xi)$ be the sample throughput of server i in $[0, T_L)$. Then its elasticity with respect to $\bar{s}_v$ is the same as (3.35). This corresponds to the following steady-state property of a closed Jackson network: the number of service completions at server i is proportional to its visit ratio y_i defined in Section 2.3.

Now, let us study $\Delta T_L/\Delta \bar{s}_v$. Consider a customer's service period at server v. During this period, customers in other servers may transfer from one server to the other. Let $\mathbf{n}(l_1)$, $\mathbf{n}(l_2)$, $\cdots$, $\mathbf{n}(l_r)$ be the states visited by the system during this duration. At the end of $\mathbf{n}(l_r)$, the customer leaves server v. The length of this customer's service duration is its service time. It is denoted as

$$s = S_{l_1} + S_{l_2} + \cdots + S_{l_r}, \tag{3.36}$$

where S_{l_i} is the time the system stays in $\mathbf{n}(l_i)$ (see Figure 3.2). Because of the change in $\bar{s}_v$, the service period of server v changes by $\Delta s = \gamma s$ (see (3.30)). This change is a perturbation obtained by server v in the service period. According to (3.30), the same ξ is used in $\Delta s = \gamma s$ for both $\bar{s}_v$ and $\bar{s}_v + \Delta \bar{s}_v$; this corresponds to the fact that in the definition of a sample derivative (see (3.26)), the same random vector is used for both θ and $\theta + \Delta\theta$.

To measure the final contribution of this perturbation to ΔT_L for any finite L, we need to modify the definition of realization. Let $T_L = t_{u,k}$ for some u and k; i.e., assume that the last transition in $[0, T_L]$ is a service completion at server u. We define

$$RI_L(\mathbf{n}, V, l) = \begin{cases} 1 & \textit{if the perturbation is propagated to server } u \textit{ at } T_L, \\ 0 & \textit{otherwise.} \end{cases}$$

Based on this definition, the contribution of the perturbation of server v at the service completion time T_{l_r} is $\gamma s RI_L(\mathbf{n}(l_r), v, l_r)$. During the service period, there is no customer transition from server v to any other server. Thus for any sample path it holds that

$$RI_L(\mathbf{n}(l_1), v, l_1) = RI_L(\mathbf{n}(l_2), v, l_2) = \cdots = RI_L(\mathbf{n}(l_r), v, l_r).$$

Therefore,

$$\begin{aligned} \gamma s RI_L(\mathbf{n}(l_r), v, l_r) &= \gamma S_{l_1} RI_L(\mathbf{n}(l_1), v, l_1) + \gamma S_{l_2} RI_L(\mathbf{n}(l_2), v, l_2) \\ &\quad + \cdots + \gamma S_{l_r} RI_L(\mathbf{n}(l_r), v, l_r). \end{aligned} \tag{3.37}$$

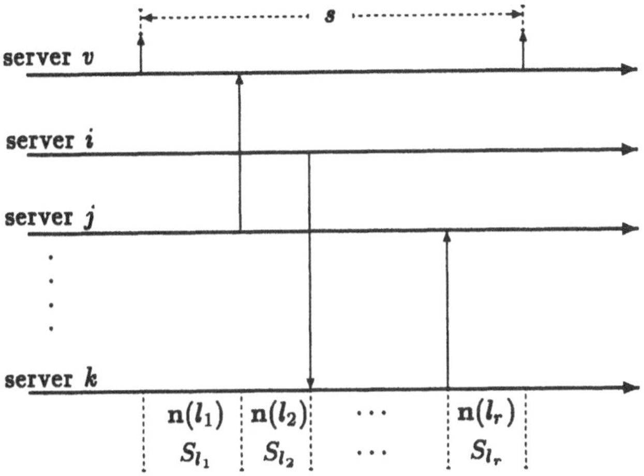

Figure 3.2: The Decomposition of a Service Period

This means that, as far as the final contribution to ΔT_L is concerned, a perturbation $\Delta s = \gamma s$ of a customer's service time is equivalent to a series of perturbations of the periods in which the system stays in states $\mathbf{n}(l_i)$, $i = 1, 2, \cdots, r$.

From the above discussion and the superposition of perturbation propagation, ΔT_L has the following value:

$$\Delta T_L = \gamma \sum_{k: t_{v,k} \leq T_L} s_{v,k} RI_L(\mathbf{n}(l_{r_k}), v, l_{r_k}),$$

where $\mathbf{n}(l_{r_k})$ is the last state in the service period of the kth customer of server v. By (3.37), this is equivalent to

$$\Delta T_L = \gamma \sum_{0 \leq l \leq L, (X_l)_v \neq 0} S_l RI_L(\mathbf{n}, v, l),$$

where $(X_l)_v$ is the vth component of X_l (i.e., the number of customers in server v at T_l). The indicies l with $(X_l)_v = 0$ are excluded in the above expression because there is no perturbation generated in idle periods of server v. However, if $(X_l)_v = 0$, then $RI_L(\mathbf{n}, v, l) = RI_L(X_l, v, l) = 0$.

Thus we can rewrite the above equation as

$$\Delta T_L = \gamma \sum_{l=0}^{L-1} S_l RI_L(\mathbf{n}, v, l). \tag{3.38}$$

Therefore,

$$\lim_{\Delta\bar{s}_v \to 0} \frac{\bar{s}_v}{\Delta\bar{s}_v} \frac{\Delta T_L}{T_L} = \lim_{\Delta\bar{s}_v \to 0} \frac{1}{\gamma} \frac{\Delta T_L}{T_L} = \frac{1}{T_L} \sum_{l=0}^{L-1} S_l RI_L(\mathbf{n}, v, l). \tag{3.39}$$

By (3.35) and (3.39), the sample elasticity of $\eta_L(\bar{s}_v, \xi)$ with respect to $\bar{s}_v$ is

$$\frac{\bar{s}_v}{\eta_L(\bar{s}_v, \xi)} \frac{\partial \eta_L(\bar{s}_v, \xi)}{\partial \bar{s}_v} = -\frac{1}{T_L} \sum_{l=0}^{L-1} S_l RI_L(\mathbf{n}, v, l). \tag{3.40}$$

The Limiting Value

In this subsection, we shall derive a formula for the limiting value of (3.40) as $L \to \infty$. First, we prove a lemma.

Lemma 3.6 *For closed Jackson networks,*

$$\lim_{L\to\infty} \frac{1}{T_L} \sum_{l=0}^{L-1} S_l RI(\mathbf{n}, v, l) = \sum_{all\ \mathbf{n}} p(\mathbf{n}) c(\mathbf{n}, v) \qquad w.p.1. \tag{3.41}$$

Proof: We first observe that whether a perturbation is realized or lost depends on the initial state $\mathbf{n}$ and the customer transitions afterwards. It does not depend on the durations the system stays in all the states, $S_l, S_{l+1}, \cdots$. Therefore, $RI(\mathbf{n}, V, l)$ is a function of $X_l, X_{l+1}, \cdots$. We denote the relation as

$$RI(\mathbf{n}, V, l) = \phi(X_l, X_{l+1}, \cdots), \qquad \textit{for } X_l = \mathbf{n}. \tag{3.42}$$

The function ϕ depends on $(\mathbf{n}, V)$. It is, however, independent of l for time-homogeneous systems, no matter whether $\mathcal{N}(t)$ is stationary or not.

Let $\chi_{\mathbf{n}}(X_l) = 1$ if $X_l = \mathbf{n}$; $\chi_{\mathbf{n}}(X_l) = 0$, otherwise. For all l with $\chi_{\mathbf{n}}(X_l) = 1$, S_l are independent and identically distributed. By the law of large numbers, we have

$$\lim_{l\to\infty} \frac{\sum_{l=0}^{L-1} S_l RI(\mathbf{n}, v, l) \chi_{\mathbf{n}}(X_l)}{\sum_{l=0}^{L-1} RI(\mathbf{n}, v, l) \chi_{\mathbf{n}}(X_l)} = E\{S_l | X_l = 1, RI(\mathbf{n}, v, l) = 1\} \quad w.p.1,$$

where $\sum_{l=0}^{L-1} RI(\mathbf{n}, v, l)\chi_{\mathbf{n}}(X_l)$ is the number of nonzero terms in the numerator. From this, we have

$$\begin{aligned}
&\lim_{L\to\infty} \frac{1}{L}\sum_{l=0}^{L-1} S_l RI(\mathbf{n}, v, l)\chi_{\mathbf{n}}(X_l) \\
&= E\{S_l|X_l = \mathbf{n}, RI(\mathbf{n}, v, l) = 1\} \times \lim_{L\to\infty} \frac{1}{L}\sum_{l=0}^{L-1} RI(\mathbf{n}, v, l)\chi_{\mathbf{n}}(X_l) \\
&= E\{S_l|X_l = \mathbf{n}, RI(\mathbf{n}, v, l) = 1\} \times \lim_{L\to\infty} \frac{1}{L}\sum_{l=0}^{L-1} Y_l,
\end{aligned}$$

where

$$Y_l = RI(\mathbf{n}, v, l)\chi_{\mathbf{n}}(X_l) = \phi(X_l, X_{l+1}, \cdots)\chi_{\mathbf{n}}(X_l) := \psi(X_l, X_{l+1}, \cdots).$$

ψ is measurable on Φ^∞ (the state space Φ contains only a finite number of points). For the ergodic process $X_l, X_{l+1}, \cdots$, the process $Y_l, Y_{l+1}, \cdots$ is also ergodic (see Theorem 2.12 or Proposition 6.31 in Breiman [9]). Thus,

$$\begin{aligned}
&\lim_{L\to\infty} \frac{1}{L}\sum_{l=0}^{L-1} Y_l = E(Y_l) \qquad\qquad w.p.1, \\
&= E[RI(\mathbf{n}, v, l)\chi_{\mathbf{n}}(X_l)] = \pi\{RI(\mathbf{n}, v, l)\chi_{\mathbf{n}}(X_l) = 1\},
\end{aligned}$$

where π is the steady-state probability measure for Markov chain $\mathbf{Y} = \{Y_0, Y_1, \cdots\}$. Therefore,

$$\begin{aligned}
&\lim_{L\to\infty} \frac{1}{L}\sum_{l=0}^{L-1} S_l RI(\mathbf{n}, v, l)\chi_{\mathbf{n}}(X_l) \\
&= E\{S_l|X_l = \mathbf{n}, RI(\mathbf{n}, v, l) = 1\} \times \pi\{X_l = \mathbf{n}, RI(\mathbf{n}, v, l) = 1\} \\
&= E\{S_l\chi_{\mathbf{n}}(X_l)RI(\mathbf{n}, v, l)\} \\
&= E\{S_l RI(\mathbf{n}, v, l)|X_l = \mathbf{n}\} \times \pi(\mathbf{n}), \qquad (3.43)
\end{aligned}$$

where $\pi(\mathbf{n})$ is the stationary probability of state $\mathbf{n}$ of the embedded chain $\mathbf{X} = \{X_l\}_{l=0}^\infty$. By the Markov property of $\mathcal{N}(t)$, given $X_l = \mathbf{n}$, the two random variables S_l and $RI(\mathbf{n}, v, l)$ are independent. Thus, by (3.23), we have

$$\begin{aligned}
&E\{S_l RI(\mathbf{n}, v, l)|X_l = \mathbf{n}\} \\
&= E(S_l|X_l = \mathbf{n}) \times E\{RI(\mathbf{n}, v, l)|X_l = \mathbf{n}\} \\
&= \frac{c(\mathbf{n}, v)}{\mu(\mathbf{n})}.
\end{aligned}$$

Finally, we have

$$\lim_{L\to\infty}\sum_{l=0}^{L-1} S_l RI(\mathbf{n},v,l)\chi_{\mathbf{n}}(X_l) = c(\mathbf{n},v)\frac{\pi(\mathbf{n})}{\mu(\mathbf{n})} \qquad w.p.1.$$

Using Equation (3.3), we obtain

$$\lim_{L\to\infty}\frac{1}{L}\sum_{l=0}^{L-1} S_l RI(\mathbf{n},v,l)\chi_{\mathbf{n}}(X_l) = \frac{1}{\eta}p(\mathbf{n})c(\mathbf{n},v) \qquad w.p.1.$$

Summing up both sides of the above equation over all $\mathbf{n}$ and noting that $\sum_{all\ \mathbf{n}} \chi_{\mathbf{n}}(X_l) = 1$, we get

$$\lim_{L\to\infty}\frac{1}{L}\sum_{l=0}^{L-1} S_l RI(\mathbf{n},v,l) = \frac{1}{\eta}\sum_{all\ \mathbf{n}} p(\mathbf{n})c(\mathbf{n},v) \qquad w.p.1. \qquad (3.44)$$

On the other hand, we can similarly prove

$$\begin{aligned}
\lim_{L\to\infty}\frac{T_L}{L} &= \lim_{L\to\infty}\frac{1}{L}\sum_{l=0}^{L-1} S_l \\
&= \lim_{l\to\infty}\frac{1}{L}\sum_{all\ \mathbf{n}}\sum_{l=0}^{L-1} S_l\chi_{\mathbf{n}}(X_l) \\
&= \sum_{all\ \mathbf{n}}\{\lim_{L\to\infty}\frac{1}{L}\sum_{l=0}^{L-1} S_l\chi_{\mathbf{n}}(X_l)\} \\
&= \sum_{all\ \mathbf{n}} E[\chi_{\mathbf{n}}(X_l)]\times E(S_l|X_l=\mathbf{n}) \qquad w.p.1 \\
&= \sum_{all\ \mathbf{n}}\frac{\pi(\mathbf{n})}{\mu(\mathbf{n})} = \frac{1}{\eta} \qquad w.p.1. \qquad (3.45)
\end{aligned}$$

Note that

$$\lim_{L\to\infty}\frac{1}{T_L}\sum_{l=0}^{L-1} S_l RI(\mathbf{n},v,l) = \{\lim_{L\to\infty}\frac{L}{T_L}\}\times\{\lim_{L\to\infty}\frac{1}{L}\sum_{l=0}^{L-1} S_l RI(\mathbf{n},v,l)\}.$$

Equation (3.41) then follows directly from Equations (3.44) and (3.45). □

Equation (3.41) is based on a single sample path of the system. However, to evaluate the left-hand side of (3.41) is somewhat inconvenient in practice. After the time period in which L customers are served, one has to continue observing the system until all the perturbations generated in that period have been either realized or lost. Sometimes this additional observation period may be too long. Besides, the left-hand side of (3.41) does not correspond to the sample derivative we have derived. Theorem 3.3 below provides a more meaningful result.

Theorem 3.3 *For closed Jackson networks,*

$$\lim_{L\to\infty} \frac{1}{T_L} \sum_{l=1}^{L} S_l RI_L(\mathbf{n}, v, l) = \sum_{all\ \mathbf{n}} p(\mathbf{n})c(\mathbf{n}, v) \qquad w.p.1. \tag{3.46}$$

Proof: By Lemma 3.6, it is sufficient to prove

$$\lim_{L\to\infty} \frac{1}{T_L} \sum_{l=0}^{L-1} S_l[RI_L(\mathbf{n}, v, l) - RI(\mathbf{n}, v, l)] = 0 \qquad w.p.1. \tag{3.47}$$

Since $\lim_{L\to\infty} \frac{L}{T_L} = \eta$, (3.47) is equivalent to

$$\lim_{L\to\infty} \frac{1}{L} \sum_{l=0}^{L-1} S_l[RI_L(\mathbf{n}, v, l) - RI(\mathbf{n}, v, l)] = 0 \qquad w.p.1. \tag{3.48}$$

Let $\mathbf{n}_1 = (N, 0, 0, \cdots, 0)$, $L_0 = 0$, and $L_k = min\{l > L_{k-1}, X_l = \mathbf{n}_1\}$. The sequence $\{T_{L_k}, \ k = 1, 2, \cdots\}$ is a delayed renewal process (Çinlar [41]), and T_{L_k} are regenerative times. Since $\mathbf{n}_1$ is a recurrent nonnull state, $E(T_{L_{k+1}} - T_{L_k})$ is finite. Next, by the perturbation propagation rules, any perturbation propagated to state $\mathbf{n}_1$ is already realized (or lost) by the network. In other words, all the perturbations previously generated will become realized (or lost) when the network enters state $\mathbf{n}_1$. Therefore, $RI_{L_k}(\mathbf{n}, v, l) = RI(\mathbf{n}, i, l)$ for any k, $\mathbf{n}$, and $l \leq L_k$. Moreover, for any L, there is an integer d such that $L_d \leq L < L_{d+1}$. Thus, $RI_L(\mathbf{n}, v, l) = RI(\mathbf{n}, v, l)$ for any $l \leq L_d$. Therefore,

$$\begin{aligned} &\lim_{L\to\infty} \frac{1}{L} \sum_{l=0}^{L-1} S_l[RI_L(\mathbf{n}, v, l) - RI(\mathbf{n}, v, l)] \\ &= \lim_{L\to\infty} \frac{1}{L} \sum_{l=L_d+1}^{L-1} S_l[RI_L(\mathbf{n}, v, l) - RI(\mathbf{n}, v, l)]. \end{aligned} \tag{3.49}$$

Note that d depends on L, and $d \to \infty$ as $L \to \infty$. Thus, Equation (3.48) holds if

$$\lim_{L\to\infty} \frac{1}{L} \sum_{l=L_d+1}^{L-1} S_l|RI_L(\mathbf{n}, v, l) - R(\mathbf{n}, v, l)| = 0 \qquad w.p.1. \tag{3.50}$$

Since $|RI_L(\mathbf{n}, v, l) - R(\mathbf{n}, v, l)| \leq 1$, (3.50) holds if

$$\lim_{L\to\infty} \frac{1}{L} \sum_{l=L_d+1}^{L-1} S_l = 0 \qquad w.p.1. \tag{3.51}$$

This is true, since

$$\begin{aligned}
\lim_{L\to\infty} \frac{1}{L} \sum_{l=L_d+1}^{L-1} S_l &\leq \lim_{L\to\infty} \frac{1}{L_d} \sum_{l=L_d+1}^{L_{d+1}} S_l \\
&= \lim_{L\to\infty} \frac{1}{L_d} \{ \sum_{l=L_1+1}^{L_{d+1}} S_l - \sum_{l=L_1+1}^{L_d} S_l \} \\
&= \lim_{L\to\infty} \frac{d}{L_d} \{ \frac{\sum_{k=1}^{d} \sum_{l=L_k+1}^{L_{k+1}} S_l}{d} - \frac{d-1}{d} \frac{\sum_{k=1}^{d-1} \sum_{l=L_k+1}^{L_{k+1}} S_l}{d-1} \} \\
&= \{ \lim_{L\to\infty} \frac{d}{L_d} \} \times \{ E(T_{L_{k+1}} - T_{L_k}) - E(T_{L_{k+1}} - T_{L_k}) \} = 0.
\end{aligned}$$

The last equation is due to $E(T_{L_{k+1}} - T_{L_k})$ being finite. □

Using the results obtained, we have the following main theorem of this subsection.

Theorem 3.4 *The sample elasticity of the system throughput in a closed Jackson network satisfies*

$$\lim_{L\to\infty} \frac{\bar{s}_v}{\eta_L(\bar{s}_v, \xi)} \frac{\partial \eta_L(\bar{s}_v, \xi)}{\partial \bar{s}_v} = -\sum_{all\ \mathbf{n}} p(\mathbf{n}) c(\mathbf{n}, v) \qquad w.p.1, \tag{3.52}$$

and

$$\lim_{L\to\infty} E\{ \frac{\bar{s}_v}{\eta_L(\bar{s}_v, \xi)} \frac{\partial \eta_L(\bar{s}_v, \xi)}{\partial \bar{s}_v} \} = -\sum_{all\ \mathbf{n}} p(\mathbf{n}) c(\mathbf{n}, v). \tag{3.53}$$

Proof: Equation (3.52) follows directly from (3.40) and Theorem 3.3. Next, from (3.40), we have

$$\left| \frac{\bar{s}_v}{\eta_L(\bar{s}_v, \xi)} \frac{\partial \eta_L(\bar{s}_v, \xi)}{\partial \bar{s}_v} \right| \leq \frac{1}{T_L} \sum_{l=0}^{L-1} S_l \leq 1.$$

Equation (3.53) then follows from (3.52) and the Lebesgue bounded convergence theorem (see Theorem 2.3 or Billingsley [5]). □

3.4 Sensitivities of Steady-State Throughputs to Mean Service Times

In this section we shall show that as the length of the sample path goes to infinity, the sample derivative of the system throughput derived in the last section in fact convereges to the derivative of the steady-state throughput with probability one. Thus, Theorem 3.4 also gives the formula for the derivative of the steady-state throughput. To this end, we first study the statistic property of $T_L(\bar{s}_v, \xi)$.

3.4.1 The Sample Performance Function $T_L(\bar{s}_v, \xi)$

The Linearity of $T_L(\bar{s}_v, \xi)$

First, from Equations (3.29) and (3.30), we observe that for any i, k such that $t_{i,k} \in [0, T_L)$, the following linearity holds:

$$\Delta_{i,k}[\alpha(\Delta\bar{s}_v)_1 + \beta(\Delta\bar{s}_v)_2] = \alpha\Delta_{i,k}[(\Delta\bar{s}_v)_1] + \beta\Delta_{i,k}[(\Delta\bar{s}_v)_2], \tag{3.54}$$

provided that the sample paths with $\Delta\bar{s}_v = (\Delta\bar{s}_v)_1$, $(\Delta\bar{s}_v)_2$, and $\alpha(\Delta\bar{s}_v)_1 + \beta(\Delta\bar{s}_v)_2$ are similar to the nominal path. In the equation, $\Delta_{i,k}[\delta]$ denotes the perturbation of $t_{i,k}$ when $\Delta\bar{s}_v = \delta$. From Lemma 3.5, this shows that in a small neighborhood of $\bar{s}_v$ (e.g., within $|\Delta\bar{s}_v| < (S_{L,min}\bar{s}_v)/T_L$), the sample function $T_L(\bar{s}_v, \xi)$ is a linear function of $\bar{s}_v$. As $\Delta\bar{s}_v > 0$ increases (in this subsection we discuss only the case of $\Delta\bar{s}_v > 0$; however, all the discussions apply to $\Delta\bar{s}_v < 0$ as well) to exceed a certain value, say $(\Delta\bar{s}_v)_0$, the linearity is violated because the sample path with $\bar{s}_v + (\Delta\bar{s}_v)_0$ is no longer similar to that of $\bar{s}_v$. It is easy to see that on the sample path with $\bar{s}_v + (\Delta\bar{s}_v)_0$, there are two transitions occurring at the same time. As $\Delta\bar{s}_v$ increases further, the sample paths form another set of similar paths. Also, as $\Delta\bar{s}_v$ increases, all the transition times move continuously. Although the Lth transition time may change from the service completion time of one server to that of the other, it does not lead to the discontinuity of T_L, because such a change happens only when the service completion times of these two servers occur simultaneously on a sample path, and the service completion time of each of these two servers is continuous. Therefore, $T_L(\bar{s}_v, \xi)$ is continuous with respect to $\bar{s}_v$ (Cao [28]). The results are summarized in the following lemma.

Lemma 3.7 *The sample function $T_L(\bar{s}_v, \xi)$ is, with probability one, a continuous and piecewise linear function of $\bar{s}_v$.*

From (3.31), we have $\Delta T_L \leq (\Delta \bar{s}_v / \bar{s}_v) T_L$. This gives a bound for the slope of the sample function at any $\bar{s}_v$:

$$\frac{\Delta T_L(\bar{s}_v, \xi)}{\Delta \bar{s}_v} \leq \frac{T_L(\bar{s}_v, \xi)}{\bar{s}_v}. \tag{3.55}$$

Remark 3.2 If $\bar{s}_v$ is in the middle of a linear segment, the sample derivatives at both the left and the right sides of $\bar{s}_v$ are equal. But if $\bar{s}_v$ is a conjunct point of two linear segments, then the left-side and the right-side derivatives are different; the sample derivative has to be viewed as a one-sided derivative.

The Unbiasedness of the Sample Derivative

Now, we are ready to prove that for any finite L, the sample derivative of $T_L(\bar{s}_v, \xi)$ with respect to $\bar{s}_v$ is an unbiased estimate for the derivative of its mean. This is formally stated in Lemma 3.8.

Lemma 3.8 *For closed Jackson networks, we have*

$$E\{\frac{\partial}{\partial \bar{s}_v}[T_L(\bar{s}_v, \xi)] | \mathbf{n}_0\} = \frac{\partial}{\partial \bar{s}_v}\{E[T_L(\bar{s}_v, \xi)] | \mathbf{n}_0\}, \tag{3.56}$$

where $\mathbf{n}_0$ *is the initial state of the sample path.*

Proof: We consider a neighborhood of $\bar{s}_v$, $[\bar{s}_v, \bar{s}_v + \Delta \bar{s}_v]$. Let

$$e(\bar{s}_v, \xi, h) = \frac{T_L(\bar{s}_v + h, \xi)] - T_L(\bar{s}_v, \xi)]}{h} - \frac{\partial[T_L(\bar{s}_v, \xi)]}{\partial \bar{s}_v}, \quad h \in [0, \Delta \bar{s}_v].$$

Then from Lemma 3.7, we have

$$\lim_{h \to 0} e(\bar{s}_v, \xi, h) = 0 \qquad w.p.1. \tag{3.57}$$

In $[\bar{s}_v, \bar{s}_v + \Delta \bar{s}_v]$, the sample function $T_L(\bar{s}_v, \xi)$ may have more than one piece of linear segment. Since $T_L(\bar{s}_v, \xi)$ is an increasing function of $\bar{s}_v$, from (3.55), all the slopes of these segments are bounded by $T_L(\bar{s}_v + \Delta \bar{s}_v, \xi)/\bar{s}_v$. Thus,

$$\frac{T_L(\bar{s}_v + h, \xi)] - T_L(\bar{s}_v, \xi)]}{h} < \frac{T_L(\bar{s}_v + \Delta \bar{s}_v, \xi)}{\bar{s}_v}, \qquad h \in [0, \Delta \bar{s}_v].$$

On the other hand, by the linearity of $T_L(\bar{s}_v, \xi)$, we have

$$\frac{\partial[T_L(\bar{s}_v, \xi)]}{\partial \bar{s}_v} < \frac{T_L(\bar{s}_v, \xi)}{\bar{s}_v} < \frac{T_L(\bar{s}_v + \Delta \bar{s}_v, \xi)}{\bar{s}_v}.$$

Thus,

$$|e(\bar{s}_v, \xi, h)| < 2\frac{T_L(\bar{s}_v + \Delta\bar{s}_v, \xi)}{\bar{s}_v}.$$

That is, $\frac{2}{\bar{s}_v}T_L(\bar{s}_v + \Delta\bar{s}_v)$ is a dominating function of $e(\bar{s}_v, \xi, h)$. Note that $E[T_L(\bar{s}_v + \Delta\bar{s}_v)]$ is finite. Applying the Lebesgue dominated convergence theorem (Billingsley [5]) to (3.57), we get

$$\lim_{h\to 0} E\{e(\bar{s}_v, \xi, h)\} = 0.$$

This leads directly to (3.56). □

3.4.2 Formulas for Steady-State Throughput Sensitivities

To study the convergence property, it is convenient to use the reciprocal of the sample throughput. Let

$$R_L(\bar{s}_v, \xi) = \frac{T_L(\bar{s}_v, \xi)}{L} = \frac{1}{\eta_L(\bar{s}_v, \xi)} \tag{3.58}$$

be the average time between two successive transitions of the state process in $[0, T_L(\bar{s}_v, \xi)]$, and let

$$R_L(\bar{s}_v|\mathbf{n}) = E\{R_L(\bar{s}_v, \xi)|\mathcal{N}(0) = \mathbf{n}\}$$

be the conditional expectation of $R_L(\bar{s}_v, \xi)$ given that the initial state is $\mathbf{n}$. Then

$$R_L(\bar{s}_v, \xi) = \frac{1}{L}\sum_{l=0}^{L-1} S_l, \tag{3.59}$$

where $S_l = T_{l+1} - T_l$. By ergodicity, we have

$$\begin{aligned} R(\bar{s}_v) &:= \lim_{L\to\infty} R_L(\bar{s}_v, \xi) \\ &= E(S_l) = \sum_{all\ \mathbf{n}} \frac{\pi(\mathbf{n})}{\mu(\mathbf{n})} \qquad w.p.1 \\ &= \frac{1}{\eta(\bar{s}_v)}. \end{aligned} \tag{3.60}$$

To study the steady-state property, we assume that the initial state distribution is the steady-state distribution $\pi(\mathbf{n})$. In this case, $E(S_l)$ are equal for all $l = 0, 1, \cdots$. From (3.59) and (3.4), we have

$$R_L(\bar{s}_v) := E[R_L(\bar{s}_v, \xi)] = \frac{1}{\eta(\bar{s}_v)}, \tag{3.61}$$

where $\eta(\bar{s}_v) = \lim_{L\to\infty} \eta_L(\bar{s}_v, \xi)$ is the steady-state throughput.

Next, from

$$R_L(\bar{s}_v|\mathbf{n}) = \frac{1}{L}\sum_{l=0}^{L-1} E[S_l|\mathcal{N}(0) = \mathbf{n}]$$

and

$$\lim_{l\to\infty} E[S_l|\mathcal{N}(0) = \mathbf{n}] = E(S_l) = R(\bar{s}_v)$$

and by the Cesaro theorem in calculus (see, e.g., Spivak [99]), we have

$$\lim_{L\to\infty} R_L(\bar{s}_v|\mathbf{n}) = R(\bar{s}_v). \tag{3.62}$$

Theorem 3.5 *The sample derivative of the system throughput in a closed Jackson network satisfies*

$$\lim_{L\to\infty} E\{\frac{\partial[R_L(\bar{s}_v,\xi)]}{\partial \bar{s}_v}\} = \frac{\partial[R(\bar{s}_v)]}{\partial \bar{s}_v}. \tag{3.63}$$

Proof: First, from (3.60) and (3.61), we have

$$R(\bar{s}_v) = E[R_L(\bar{s}_v,\xi)] = E\{E[R_L(\bar{s}_v,\xi)|\mathcal{N}(0) = \mathbf{n})]\}.$$

The first expectation on the right-hand side is taken over the stationary distribution $\pi(\mathbf{n})$. The derivative of $R(\bar{s}_v)$ is then:

$$\begin{aligned}
\frac{\partial[R(\bar{s}_v)]}{\partial \bar{s}_v} &= \frac{\partial}{\partial \bar{s}_v} E\{E[R_L(\bar{s}_v,\xi)|\mathbf{N}(0) = \mathbf{n}]\} \\
&= \frac{\partial}{\partial \bar{s}_v}\{\sum_{all\ \mathbf{n}} E[R_L(\bar{s}_v,\xi)|\mathcal{N}(0) = \mathbf{n}]\pi(\mathbf{n})\} \\
&= \sum_{all\ \mathbf{n}} \frac{\partial}{\partial \bar{s}_v}\{E[R_L(\bar{s}_v,\xi)|\mathcal{N}(0) = \mathbf{n}]\pi(\mathbf{n})\}.
\end{aligned}$$

The last equation is due to the fact that the number of states is finite. Thus,

$$\begin{aligned}
\frac{\partial[R(\bar{s}_v)]}{\partial \bar{s}_v} &= \sum_{all\ \mathbf{n}} \frac{\partial}{\partial \bar{s}_v}\{E[R_L(\bar{s}_v,\xi)|\mathcal{N}(0) = \mathbf{n}]\}\pi(\mathbf{n}) \\
&+ \sum_{all\ \mathbf{n}} \{E[R_L(\bar{s}_v,\xi)|\mathcal{N}(0) = \mathbf{n}]\}\frac{\partial \pi(\mathbf{n})}{\partial \bar{s}_v}.
\end{aligned}$$

Applying Lemma 3.8 to the first term on the right-hand side of the above equation, we get

$$\begin{aligned}
\frac{\partial[R(\bar{s}_v)]}{\partial \bar{s}_v} &= \sum_{all\ \mathbf{n}} E\{\frac{\partial R_L(\bar{s}_v,\xi)}{\partial \bar{s}_v}|\mathbf{N}(0)=\mathbf{n}\}\pi(\mathbf{n}) \\
&+ \sum_{all\ \mathbf{n}} \{E[R_L(\bar{s}_v,\xi)|\mathcal{N}(0)=\mathbf{n}]\}\frac{\partial \pi(\mathbf{n})}{\partial \bar{s}_v} \\
&= E\{E[\frac{\partial R_L(\bar{s}_v,\xi)}{\partial \bar{s}_v}|\mathcal{N}(0)=\mathbf{n}]\} + \sum_{all\ \mathbf{n}} \{R_L(\bar{s}_v|\mathbf{n})\}\frac{\partial \pi(\mathbf{n})}{\partial \bar{s}_v} \\
&= E\{\frac{\partial R_L(\bar{s}_v,\xi)}{\partial \bar{s}_v}\} + \sum_{all\ \mathbf{n}} \{R_L(\bar{s}_v|\mathbf{n})\}\frac{\partial \pi(\mathbf{n})}{\partial \bar{s}_v}. \qquad (3.64)
\end{aligned}$$

The second term on the right-hand side of (3.64) is the bias caused by the change in the probability distribution of the initial state. Letting $L \to \infty$ on both sides of (3.64), we get

$$\frac{\partial[R(\bar{s}_v)]}{\partial \bar{s}_v} = \lim_{L\to\infty} E\{\frac{\partial R_L(\bar{s}_v,\xi)}{\partial \bar{s}_v}\} + \sum_{all\ \mathbf{n}} \{\lim_{L\to\infty} R_L(\bar{s}_v|\mathbf{n})\}\frac{\partial \pi(\mathbf{n})}{\partial \bar{s}_v}. \qquad (3.65)$$

Equations (3.62), (3.65) and

$$\sum_{all\ \mathbf{n}} \frac{\partial \pi(\mathbf{n})}{\partial \bar{s}_v} = \frac{\partial}{\partial \bar{s}_v}\{\sum_{all\ \mathbf{n}} \pi(\mathbf{n})\} = 0$$

lead directly to Equation (3.63). □

Theorem 3.6 *In an irreducible closed Jackson network, the sample elasticity of the throughput converges in mean to the elasticity of the steady-state throughput as the length of the sample path goes to infinity. That is,*

$$\lim_{L\to\infty} E\{\frac{\bar{s}_v}{\eta_L(\bar{s}_v,\xi)}\frac{\partial \eta_L(\bar{s}_v,\xi)}{\partial \bar{s}_v}\} = \frac{\bar{s}_v}{\eta(\bar{s}_v)}\frac{\partial \eta(\bar{s}_v)}{\partial \bar{s}_v}. \qquad (3.66)$$

Proof: From (3.58) and (3.60), the theorem is equivalent to

$$\lim_{L\to\infty} E\{\frac{\bar{s}_v}{R_L(\bar{s}_v,\xi)}\frac{\partial R_L(\bar{s}_v,\xi)}{\partial \bar{s}_v}\} = \frac{\bar{s}_v}{R(\bar{s}_v)}\frac{\partial R(\bar{s}_v)}{\partial \bar{s}_v}.$$

By (3.40) and (3.58), we have

$$|\frac{\bar{s}_v}{R_L(\bar{s}_v,\xi)}\frac{\partial R_L(\bar{s}_v,\xi)}{\partial \bar{s}_v}| = |\frac{\bar{s}_v}{T_L(\bar{s}_v,\xi)}\frac{\partial T_L(\bar{s}_v,\xi)}{\partial \bar{s}_v}| \leq 1$$

for all $L > 0$. Let

$$\begin{aligned} f_L(\bar{s}_v,\xi) &= \frac{\bar{s}_v}{R_L(\bar{s}_v,\xi)}\frac{\partial R_L(\bar{s}_v,\xi)}{\partial \bar{s}_v} - \frac{\bar{s}_v}{R(\bar{s}_v)}\frac{\partial R_L(\bar{s}_v,\xi)}{\partial \bar{s}_v} \\ &= \{1-\frac{R_L(\bar{s}_v,\xi)}{R(\bar{s}_v)}\}\times\frac{\bar{s}_v}{R_L(\bar{s}_v,\xi)}\frac{\partial R_L(\bar{s}_v,\xi)}{\partial \bar{s}_v}. \end{aligned}$$

From Equation (3.60), we have

$$\lim_{L\to\infty}\frac{R_L(\bar{s}_v,\xi)}{R(\bar{s}_v)} = 1 \qquad w.p.1.$$

Thus,

$$\lim_{L\to\infty} f_L(\bar{s}_v,\xi) = 0 \qquad w.p.1.$$

Also,

$$|f_L(\bar{s}_v,\xi)| \le 1 + \frac{R_L(\bar{s}_v,\xi)}{R(\bar{s}_v)}.$$

Next, Equation (3.61) can be written as

$$\lim_{L\to\infty} E\{R_L(\bar{s}_v,\xi)\} = R(\bar{s}_v).$$

From this equation, (3.60), and the fact that $R_L(\bar{s}_v,\xi) > 0$, we conclude that $R_L(\bar{s}_v,\xi)$ are uniformly integrable on $L \in \{1,2,\cdots\}$ (Theorem 2.2). Thus, $1 + R_L(\bar{s}_v,\xi)/R(\bar{s}_v)$ are also uniformly integrable (see, e.g., Problem 16.17 in Billingsley [5]). By Theorem 2.1, $f_L(\bar{s}_v,\xi)$ are also uniformly integrable. Therefore (see Theorem 2.4 or Theorem 16.13 in Billingsley [5]),

$$\lim_{L\to\infty} E\{f_L(\bar{s}_v,\xi)\} = E\{\lim_{L\to\infty} f_L(\bar{s}_v,\xi)\} = 0.$$

This is equivalent to

$$\begin{aligned} &\lim_{L\to\infty} E\{\frac{\bar{s}_v}{R_L(\bar{s}_v,\xi)}\frac{\partial R_L(\bar{s}_v,\xi)}{\partial \bar{s}_v}\} \\ &= \frac{\bar{s}_v}{R(\bar{s}_v)}\lim_{L\to\infty} E\{\frac{\partial R_L(\bar{s}_v,\xi)}{\partial \bar{s}_v}\} = \frac{\bar{s}_v}{R(\bar{s}_v)}\frac{\partial R(\bar{s}_v)}{\partial \bar{s}_v}. \end{aligned}$$

The last equation is due to Theorem 3.5. □

Theorems 3.4 and 3.6 lead directly to the formula for the elasticity of the system throughput, which is stated in the following theorem.

Theorem 3.7 *For closed Jackson networks, it holds that*

$$\frac{\bar{s}_v}{\eta}\frac{\partial \eta}{\partial \bar{s}_v} = -\sum_{all\ \mathbf{n}} p(\mathbf{n})c(\mathbf{n},v) \qquad v \in \Gamma. \tag{3.67}$$

Summing up both sides of (3.67) and using Lemma 3.3, we have

$$\sum_{v=1}^{M} \frac{\bar{s}_v}{\eta} \frac{\partial \eta}{\partial \bar{s}_v} = -1. \quad (3.68)$$

An intuitive explanation of this equation is as follows: if the mean service time of each server in the network changes by the same proportion $\gamma = \Delta\bar{s}_i/\bar{s}_i$, $i = 1, 2, \cdots, M$, then the effect will be as if the time scale is changed by $1 + \gamma$ times. Therefore, the time needed to complete the same number of customers will change by the same proportion $1 + \gamma$; i.e., $\Delta T/T = -\gamma$. (3.68) then follows directly from (3.35).

The following corollaries ensue directly from Theorems 3.7 and 3.3 and Lemma 3.6.

Corollary 3.1

$$\lim_{L\to\infty} \frac{1}{T_L} \sum_{l=1}^{L} S_l RI(\mathbf{n}, v, l) = -\frac{\bar{s}_v}{\eta} \frac{\partial \eta}{\partial \bar{s}_v} \qquad w.p.1, \quad (3.69)$$

or

$$\lim_{L\to\infty} \frac{1}{T_L} \sum_{k:t_{v,k}\le T_L} s_{v,k} RI(\mathcal{N}(t_{v,k}), v, t_{v,k}) = -\frac{\bar{s}_v}{\eta} \frac{\partial \eta}{\partial \bar{s}_v} \qquad w.p.1. \quad (3.70)$$

As observed before, to use the results stated in the next Corollary is more convenient in practice.

Corollary 3.2

$$\lim_{L\to\infty} \frac{1}{T_L} \sum_{l=1}^{L} S_l RI_L(\mathbf{n}, v, l) = -\frac{\bar{s}_v}{\eta} \frac{\partial \eta}{\partial \bar{s}_v} \qquad w.p.1, \quad (3.71)$$

or

$$\lim_{L\to\infty} \frac{1}{T_L} \sum_{k:t_{v,k}\le T_L} s_{v,k} RI_L(\mathcal{N}(t_{v,k}), v, t_{v,k}) = -\frac{\bar{s}_v}{\eta} \frac{\partial \eta}{\partial \bar{s}_v} \qquad w.p.1. \quad (3.72)$$

The left-hand side of (3.71) and (3.72) for a finite L can be obtained by applying the perturbation analysis algorithms developed in Ho and Cao [70] to a single sample path. Theorem 3.3 indicates that the perturbation analysis algorithms in fact can be viewed as an efficient way of estimating $\sum_{all\ \mathbf{n}} p(\mathbf{n})c(\mathbf{n}, v)$. Thus, the corollary provides a theoretical background for practical algorithms using perturbation analysis to estimate the elasticities.

Example 3.3 Consider the same queuing network as that in Example 3.1. From (3.21) and Theorem 3.7, we have

$$\frac{\bar{s}_v}{\eta}\frac{\partial\eta}{\partial\bar{s}_v} = -\sum_{all\ \mathbf{n}} \frac{n_v}{N} p(\mathbf{n}) = -\sum_{n_v} \{\frac{n_v}{N} \sum_{n_j: j\neq v} p(\mathbf{n})\}$$

$$= -\sum_{n=1}^{N} \frac{n}{N} p(n_k = n) = -\frac{1}{N}\bar{n}_v(N),$$

where $\bar{n}_v(N)$ is the average queue length of server v in a network with a total of N customers.

In our example, $\mu_i q_{i,j} = \mu_j q_{j,i}$. It is easy to prove that $x_1 = x_2 = \cdots = x_M = 1$ in the product-form solution (2.45). Thus,

$$G(N) = \binom{M+N-1}{N}.$$

The probabilities $p(\mathbf{n})$ for all $\mathbf{n}$ have the same value $1/G(N)$. After some calculations, we obtain

$$p(n_v = n) = \frac{\binom{M+N-n-2}{M-2}}{\binom{M+N-1}{N}}.$$

From this we get $\bar{n}_v(N) = N/M$ and

$$\frac{\bar{s}_v}{\eta}\frac{\partial\eta}{\partial\bar{s}_v} = -\frac{1}{M}.$$

This agrees exactly with (2.53), which is derived from the product-form solution.

Example 3.4 In this example, we choose $M = 3$, $N = 8$, $\bar{s}_1 = 5$, $\bar{s}_2 = 10$, and $\bar{s}_3 = 12$. The routing probability matrix is

$$Q = \begin{bmatrix} 0 & 0.5 & 0.5 \\ 0.7 & 0 & 0.3 \\ 0.4 & 0.6 & 0 \end{bmatrix}.$$

Equations (3.14)-(3.19) are solved numerically. The elasticities calculated by Equation (3.67) are -0.0365, -0.5133, and -0.4502, which are exactly the same as those given by (2.53). These values satisfy (3.68).

3.5 Other Sensitivities

In the last section, we showed how the formulas for the steady-state sensitivities of the system throughput with respect to mean service times can be derived by using realization probabilities, which reflect the dynamic property of the system. In this section, we shall show that the realization probability provides more information than the parametric sensitivity and can be used to derive other sensitivity formulas. In fact, the realization probability can be viewed as the fundamental "building block" of sensitivity analysis in the sense that all the performace sensitivities can be decomposed into the sum of the effects of all the small perturbations. The cases studied in this section will demonstrate how such decomposition can be accomplished.

3.5.1 Service Rates Change in a Set of States

The realization probability $c(\mathbf{n}, v)$ characterizes the sensitivity of the system throughput with respect to the service rate of server v when the system state is $\mathbf{n}$. Let $\mu_{i,\mathbf{n}}$ be the service rate of server i when the system state is $\mathbf{n}$. Similar to Equation (3.46) is

$$\lim_{L\to\infty} \frac{1}{T_L} \sum_{l=1}^{L} S_l RI_L(\mathbf{n}, v, l)\chi_{\mathbf{n}}(X_l) = p(\mathbf{n})c(\mathbf{n}, v) \qquad w.p.1.$$

By the same analysis as that in Section 3.4, we can prove

$$\left\{\frac{\mu_{v,\mathbf{n}}}{\eta}\frac{\partial \eta}{\partial \mu_{v,\mathbf{n}}}\right\}_{\mu_{v,\mathbf{n}}=\mu_v} = p(\mathbf{n})c(\mathbf{n}, v).$$

The subscript ($\mu_{v,\mathbf{n}} = \mu_v$) indicates that the value in the braces in the above equation is taken at the point $\mu_{v,\mathbf{n}} = \mu_v$. The derivative $\partial\eta/\partial\mu_{v,\mathbf{n}}$ measures the rate of change in η when at state $\mathbf{n}$ the service rate of server v changes from μ_v to $\mu_v + \Delta\mu_{v,\mathbf{n}}$ and all the other service rates remain μ_v. The perturbed system is no longer a state-independent Jackson networks.

Similarly, let μ_{i,n_i} be the service rate of server i when there are n_i customers in it; then

$$\left\{\frac{\mu_{v,j}}{\eta}\frac{\partial \eta}{\partial \mu_{v,j}}\right\}_{\mu_{v,j}=\mu_v} = \sum_{n_v=j} p(\mathbf{n})c(\mathbf{n}, v). \qquad (3.73)$$

Example 3.5 The routing probabilities are the same as those in Example 3.4. $N = 5$, $\bar{s}_1 = 8$, $\bar{s}_2 = 10$, and $\bar{s}_3 = 15$. The realization probabilities and the steady-state probabilities are numerically calculated. The elasticities of the throughput are 0.1655, 0.2910, and 0.5435. We also calculated the elasticities (3.73) for $v = 1$. Their values are

j :	1	2	3	4	5
	0.0343	0.0446	0.0410	0.0300	0.0155

The above results can be further generalized. Let $\Sigma = \Phi \times \Gamma$ be the set of all $(\mathbf{n}, i)$, $\mathbf{n} \in \Phi$ and $i \in \Gamma$. Let $\Sigma_0 \subseteq \Sigma$ be any subset. Suppose that if $(\mathbf{n}, i) \in \Sigma_0$, then the service rates of server i change according to the same rate α_{Σ_0}; i.e., if $(\mathbf{n}, i) \in \Sigma_0$, then the service rate of server i changes from μ_i to $\mu_i + \alpha_{\Sigma_0}\mu_i$. In this case, the sensitivity is

$$\left\{\frac{1}{\eta}\frac{\partial \eta}{\partial \alpha_{\Sigma_0}}\right\}_{\alpha_{\Sigma_0}=0} = \sum_{(i,\mathbf{n})\in\Sigma_0} p(\mathbf{n})c(\mathbf{n}, i).$$

3.5.2 Sensitivities to the Number of Customers

In a slightly different form, the equation derived in Suri [100] can be written as

$$\left\{\frac{\mu_{i,j}}{\eta}\frac{\partial \eta}{\partial \mu_{i,j}}\right\}_{\mu_{i,j}=\mu_i} = p_{N-1}(n_i \geq j) - p_N(n_i \geq j), \qquad (3.74)$$

where $p_N(n_i \geq j)$ is the steady-state probability of $(n_i \geq j)$ in a closed Jackson network with a total of N customers. From Equations (3.73) and (3.74), we have

$$\sum_{n_i=j} p(\mathbf{n})c(\mathbf{n}, i) = p_N(n_i \geq j) - p_{N-1}(n_i \geq j). \qquad (3.75)$$

Note that $\sum_{j=1}^{N} p_N(n_i \geq j) = \bar{n}_i(N)$. From (3.75), we have

$$\sum_{all\ \mathbf{n}} p(\mathbf{n})c(\mathbf{n}, i) = \bar{n}_i(N) - \bar{n}_i(N-1). \qquad (3.76)$$

Let $\eta_N(i) = p_N(n_i \geq 1)\mu_i$ be the steady-state throughput of server i in a closed Jackson network with N total customers. Letting $j = 1$ in (3.75), we get

$$\Delta\eta = \eta_N(i) - \eta_{N-1}(i) = \mu_i\{\sum_{n_i=1} p(\mathbf{n})c(\mathbf{n}, i)\} \qquad (3.77)$$

where $\Delta\eta$ is the increment of the throughput of server i when the number of customers in a Jackson network increases from $N-1$ to N. An interesting feature of this equation is that it establishes a relationship between the realization probability, which measures the sensitivity with respect to infinitesimal changes of a continuous parameter, and the throughput increment because of the change of a discrete parameter.

Example 3.6 Consider the same symmetric network as that in Example 3.1. We have $c(\mathbf{n}, i) = n_i/N$. Also, as calculated in Example 3.3,

$$p_N(n_i = 1) = \frac{N(M-1)}{(M+N-1)(M+N-2)}.$$

Thus, according to (3.77), we have

$$\begin{aligned} \eta_N(i) - \eta_{N-1}(i) &= \mu_i\{\sum_{n_1=1} p(\mathbf{n})c(\mathbf{n}, i)\} \\ &= \mu_i\{\frac{1}{N}\sum_{n_1=1} p(\mathbf{n})\} = \mu_i\{\frac{1}{N}p_N(n_i = 1)\} \\ &= \mu_i\{\frac{(M-1)}{(M+N-1)(M+N-2)}\}. \end{aligned} \quad (3.78)$$

On the other hand, by the product-form formula we have $\eta_N(i) = \mu_i[1 - p_N(n_i = 0)] = \mu_i(\frac{N}{M+N-1})$. The $\Delta\eta$ in (3.78) obtained by the realization probability is exactly the same as $\eta_N(i) - \eta_{N-1}(i)$ given by the product-form formula.

Example 3.7 Consider the same system as in Example 3.2. Using Equation (3.77) and the data in Table 3.1, we get $\Delta\eta(1) = 0.00417$. The throughputs of the system and another system with $N - 1 = 4$ customers are also calculated by using the product-form formula. The values are $\eta_5(1) = 0.08954$ and $\eta_4(1) = 0.08537$. It is clear that $\Delta\eta(1) = \eta_5(1) - \eta_4(1)$.

Our last remark is that using (3.77), one does not need to use the product-form formula for the system with $N - 1$ customers. The practical significance is, the information of $\Delta\eta$ is completely included in a sample path of the system with N customers. One can obtain both $p(\mathbf{n})$ and $c(\mathbf{n}, i)$ by analyzing a single sample path of the system with N customers.

3.5.3 Sensitivities to Any Changes in Service Time Distributions

As discussed in Section 1.2, the theory of the realization probability is similar to the linearization of nonlinear continuous systems. In the latter case, the system trajectory may be perturbed in any pattern. This hints that realization probabilities can be used to study the sensitivities when the perturbed system is not a Jackson network. This is the subject of this section.

We have shown that if the nominal and the perturbed systems are both Jackson networks, then the formulas based on the realization probability yield the same values for throughput derivatives as those obtained by taking the derivatives of the throughputs given by the product-form formulas. However, if the service distributions are not exponential in the perturbed system, then the product-form formulas do not apply to the perturbed system. One can not obtain the sensitivity by taking the derivative. Nevertheless, we shall see that by using the realization probability, one can still obtain formulas for the sensitivities.

Suppose that in the perturbed path the service times of server v are determined by $F_v(\bar{s}_v + \Delta\bar{s}_v, s)$, with means $\bar{s}_v + \Delta\bar{s}_v$, $v \in \Gamma$, as follows:

$$s_{v,k} + \Delta s_{v,k} = F_v^{-1}(\bar{s}_v + \Delta\bar{s}_v, \eta_{v,k}), \qquad v \in \Gamma,\ k = 1, 2, \cdots \tag{3.79}$$

where $\Delta s_{v,k}$ denotes the difference between the perturbed and the nominal service times. $E(\Delta s_{v,k}) = E(s_{v,k} + \Delta s_{v,k}) - E(s_{v,k}) = \Delta\bar{s}_v$. Assume that for any $L < \infty$ and any preassigned positive number $\epsilon > 0$, we can always (with probability one) choose a $\Delta\bar{s}_v$ small enough such that all $|\Delta s_{v,k}|$, determined by (3.79) and satisfying $t_{v,k} < T_L$, are smaller than ϵ. Such a distribution and the nominal distribution are called two *neighboring distributions*. For two neighboring distributions, one can always choose a $\Delta\bar{s}_v$ so small that $\Delta_{i,k} < S_{L,min}$ for all $t_{i,k} \le T_L$, $i = 1, 2, \cdots, M$. That is, the nominal and the perturbed paths are similar in $[0, T_L]$. Thus, the propagation of the perturbations generated according to (3.79) obeys the propagation rules (3.11). Since $\Delta s_{v,k}$ is no longer proportional to $s_{v,k}$, the decomposition (3.36) does not lead to the equivalence of the final contribution of the perturbations to ΔT_L. (Cf. the discussion on perturbation generation for general service-time distributions in Chapter 5.) That is, we cannot replace the perturbation $\Delta s_{v,k}$ by a series of perturbations that are proportional to the lengths of the periods in which the system stays in the same state. The perturbation $\Delta s_{v,k}$ should be generated at the service completion times $t_{v,k}$ and then it will be propagated. Define $C_i = \{t_{i,k}\}_{k=1}^{\infty}$; then $\{T_l\}_{l=0}^{\infty} = \cup_{i=1}^{M} C_i$. By an analysis similar to Corollary 3.2, we obtain

$$\frac{\bar{s}_v}{\eta}\frac{\partial\eta}{\partial\bar{s}_v} = -\lim_{L\to\infty}\{\frac{1}{T_L}\sum_{l=1}^{L}\Delta s_{v,k} RI_t(\mathrm{n}, v, l)\chi_{C_v}(T_l)\}\frac{\bar{s}_v}{\Delta\bar{s}_v} \qquad w.p.1, \tag{3.80}$$

where $\chi_{C_v}(T_l) = 1$ if $T_l \in C_v$, $\chi_{C_v}(T_l) = 0$ if it is otherwise. In (3.80), $\Delta s_{v,k}$ is the perturbation of the service time $s_{v,k}$ of a customer whose service completion time is $t_{v,k} = T_l$. $RI_t(\mathbf{n}, v, l)$ is the realization index of a perturbation at the time instant $t_{v,k} - 0$. The above partial derivative is with respect to $\bar{s}_v$ in a sense that service times change according to (3.79). Similar to (3.43) is

$$\frac{\bar{s}_v}{\Delta \bar{s}_v} \lim_{L\to\infty} \{\frac{1}{L}\sum_{l=1}^{L} \Delta s_{v,k} RI_t(\mathbf{n}, v, l)\chi_{C_v}(T_l)\chi_{\mathbf{n}}(X_l)\}$$

$$= E\{\Delta s_{v,k}\chi_{C_v}(T_l)\chi_{\mathbf{n}}(X_l)\}\ c_t(\mathbf{n}, v)\ \frac{\bar{s}_v}{\Delta \bar{s}_v} \qquad w.p.1, \tag{3.81}$$

where $c_t(\mathbf{n}, v) = E\{RI_t(\mathbf{n}, v, l)\}$ is the realization probability of a perturbation at server v's completion time and when the system is in state $\mathbf{n}$ before this time instant.

To continue the analysis, we consider the particular distribution function $F_v(\bar{s}_v + \Delta \bar{s}_v, \xi)$ in which $\Delta s_{v,k} \equiv \Delta \bar{s}_v$ for all k. Under this assumption we have

$$\begin{aligned} E\{\Delta s_{v,k}\chi_{\mathbf{n}}(X_l)\chi_{C_v}(T_l)\} &= E(\Delta s_{v,k})E\{\chi_{\mathbf{n}}(X_l)\chi_{C_v}(T_l)\} \\ &= \Delta \bar{s}_v\ E(l \in C_v, X_l = \mathbf{n}), \end{aligned}$$

where E denotes the steady-state expected value. Thus,

$$\lim_{L\to\infty} \{\frac{1}{T_L}\sum_{l=1}^{L} \Delta_{v,k} RI_t(\mathbf{n}, v, l)\chi_{C_v}(T_l)\chi_{\mathbf{n}}(X_l)\}\frac{\bar{s}_v}{\Delta \bar{s}_v}$$

$$= \bar{s}_v E(l \in C_v, X_l = \mathbf{n})c_t(\mathbf{n}, v) \lim_{L\to\infty} \frac{L}{T_L}. \tag{3.82}$$

From (3.45), we have $\lim_{L\to\infty} L/T_L = \eta$. On the other hand,

$$\begin{aligned} E(l \in C_v, X_l = \mathbf{n}) &= P(l \in C_v | X_l = \mathbf{n})\pi(\mathbf{n}) \\ &= \sum_{j=1}^{M} P(X_{l+1} = \mathbf{n}_{v,j} | X_l = \mathbf{n})\pi(\mathbf{n}) \\ &= \frac{\pi(\mathbf{n})}{\mu(\mathbf{n})}\mu_v. \end{aligned}$$

This probability can also be obtained by the arrival theorem (see Sevcik and Mitranis [95]). Thus,

$$\lim_{L\to\infty} \{\frac{1}{T_L}\sum_{l=1}^{L} \Delta_{v,k} RI_t(\mathbf{n}, v, l)\chi_{C_v}(T_l)\chi_{\mathbf{n}}(X_l)\}\frac{\bar{s}_v}{\Delta \bar{s}_v}$$

$$= \eta \bar{s}_v \mu_v \frac{\pi(\mathbf{n})}{\mu(\mathbf{n})} c_t(\mathbf{n}, v) = p(\mathbf{n})c_t(\mathbf{n}, v). \tag{3.83}$$

The last equation is due to (3.3). From (3.80) and (3.83), we have

$$\frac{\bar{s}_v}{\eta}\frac{\partial \eta}{\partial \bar{s}_v} = -\sum_{all\ \mathbf{n}} p(\mathbf{n})c_t(\mathbf{n}, v). \tag{3.84}$$

Note that $c_t(\mathbf{n}, i)$ usually does not equal $c(\mathbf{n}, i)$. At the service completion time of server i, a customer will definitely leave the server and will enter server j with probability $q_{i,j}$; i.e., state $\mathbf{n}$ will definitely transfer to $\mathbf{n}_{i,j}$ for some j. Therefore,

$$c_t(\mathbf{n}, i) = \sum_{j=1}^{M} q_{i,j}c(\mathbf{n}_{i,j}, i) + \sum_{j=1}^{M}[1 - \epsilon(n_j)]q_{i,j}c(\mathbf{n}_{i,j}, j). \tag{3.85}$$

We also make the convention that if $n_i = 0$ (there can not be any service completion of server i at state $\mathbf{n}$), then $c_t(\mathbf{n}, i) = 0$. Using (3.84) and (3.85), we can compute the sensitivity of the throughput with respect to the mean service time in the sense that each customer's service time increases by the same amount, $\Delta\bar{s}_v$.

Example 3.8 The system is the same as in Example 3.5. The values of $c_t(\mathbf{n}, i)$ are calculated by (3.85). Equation (3.84) gives the elasticities with respect to μ_i, $i = 1, 2, 3$, as follows:

0.1104 0.2062 0.4224.

These are the elasticities of the system throughput if the service time of each customer decreases by the same amount. Note that the sum of these values is not 1. These values are smaller than their counterparts listed in Example 3.5. This is intuitively reasonable. In Example 3.5 the perturbation is proportional to the service time. Longer service times acquire larger perturbations. These larger perturbations have more opportunity to be realized because at the completion times of long service times there tend to be more customers present at that server, and thus it has more opportunity to terminate the idle periods of other servers.

The above results (3.84) and (3.85) can be extended to the case where $\Delta s_{v,k}$ is independent of X_l. In this case, (3.82) still holds, and the same argument leads to (3.84). To show this precisely, let the perturbed service time be determined by

$$s = -\bar{s}_v ln(1 - \xi) + \Delta\bar{s}_v + g(\xi'), \tag{3.86}$$

where ξ, ξ' are two independent random variables which are uniformly distributed on $[0, 1)$, and $g(\xi')$ is a measurable function on $[0, 1)$ such that $g(\xi') + \Delta\bar{s}_v > 0$ w.p.1. and $E[g(\xi')] = 0$. The form (3.86) is quite general. For example, if the service time distribution changes slightly from an exponential distribution $-\bar{s}_v ln(1 - \xi)$ to a Cox distribution (see Cox [45] or Kleinrock [76]), then we can write $-\bar{s}_v ln(1 - \xi)$ as the first stage of the Cox distribution and $f(\xi') = \Delta\bar{s}_v + g(\xi')$ as its second stage. The above discussion claims that the throughput sensitivity of a system whose service distribution changes according to (3.86) can also be calculated by (3.84) and (3.85).

Finally, Ho and Cao [69] proposed a method for the performance sensitivity with respect to the changes in routing probabilities in closed queuing networks.

Chapter 4

Realization Factors in State-Dependent Networks

In this chapter, we shall generalize the results in Chapter 3 to state-dependent queuing networks and general performance measures. A *state-dependent queuing network* here means a network in which the service requirements are exponentially distributed, but the service rates may depend on the system state (i.e., the numbers of customers in all servers). We study both open and closed networks. The performance measure can take two general forms, customer average and time average. We shall establish results parallel to those in Chapter 3. Especially, we shall prove that the performance sensitivity can be expressed in terms of the expected value of the *realization factor*, a generalization of the realization probability. We shall first study closed queuing networks and then extend the results to open networks. This chapter covers the material presented in Cao [15], [17], [19], Cao and Ma [34], and Glasserman [56].

4.1 Realization Factors and Their Equations

4.1.1 Perturbation Evolution in State-Dependent Networks

The System

Consider a closed queuing network consisting of N single-class customers and M single FCFS server nodes; each has a buffer of an infinite capacity. Customers circulate among servers according to routing probabilities $q_{i,j}$, $i, j = 1, 2, \cdots, M$. The service requirements of all customers at all servers are exponentially distributed, and the service time of a customer at a server is the customer's service requirement divided by the service rate of that server. Without loss of generality, we assume that the mean service requirements are 1 for customers from all the servers. The service rate of any server in the network, however, is assumed to depend on the numbers of customers in all the servers of the network. Let $\mathbf{n} = (n_1, n_2, \cdots, n_M)$, where n_i is the number of customers in server i, $i = 1, 2, \cdots, M$, be the system state and let Φ be the state space; let the service rates of server i at state $\mathbf{n}$ be denoted by $\mu_{i,\mathbf{n}}$. We assume $0 < \mu_{i,\mathbf{n}} < \infty$ for all $i = 1, 2, \cdots, M$, and $\mathbf{n} \in \Phi$.

Similar to the case of closed Jackson networks, a sample path of such a network can be represented by a random vector $\xi = (\xi_1, \xi_2, \cdots, \xi_M; \zeta_1, \zeta_2, \cdots, \zeta_M; \mathbf{n}_0)$, where $\mathbf{n}_0$ is the initial state, ξ_i and ζ_i, $i = 1, 2, \cdots, M$, are random vectors that determine the service times and the customer routes. In particular, $\xi_i = (\xi_{1,1}, \xi_{1,2}, \cdots, \xi_{i,k}, \cdots)$, with $\xi_{i,k}$ being a random variable uniformly distributed on $[0, 1)$; the service time of the kth customer in server i if the system is in state $\mathbf{n}$ can be determined by

$$-\frac{1}{\mu_{i,\mathbf{n}}} ln(1 - \xi_{i,k}) \qquad k = 1, 2, \cdots .$$

Let $(\Omega, \mathcal{F}, P)$ be the underlying probability space. Then the state process is a Markov process defined on Ω. We denote the Markov process as $\mathcal{N}(t, \xi)$ and its embedded Markov chain as $\mathbf{X} = \{X_l\}_{l=0}^{\infty}$.

We assume that the closed queuing network is irreducible and its state process is ergodic (see Section 2.2).

Perturbation Evolution

In a state-dependent network, a perturbation evolves on a sample path in a manner different from that described in Chapter 3. The main difference is

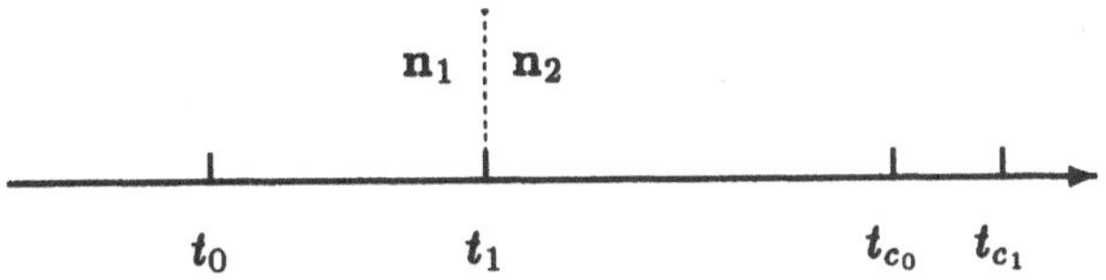

Figure 4.1: The Change of the Service Completion Time

that in a state-dependent network, a perturbation at any server may affect the service completion time of other servers even when no idle periods are involved. The intuitive reason for this is, if the service completion time of server i, $t_{i,k}$, is delayed to $t_{i,k} + \Delta t_{i,k}$, and if at this service completion time the customer in server i moves to server j, then in the time period $[t_{i,k}, t_{i,k}+\Delta t_{i,k})$ the system state changes from $\mathbf{n}$ to $\mathbf{n}_{i,j}$. This change in the system state changes the service rates of other servers, and hence induces perturbations of the service completion times of other servers. Although the nature of obtaining these perturbations is different from perturbation propagation through idle periods, for terminology simplicity we still call them perturbations propagated from server i. The sizes of these perturbations usually do not equal the original perturbation.

To see how a perturbation of a service completion time can be generated, let us suppose that a server, say server v, starts serving a customer at time t_0 at rate $\mu_{v,\mathbf{n}_1}$. Server v has a predicted service completion time t_{c0} with $t_{c0} = t_0 + r/\mu_{v,\mathbf{n}_1}$; r is the service requirement of the customer from server v. If the system jumps to state $\mathbf{n}_2$ at t_1 before t_{c0} because of a customer transition (which may be from any server, see Figure 4.1), then the service rate changes from $\mu_{v,\mathbf{n}_1}$ to $\mu_{v,\mathbf{n}_2}$ and the predicted service completion time t_{c1} at t_1 becomes

$$t_{c1} = t_1 + \frac{\mu_{v,\mathbf{n}_1}}{\mu_{v,\mathbf{n}_2}}(t_{c0} - t_1) = (1-\kappa)t_1 + \kappa t_{c0}$$

with $\kappa = \frac{\mu_{v,\mathbf{n}_1}}{\mu_{v,\mathbf{n}_2}}$. From this equation, we can obtain the change of the service completion time t_{c1}, Δt_{c1}, as follows:

$$\Delta t_{c1} = (1-\kappa)\Delta t_1 + (t_{c0} - t_1)\Delta\kappa + \kappa \Delta t_{c0}. \tag{4.1}$$

Denoting the residual service requirement of the customer from server v at time t_1 by $r_1 = \mu_{v,\mathbf{n}}(t_{c0} - t_1)$, we can rewrite Equation (4.1) in the form

$$\Delta t_{c1} = (1-\kappa)\Delta t_1 + \kappa\Delta t_0 - \kappa(t_1 - t_0)\frac{\Delta\mu_{v,\mathbf{n}_1}}{\mu_{v,\mathbf{n}_1}} - (\frac{r_1}{\mu_{v,\mathbf{n}_2}})\frac{\Delta\mu_{v,\mathbf{n}_2}}{\mu_{v,\mathbf{n}_2}}, \tag{4.2}$$

where the equality

$$\Delta\kappa = \frac{\mu_{v,\mathbf{n}_2}\Delta\mu_{v,\mathbf{n}_1} - \mu_{v,\mathbf{n}_1}\Delta\mu_{v,\mathbf{n}_2}}{\mu_{v,\mathbf{n}_2}{}^2}$$

is used. The quantity Δt_{c1} is the perturbation of server v at t_1; it represents the perturbation of the predicted service completion time of the customer in server v at t_1.

Equation (4.2) describes the perturbation evolution on a sample path. The first two terms on the right-hand side of (4.2) yield the perturbation propagation rule. Δt_1 is the perturbation of the state transition time t_1, or the perturbation of the server that triggers this transition; thus the term $(1-\kappa)\Delta t_1$ reflects the perturbation propagated from that server. In the second term, Δt_0 comes from the perturbation of the starting time of the service: if server i is not idle before t_0, Δt_0 is the perturbation propagated from the previous service completion time of server v itself; otherwise, it is the perturbation of the service completion time of the customer that, coming from another server, terminates the idle period. In other words, Δt_0 obeys the propagation rules on page 81.

The other two terms of (4.2) describe the perturbation generation. The term $-\kappa(t_1 - t_0)\frac{\Delta\mu_{v,\mathbf{n}_1}}{\mu_{v,\mathbf{n}_1}}$ is the perturbation generated in $[t_0, t_1]$, which is proportional to $\Delta\mu_{v,\mathbf{n}_1}$ and $(t_1 - t_0)$, whereas the last term $-(\frac{r_1}{\mu_{v,\mathbf{n}_2}})\frac{\Delta\mu_{v,\mathbf{n}_2}}{\mu_{v,\mathbf{n}_2}}$ is simply the perturbation generated after t_1. Note that in this case, $t_1 - t_0$ is simply the duration of two successive transitions.

It is important that $t_1 + \Delta t_1 > t_0$ holds in (4.1) so that the transitions of the perturbed path and the nominal path occur in the same order. This requires that $\Delta\mu_{v,\mathbf{n}}$ be chosen small enough.

From the above discussion, we conclude that if $X_l = \mathbf{n}$, then the perturbation generated in $[T_l, T_{l+1})$ at server i is

$$\Delta S_l = -(T_l - T_{l-1})\frac{\Delta\mu_{v,\mathbf{n}}}{\mu_{v,\mathbf{n}}} = -\gamma S_l, \tag{4.3}$$

where $\gamma = \Delta\mu_{v,\mathbf{n}}/\mu_{v,\mathbf{n}}$. Equation (4.3) describes the generation of perturbations and is a counterpart of (3.30).

Suppose that at transition time T_l, the hth customer of server j transfers to server m. Based on (4.2) and the above discussion, the propagation rules can be described as follows:

$$\Delta_{i,k} = \begin{cases} \Delta_{j,k-1} & \textit{if } i = j, \\ \Delta_{j,h} & \textit{if } i = m \textit{ and } n_i(T_l-) = 0, \\ (1-\kappa_j)\Delta_{j,h} + \kappa_j \times \Delta_{j,h} & \textit{otherwise}, \end{cases} \tag{4.4}$$

where

$$\kappa_j = \frac{\mu_{j,X(T_{l-1})}}{\mu_{j,X(T_l)}}. \tag{4.5}$$

Equation (4.4) is a counterpart of (3.32).

From (4.4), at any transition time only the perturbation of the transition server remains unchanged; the idle server obtains a perturbation that equals that of the server terminating the idle period; all the other servers obtain perturbations whose sizes are specified in the last equation of (4.4).

It is clear that the superposition holds for perturbation propagation.

4.1.2 General Performance Measures and Realization Factors

In this section, we study the final effect of a perturbation on a general performance measure.

General Performance Measures

Let f be a mapping from Φ to $R = (-\infty, \infty)$. The sample performance measure of interest is defined as

$$\eta_L^{(f)} = \frac{1}{L}\int_0^{T_L} f(\mathbf{N}(t))dt = \frac{1}{L}F_L \qquad L = 1, 2, \cdots \tag{4.6}$$

or

$$\eta_{T_L}^{(f)} = \frac{1}{T_L}\int_0^{T_L} f(\mathbf{N}(t))dt = \frac{1}{T_L}F_L \qquad L = 1, 2, \cdots. \tag{4.7}$$

For example, if $f = I \equiv 1$, then

$$\eta_L^{(I)} = \frac{1}{\eta_L} = \frac{T_L}{L} \qquad L = 1, 2, \cdots,$$

and η_L is the sample throughput defined in (3.33). If $f(\mathbf{n}) = n_i$, then $\eta_L^{(f)}$ is the average response time of server i. Note that the relations

$$\eta_{T_L}^{(f)} = \eta_L \eta_L^{(f)} = \frac{\eta_L^{(f)}}{\eta_L^{(I)}} \qquad L = 1, 2, \cdots \tag{4.8}$$

hold. Equations (4.6) and (4.7) are in the forms of *customer average* and *time average*, respectively.

To indicate the dependency of the sample performance on the sample path ξ and on the parameter $\mu_{i,\mathbf{n}}$, we explicitly write $(\mu_{i,\mathbf{n}}, \xi)$ as the variables in the sample performance functions.

In (4.6) and (4.7), F_L has the form

$$\begin{aligned} F_L &= \int_0^{T_L} f(\mathbf{N}(t))dt \\ &= \sum_{l=0}^{L-1} f(X_l)(T_{l+1} - T_l) = \sum_{l=0}^{L-1} f(X_l)S_l, \qquad L = 1, 2, \cdots. \end{aligned}$$

From this, we have

$$\begin{aligned} \Delta F &= F_L(\mu_{i,\mathbf{n}} + \Delta\mu_{i,\mathbf{n}}, \xi) - F_L(\mu_{i,\mathbf{n}}, \xi) \\ &= \sum_{l=0}^{L-1} f(X_l)\Delta S_l \qquad L = 1, 2, \cdots. \end{aligned} \tag{4.9}$$

Realization Factors

Now, we study the evolution of perturbations on a sample path. Since superposition holds for the propagation rules in state-dependent networks, we need to study only a single perturbation at a server. Suppose that at time $t_0 = 0$, a server obtains a perturbation Δ; this perturbation will be propagated according to (4.4). Since $0 < \mu_{i,\mathbf{n}} < \infty$, all the κ_j's defined in (4.5) are finite. Thus, for any finite L, we can always choose a Δ small enough so that $\Delta_{i,k} < S_{L,min}$ for all $t_{i,k} \leq T_L$. That is, the nominal and the perturbed paths are similar, and (4.4) applies.

In the state-independent Jackson network case, other servers obtain a perturbation with the same size Δ. In state-dependent networks, the size of the perturbation may change during the evolution, because of the perturbation induced at other servers (the last equation of (4.4)). Thus, the definition of perturbation realization has to be modified.

Definition 4.1 *A perturbation Δ of a server in a network is said to be realized with a factor $c(\xi)$ by the network along a sample path ξ if, after being propagated along the sample path, every server in the network finally gets a perturbation with the same size $c(\xi)\Delta$.*

By a method similar to that for Theorem 3.1, we can prove that for an irreducible closed state-dependent network, a perturbation will be realized with probability one. That is, for every sample path ξ, there always (w.p.1) exists an $L^* < \infty$ (depending on ξ) such that

$$\Delta T_L = c(\xi)\Delta \qquad L > L^*. \tag{4.10}$$

The Jackson network discussed in Chapter 3 becomes a special case where $c(\xi) = 1$ or 0. We say a perturbation is lost on a sample path when $c(\xi) = 0$. The factor $c(\xi)$ is called a *sample realization factor*.

After a perturbation Δ is realized by the network with a factor $c(\xi)$, the perturbed path $\{T_L + \Delta T_L\}$ (for $L > L^*$) looks exactly the same as the nominal path $\{T_L\}$ except that every transition time is shifted by the same amount of $c(\xi)\Delta$. Both c and L^* are random variables defined on $(\Omega, \mathcal{F}, P)$. The integer-valued random variable L^* is finite with probability one and is less than or equal to the first stopping time that the system reaches state $(N, 0, ..., 0)$.

Next, we consider a single perturbation Δ of server i at $t_0 = T_l$. Let $\{\Delta S_k\}$ and $\{\Delta F_k\}$, $k \geq l$, be the perturbations of S_k and F_k, respectively, due to this perturbation. It follows from (4.10) that $\Delta S_L = 0$ for all $L > L^*$; consequently, $\Delta F_L = \sum_{k=0}^{L^*-1} f(X_k)\Delta S_k$ are the same for all $L > L^*$. This is stated as a corollary.

Corollary 4.1 *Suppose that at transition time T_l the system state is* $\mathbf{n}$ *and server i obtains a perturbation Δ. Then there exists an integer-valued random variable $L^*(\mathbf{n}, i, l)$ which is finite with probability one, such that ΔF_L are the same for all $L > L^*(\mathbf{n}, i, l)$.*

From this corollary, we can write $\Delta F_L := c^{(f)}(\xi)\Delta$ for $L > L^*$. To emphasize $\mathbf{n}$, i and T_l, we rewrite it as

$$\Delta F_L := c^{(f)}(\mathbf{n}, i, l)\Delta \qquad \textit{for all } L > L^*(\mathbf{n}, i, l), \tag{4.11}$$

where l indicates that the original perturbation is at the lth transition of a sample path ξ. $c^{(f)}(\mathbf{n}, i, l)$ is called the *sample realization factor of the perturbation with respect to f*.

Since the propagation rules depend only on customer transitions and are independent of S_k, $k = l, l+1, \cdots$, the sample realization factor $c^{(f)}$ depends only on the random variables $X_l, X_{l+1}, ...$ with $X_l = \mathbf{n}$. Thus,

$$c^{(f)}(\mathbf{n}, i, l) = \psi(X_l, X_{l+1}, ...) \qquad \textit{with } X_l = \mathbf{n}, \tag{4.12}$$

for some mapping ψ from Φ^∞ to R, independent of l.

For state-independent networks and $f = I \equiv 1$, the factor $c^{(f)}(\mathbf{n}, i, l) = c^{(I)}(\mathbf{n}, i, l)$ reduces to the realization index $RI(\mathbf{n}, i, l)$ (see Definition 3.3). While the realization index measures the final effect of a perturbation on the transition times $\{T_L\}$, the sample realization factor $c^{(f)}(\mathbf{n}, i, l)$ measures the final effect of a perturbation on the sample performance measure $\{F_L\}$.

Definition 4.2

$$c^{(f)}(\mathbf{n}, i) := E[c^{(f)}(\mathbf{n}, i, l)|X_l = \mathbf{n}], \qquad l = 0, 1, \cdots, \tag{4.13}$$

is called the realization factor of the perturbation of server i at state $\mathbf{n}$ with respect to f.

Note that by virtue of (4.12), for time-homogenous systems, the right-hand side of (4.13) is independent of l. For state-independent networks and when $f = I \equiv 1$, we have $c^{(I)}(\mathbf{n}, i) = c(\mathbf{n}, i)$; the realization factor reduces to the realization probability.

4.1.3 Equations for Realization Factors

The State-Dependent Networks

Let V be a subset of $\Gamma = \{1, 2, \cdots, M\}$. Suppose that all the servers in V have a perturbation with the same size Δ and that the system state is $\mathbf{n}$. Denote by $(\mathbf{n}, V)$ the set of perturbations. The perturbation in each server, $(\mathbf{n}, i)$, $i \in V$, will have a final effect on the sample performance $\{F_L\}$ with probability one. Because of the superposition property, the total set of the perturbations $(\mathbf{n}, V)$ will have an aggregated final effect (perturbation) on the sample performance $\{F_L\}$ with probability one. Let $c^{(f)}(\mathbf{n}, V)$ be the realization factor of the perturbation in $(\mathbf{n}, V)$, i.e., the expected value of the finally realized perturbation of the sample performance $\{F_L\}$, corresponding to the perturbations Δ of $(\mathbf{n}, V)$.

The realization factors satisfy the following properties.

Lemma 4.1

$$If\ n_i = 0,\ then\ c^{(f)}(\mathbf{n}, i) = 0; \tag{4.14}$$

$$c^{(f)}(\mathbf{n}, \Gamma) = f(\mathbf{n}); \tag{4.15}$$

If $V_1 \cap V_2 = \emptyset$ and $V_1 \cup V_2 = V_3$, then

$$c^{(f)}(\mathbf{n}, V_1) + c^{(f)}(\mathbf{n}, V_2) = c^{(f)}(\mathbf{n}, V_3); \tag{4.16}$$

and

$$\sum_{i=1}^{M} c^{(f)}(\mathbf{n}, i) = f(\mathbf{n}). \tag{4.17}$$

Proof: We prove only (4.15). Observe that if all servers have the same perturbation at T_l, with $X_l = \mathbf{n}$, then the perturbed path and the nominal path look exactly the same except that on the perturbed path all the transition times from the $(l+1)$st transition onward are delayed (or advanced) by Δ. This implies that $\Delta S_l = \Delta$ and that $\Delta S_k = 0$ for $k \neq l$. Thus, from (4.9), $\Delta F_L = f(\mathbf{n})\Delta$ for $L > l$, and (4.15) is proved. □

Theorem 4.1 *For a state-dependent closed network and any performance function f, the following equations hold:*

$$
\begin{aligned}
&\{\sum_{i=1}^{M} \epsilon(n_i)\mu_{i,\mathbf{n}}\} c^{(f)}(\mathbf{n},k) \\
= \quad & \sum_{i=1,i\neq k}^{M} \sum_{j=1}^{M} \epsilon(n_i)\mu_{i,\mathbf{n}} q_{i,j} \kappa_k(\mathbf{n},\mathbf{n}_{i,j}) c^{(f)}(\mathbf{n}_{i,j},k) \\
+ \quad & \sum_{j=1}^{M} \mu_{k,\mathbf{n}} q_{k,j} \{ c^{(f)}(\mathbf{n}_{k,j},k) + \sum_{i=1,i\neq k}^{M} [1-\kappa_i(\mathbf{n},\mathbf{n}_{k,j})]\epsilon(n_i) c^{(f)}(\mathbf{n}_{k,j},i) \\
+ \quad & [1-\epsilon(n_j)] c^{(f)}(\mathbf{n}_{k,j},j) + f(\mathbf{n}) - f(\mathbf{n}_{k,j}) \}, \\
& \qquad\qquad n_k > 0, \; k \in \Gamma, \qquad\qquad (4.18)
\end{aligned}
$$

where $\kappa_i(\mathbf{n},\mathbf{n}_{k,j}) = \mu_{i,\mathbf{n}}/\mu_{i,\mathbf{n}_{k,j}}$, $\kappa_k(\mathbf{n},\mathbf{n}_{i,j}) = \mu_{k,\mathbf{n}}/\mu_{k,\mathbf{n}_{i,j}}$.

Proof: Suppose that the system is in state, say $X_l = \mathbf{n}$, and that server k has a perturbation Δ. The realization factor of this perturbation with respect to f, $c^{(f)}(\mathbf{n},k)$, is the final effect of the perturbation Δ; this final effect is contributed by various possible perturbations after the next transition X_{l+1} according to the propagation rules. Given the current state $X_l = \mathbf{n}$, the system has a probability $[\epsilon(n_i)\mu_{i,\mathbf{n}}/\sum_{i=1}^{M}\epsilon(n_i)\mu_{i,\mathbf{n}}]q_{i,j}$ of transferring to state $\mathbf{n}_{i,j}$. Two cases are possible. First, the transition does not happen at server k. In this case, according to the updating rule, the perturbation of server k has to be modified by a weighting factor $\kappa_k(\mathbf{n},\mathbf{n}_{i,j})$, and all the other servers remain unperturbed. This explains the first summation on the right-hand side of the equation in (4.18). Second, the transition is due to a customer from server k to server j. In this case, there are four different situations, corresponding to the four different terms in Equation (4.18): (a) Server k keeps the same perturbation Δ; this results in the first term in the second summation on the right-hand side of the equation (in the braces). (b) All other servers obtain a perturbation of a size $(1-\kappa_i(\mathbf{n},\mathbf{n}_{k,j}))$, $i \neq k$; this corresponds to the second term in the summation. (c) If server j is idle before the transition, server j will have a perturbation Δ after the transition; this is expressed in the third term. (d) The last term $f(\mathbf{n}) - f(\mathbf{n}_{k,j})$

comes from the fact that if the departure from k to j is delayed by Δ then the current transition duration at state $\mathbf{n}$ will increase by Δ, and the next transition duration at state $\mathbf{n}_{k,j}$ will decrease by Δ. By the properties of superposition and conditional expectation, Equation (4.18) follows. □

Special Cases

A special case of a state-dependent server is a load-dependent server, whose service rate depends solely on the number of customers in the server. That is, $\mu_{i,\mathbf{n}} = \mu_{i,n_i}$. Equation (4.18) for a system with all load-dependent servers becomes

$$\begin{aligned}
\{\sum_{i=1}^{M} \epsilon(n_i)\mu_{i,n_i}\}c^{(f)}(\mathbf{n},k) &= \sum_{i\neq k}\sum_{j\neq k} \epsilon(n_i)\mu_{i,n_i}q_{i,j}c^{(f)}(\mathbf{n}_{i,j},k) \\
&+\sum_{i\neq k} \epsilon(n_i)\mu_{i,n_i}q_{i,k}\frac{\mu_{k,n_k}}{\mu_{k,n_k+1}}c^{(f)}(\mathbf{n}_{i,k},k) + \mu_{k,n_k}q_{k,k}c^{(f)}(\mathbf{n},k) \\
&+\sum_{j\neq k} \mu_{k,n_k}q_{k,j}\{(1-\frac{\mu_{j,n_j}}{\mu_{j,n_j+1}})\epsilon(n_j)c^{(f)}(\mathbf{n}_{k,j},j) \\
&+(1-\epsilon(n_j))c^{(f)}(\mathbf{n}_{k,j},j) + f(\mathbf{n}) - f(\mathbf{n}_{k,j})\} \\
&\qquad n_k > 0, \quad k \in \Gamma.
\end{aligned} \tag{4.19}$$

If all the service rates are load-independent; i.e., $\mu_{i,n_i} = \mu_i$ for all i and n_i, then $\mu_{i,n_i}/\mu_{i,n_i+1} = 1$, and (4.19) reduces to

$$\begin{aligned}
\{\sum_{i=1}^{M} \epsilon(n_i)\mu_i\}c^{(f)}(\mathbf{n},k) &= \sum_{i=1}^{M}\sum_{j=1}^{M} \epsilon(n_i)\mu_i q_{i,j}c^{(f)}(\mathbf{n}_{i,j},k) \\
&+\sum_{j=1}^{M} \mu_k q_{k,j}\{(1-\epsilon(n_j))c^{(f)}(\mathbf{n}_{k,j},j) + f(\mathbf{n}) - f(\mathbf{n}_{k,j})\} \\
&\qquad n_k > 0, \quad k \in \Gamma.
\end{aligned} \tag{4.20}$$

For $f = I \equiv 1$, this equation becomes the realization probability equation for closed Jackson networks (Equation (3.19)), with $c^{(I)}(\mathbf{n},k)$ corresponding to the realization probability $c(\mathbf{n},k)$.

Networks with Zero Service Rates

We have assumed that the service rates are positive; thus, there is only one closed set of the states, provided that the routing probability matrix Q is irreducible. Suppose some service rates may be zero. Then the state space Φ may be reducible even if Q is irreducible. Some of the states may be transient, and there may be several irreducible subsets in Φ. A network starting with a transient state will eventually reach one of the recurrent subsets.

Let $\Phi_0 \subset \Phi$ be one of the irreducible recurrent subsets. There must be some states $\mathbf{n} \in \Phi_0$ and some $i \in \Gamma$ for which $\mu_{i,\mathbf{n}} = 0$. It is easy to see that if $\mu_{i,\mathbf{n}} = 0$, then the perturbation generated at server i in the period when the state is $\mathbf{n}$ will be lost; this is similar to the idle period. Thus, (4.14) becomes

$$If\ \mu_{i,\mathbf{n}} = 0,\ then\ c^{(f)}(\mathbf{n}, i) = 0. \tag{4.21}$$

It is also easy to verify that Equations (4.15)-(4.18) holds for all $\mathbf{n} \in \Phi_0$.

Now, consider load-dependent networks. Suppose $N \geq n_i^* \geq 0$ and $\mu_{i,n_i^*} = 0$. All the states with $n_i < n_i^*$ are transient states. There may be more than one recurrent subset. For example, in an N-customer network if $\mu_{i,N} = 0$ for all $i \in \Gamma$, then there are N recurrent subsets, each consisting a single state whith $n_i = N$ for some i. A network starting with any transient state with reach one of these absorbing states and stays there. For another example, consider a network consisting of 3 servers and 5 customers. The routing matrix is irreducible. Suppose that $\mu_{1,3} = \mu_{2,3} = 0$ and that all the other service rates are positive. Then there are two recurrent subsets, $\Phi_0 = \{\mathbf{n}:\ n_2 \geq 3,\}$ and $\Phi_0' = \{\mathbf{n}:\ n_3 \geq 3\}$. (4.21) becomes

$$If\ n_i = n_i^*,\ then\ c^{(f)}(\mathbf{n}, i) = 0. \tag{4.22}$$

Examples

Example 4.1 Consider a cyclic queuing network consisting of two servers and two customers. The system parameters are $q_{1,2} = q_{2,1} = 1$, $q_{1,1} = q_{2,2,} = 0$, $N = M = 2$. The states of the system can be denoted as (1,1), (2,0), and (0,2), where the numbers in the parenthesis represent the numbers of customers in server 1 and 2, respectively. The service rates are state-dependent, with $\mu_{1,(1,1)} = 10$, $\mu_{2,(1,1)} = \mu$, $\mu_{1,(2,0)} = 1$, and $\mu_{2,(0,2)} = 1$. We first consider $f = I \equiv 1$, i.e., the throughput case. From Lemma 4.1, we have

$$c[(0,2),1] = c[(2,0),2] = 0.$$

$$c[(0,2),2] = c[(2,0),1] = 1.$$

From Theorem 4.1, we have

$$(\mu_{1,(1,1)}+\mu_{2,(1,1)})c[(1,1),1]=\mu_{2,(1,1)}\frac{\mu_{1,(1,1)}}{\mu_{1,(2,0)}}c[(2,0),1]$$
$$+\mu_{1,(1,1)}\{c[(0,2),1]+(1-\frac{\mu_{2,(1,1)}}{\mu_{2,(0,2)}})c[(0,2),2]\},$$

and

$$(\mu_{1,(1,1)}+\mu_{2,(1,1)})c[(1,1),2]=\mu_{1,(1,1)}\frac{\mu_{2,(1,1)}}{\mu_{2,(0,2)}}c[(0,2),2]$$
$$+\mu_{2,(1,1)}\{c[(2,0),2]+(1-\frac{\mu_{1,(1,1)}}{\mu_{1,(2,0)}})c[(2,0),1]\}.$$

Replacing $\mu_{i,\mathbf{n}}$ with their corresponding values, we get

$$(10+\mu)c[(1,1),1]=10\mu c[(2,0),1]+10\{c[(0,2),1]+(1-\mu)c[(0,2),2],$$

and

$$(10+\mu)c[(1,1),2]=10\mu c[(0,2),2]+10\{c[(2,0),2]+(1-10)c[(2,0),1].$$

Solving these equations, we obtain

$$c[(1,1),1]=\frac{10}{10+\mu},$$

and

$$c[(1,1),2]=\frac{\mu}{10+\mu}.$$

Next, we let $f(\mathbf{n})=f((n_1,n_2))=n_1$; the performance measure $\eta^{(f)}$ is the steady-state mean response time of server 1. We have

$$c^{(f)}((0,2),1)=c^{(f)}((2,0),2)=c^{(f)}((0,2),2)=0,$$
$$c^{(f)}((2,0),1)=2,$$

$$(10+\mu)c^{(f)}((1,1),1)=10\mu c^{(f)}((2,0),1)$$
$$+10\{c^{(f)}((0,2),1)+(1-\mu)c^{(f)}((0,2),2)+f((1,1))-f((0,2))\},$$

and

$$(10+\mu)c^{(f)}((1,1),2)=10\mu c^{(f)}((0,2),2)$$
$$+10\{c^{(f)}((2,0),2)+(1-10)c^{(f)}((2,0),1)+f((1,1))-f((2,0))\}.$$

Solving these equations, we obtain

$$c^{(f)}((1,1),1) = \frac{10+20\mu}{10+\mu},$$

$$c^{(f)}((1,1),2) = -\frac{19\mu}{10+\mu}.$$

These realization factors $c = c^{(I)}$ and $c^{(f)}$ characterize the effect of a perturbation on the throughput and the mean response time, respectively, and will be used in the next section to calculate their sensitivities with respect to μ.

Example 4.2 In this example, we consider a queuing network with 3 servers and 4 customers. The system has a total of 15 states and 45 realization factors. The routing probabilities are $q_{1,1} = 0.3$, $q_{1,2} = 0.3$, $q_{1,3} = 0.4$, $q_{2,1} = 0.5$, $q_{2,2} = 0.1$, $q_{2,3} = 0.4$, $q_{3,1} = 0.1$, $q_{3,2} = 0.2$, and $q_{3,3} = 0.7$. The state-dependent service rates are listed in Tables 4.1. The performance measure of interest is again taken as the steady-state mean response time of server 1; i.e., $f(\mathbf{n}) = f((n_1, n_2, n_3)) = n_1$.

A program was written to find the values for the 45 realization factors of the system by solving the equations in Theorem 4.1 and Lemma 4.1. The values thus obtained are listed in Table 4.2. The table also gives the value of steady-state probabilities, which are obtained by solving the Markov equations for these probabilities. Table 4.2 shows that the realization factor may be negative or greater than 1.

4.1.4 The Uniqueness of the Solutions

To prove the uniqueness of the solutions to the set of equations in Theorem 4.1 and Lemma 4.1, we use an augmented Markov chain. On the embedded Markov chain $\mathbf{X}$, we define $\mathbf{Z} = \{Z_l\}_{l=0}^{\infty}$, with $Z_l = \{X_l, U_l\}$, where $U_l = (u_{l,1}, u_{l,2}, ..., u_{l,M})$ is an M-dimensional real vector, $-\infty < u_{l,i} < \infty$, $i = 1, 2, ..., M$. If $u_{l,i} = 0$ or 1 for all i, then U_l is the vector defined in Section 3.2 that consists of 0's and 1's indicating which servers have obtained the perturbation Δ; in this case, $\mathbf{Z}$ is the same as the augmented Markov chain discussed in Remark 3.1. Δ can be thought to be a unit of the perturbations. In general, $u_{l,i}$ indicates that the perturbation of server i at T_l is $u_{l,i}\Delta$. U_l is called a *perturbation vector*.

The state transition probabilities of the augmented Markov chain $\mathbf{Z}$ can be determined according to the perturbation propagation rules. Namely,

n	$\mu_{1,n}$	$\mu_{2,n}$	$\mu_{3,n}$
(4,0,0)	1.5	0	0
(3,1,0)	2.0	0.5	0
(3,0,1)	1.5	0	0.6
(2,2,0)	1.0	1.0	0
(2,1,1)	0.8	0.6	1.2
(2,0,2)	1.5	0	1.0
(1,3,0)	0.7	0.8	0
(1,2,1)	0.2	0.6	1.0
(1,1,2)	0.4	0.4	0.6
(1,0,3)	1.0	0	2.0
(0,4,0)	0	1.6	0
(0,3,1)	0	1.5	0.6
(0,2,2)	0	1.0	1.0
(0,1,3)	0	0.5	1.4
(0,0,4)	0	0	2.0

Table 4.1: The Service Rates in Example 3.2

n	$p(n)$	$c^{(f)}(n,1)$	$c^{(f)}(n,2)$	$c^{(f)}(n,3)$
(4,0,0)	0.0042	4.0000	0.0000	0.0000
(3,1,0)	0.0123	3.2181	-0.2181	0.0000
(3,0,1)	0.0226	3.1089	0.0000	-0.1089
(2,2,0)	0.0233	1.7442	0.2558	0.0000
(2,1,1)	0.0538	1.5113	0.4982	-0.0095
(2,0,2)	0.0666	2.3949	0.0000	-0.3949
(1,3,0)	0.0229	1.2450	-0.2450	0.0000
(1,2,1)	0.0780	0.4703	0.3921	0.1377
(1,1,2)	0.2085	0.7960	0.3598	-0.1558
(1,0,3)	0.1090	1.4619	0.0000	-0.4619
(0,4,0)	0.0049	0.0000	0.0000	0.0000
(0,3,1)	0.0192	0.0000	-0.0093	0.0093
(0,2,2)	0.0753	0.0000	0.2090	-0.2090
(0,1,3)	0.1701	0.0000	0.1989	-0.1989
(0,0,4)	0.1294	0.0000	0.0000	0.0000

Table 4.2: A Numerical Example of Realization Factors

if $\mathbf{n}' = \mathbf{n}_{i,j}$ and U' is the perturbation vector representing the sizes of the perturbations of all the servers after the transition from state $\mathbf{n}$ and a perturbation vector U, then

$$Q[(\mathbf{n}, U), (\mathbf{n}', U')] = \frac{\epsilon(n_i)\mu_{i,\mathbf{n}}}{\sum_{i=1}^{M} \epsilon(n_i)\mu_{i,\mathbf{n}}} q_{i,j}; \tag{4.23}$$

otherwise,

$$Q[(\mathbf{n}, U), (\mathbf{n}', U')] = 0.$$

Let $\Pi := \Phi \times R^M$ be the state space of the augmented chain $\mathbf{Z}$, and let

$$\Pi_0 = \{(\mathbf{n}, U) \ : \ u_i \equiv u \ \textit{for all } i \textit{ with } n_i > 0\};$$

in other words, each element of Π_0 is a perturbation vector representing a realized perturbation. The set Π_0 is an absorbing set of the augmented chain $\mathbf{Z}$. Furthermore, by the irreducibility of the network and Corollary 4.1, all the states not in Π_0 are transient.

Now, we are ready to prove the uniqueness theorem.

Theorem 4.2 *There exists a unique solution to the set of equations in Theorem 4.1 and Lemma 4.1.*

Proof: The existence is immediate from the definition of the realization factor and the derivation of Theorem 4.1 and Lemma 4.1. From (4.14), (4.17), and (4.18), we have a total of $a = (M+1)K$ equations for $b = MK$ unknown variables (i.e., the realization factors), where $K = \frac{(M+N-1)!}{(M-1)!N!}$ is the number of states. These equations can be rewritten in a matrix form:

$$\mathbf{A}\mathbf{c}^{(f)} = \mathbf{g}^{(f)}, \tag{4.24}$$

where $\mathbf{A}$ is an $a \times b$ matrix, $\mathbf{c}^{(f)}$ a $b \times 1$ vector, and $\mathbf{g}^{(f)}$ an $a \times 1$ vector. The vector $\mathbf{c}^{(f)}$ represents the unknown realization factors. The components of $\mathbf{A}$ are uniquely determined by the coefficients in Equations (4.14), (4.17), and (4.18); these coefficients are independent of f. The vector $\mathbf{g}^{(f)}$ collects all the constants in these equations; it depends on f through (4.17) and (4.18). It is easy to see that if (4.24) has a unique solution for a particular f, then it has a unique solution for any function. Thus, in the following we shall choose $f \equiv 1$; i.e., we shall consider the perturbation on T_l.

The state space Π of the augmented chain $\mathbf{Z}$ is uncountable. However, we can restrict $\mathbf{Z}$ to a countable subspace of Π. Let $\tilde{\Pi}$ be the set of all the $\{\mathbf{n}, U\}'s$ which are reachable from one of the $\{\mathbf{n}, V\}'s$ by perturbation propagation, $\mathbf{n} \in \Phi$, V is a vector of 0's and 1's. $\tilde{\Pi}$ is countable and closed under perturbation propagation; i.e., all the states reachable from a state in $\tilde{\Pi}$ by perturbation propagation belong to $\tilde{\Pi}$. Thus, we can construct the augmented chain in such a way that $Z_l \in \tilde{\Pi}$ and the transition probabilities are the same as those defined in (4.23).

Let Q be the matrix of the transition probabilities of $\mathbf{Z}$, and let $c(\mathbf{n}, U) := c^{(l)}(\mathbf{n}, U)$ be a function defined on Π. Consider the following equations:

$$c(\mathbf{n}, U) = \sum_{i=1}^{M}\sum_{j=1}^{M} \sum_{(\mathbf{n}_{i,j},U')\in\tilde{\Pi}} p[(\mathbf{n}, U), (\mathbf{n}_{i,j}, U')]c(\mathbf{n}_{i,j}, U') \qquad \textit{all } U. \tag{4.25}$$

This can be written in a matrix form:

$$\mathbf{c} = Q\mathbf{c}, \tag{4.26}$$

where $\mathbf{c}$ is a vector whose components consist of all $(\mathbf{n}, U)'s$.

We claim that (4.26) is equivalent to (4.18) for $f \equiv 1$ plus the following linear equation:

$$c(\mathbf{n}, \alpha U^{(1)} + \beta U^{(2)}) = \alpha c(\mathbf{n}, U^{(1)}) + \beta c(\mathbf{n}, U^{(2)}), \tag{4.27}$$

where $U^{(1)}$ and $U^{(2)}$ are two perturbation vectors.

First, we observe that (4.18) is a subset of (4.25) with $U = e^k$, $k = 1, 2, ..., M$, where e^k is a standard unit vector whose kth component is one and whose other components are all zeros. This can be verified by using the perturbation propagation rules, the definition of Q, and the meaning of $c(\mathbf{n}, k)$ in (4.18). In fact, if we denote $U = e^k$, then $c(\mathbf{n}, k) = c(\mathbf{n}, e^k)$, and after a customer transition from server i to server j, $i \neq k$, $U' := e^k_{i,j} = \kappa_k(\mathbf{n}, \mathbf{n}_{i,j})e^k$. By (4.27), $c(\mathbf{n}_{i,j}, U') = \kappa_k(\mathbf{n}, \mathbf{n}_{i,j})c(\mathbf{n}_{i,j}, e^k)$. Similarly, after a transition from server k to server j, the perturbation vector $U = e^k$ changes to

$$U' \equiv e^k_{k,j} = e^k + \sum_{i\neq k}^{M}[(1 - \kappa_i(\mathbf{n}, \mathbf{n}_{k,j})]\epsilon(n_i)e^i + (1 - \epsilon(n_j))e^j.$$

Thus,

$$\begin{aligned} c(\mathbf{n}_{k,j}, U') &\equiv c(\mathbf{n}_{k,j}, e^k_{k,j}) \\ &= c(\mathbf{n}_{k,j}, e^k) + \sum_{i\neq k}^{M}[(1 - \kappa_i(\mathbf{n}, \mathbf{n}_{k,j})]\epsilon(n_i)c(\mathbf{n}_{k,j}, e^i) \\ &\quad +(1 - \epsilon(n_j))c(\mathbf{n}_{k,j}, e^j). \end{aligned}$$

This is the same as the term in the braces on the right-hand side of (4.18) (with $f \equiv 1$). On the other hand, dividing both sides of (4.18) by $\sum_{i=1}^{M} \epsilon(n_i)\mu_{i,\mathbf{n}}$ and using (4.23), we can write (4.18) in the same form as (4.25):

$$\begin{aligned} c(\mathbf{n}, e^k) &= \sum_{i \neq k}^{M} \sum_{j=1}^{M} p[(\mathbf{n}, e^k), (\mathbf{n}_{i,j}, e^k_{i,j})] c(\mathbf{n}_{i,j}, e^k_{i,j}) \\ &+ \sum_{j=1}^{M} p[(\mathbf{n}, e^k), (\mathbf{n}_{k,j}, e^k_{k,j})] c(\mathbf{n}_{k,j}, e^k_{k,j}) \qquad k \in \Gamma. \end{aligned} \tag{4.28}$$

Next, by the linearity and superposition of the propagation rules, the solution to (4.26), $c(\mathbf{n}, U)$, is a linear function of U; i.e., (4.27) holds for the solution of (4.26). Therefore, for any $U = \sum_{k=1}^{M} \alpha_k e^k$, the corresponding equation in (4.25) is equivalent to (4.28) and (4.27).

Finally, Equations (4.14), (4.15), and (4.27) specify the values of $c(\mathbf{n}, U)$ in the absorbing set Π_0. By Theorem 8.41, p.209, of Kemeny et al. [75], we conclude that there is a unique bounded function $c(\mathbf{n}, U)$ on Φ, satisfying (4.25) and with a restriction on Π_0 specified by Equations (4.14), (4.15), and (4.27). This means that there is a unique bounded solution to (4.25), or equivalently, there is a unique bounded solution to the equations in Theorem 4.1 and Lemma 4.1. □

4.2 Sensitivity Formulas for Closed Networks

To derive the steady-state performance sensitivity formulas for state-dependent networks, we first study the convergence property of the sample derivatives and then prove that the limiting values are in fact equal to the steady-state sensitivities. Most results are similar to those in Sections 3.3 and 3.4; we shall focus on new results and new conditions for the results to hold.

4.2.1 The Convergence of Sample Sensitivities

Suppose that one service rate, say $\mu_{v,\mathbf{n}}$, changes to $\mu_{v,\mathbf{n}} + \Delta\mu_{v,\mathbf{n}}$. The sample derivative of $\eta_L^{(f)}$ with respect to $\mu_{v,\mathbf{n}}$ is given by

$$\begin{aligned} \frac{\partial \eta_L^{(f)}}{\partial \mu_{v,\mathbf{n}}} &= \lim_{\Delta\mu_{v,\mathbf{n}} \to 0} \frac{\eta_L^{(f)}(\mu_{v,\mathbf{n}} + \Delta\mu_{v,\mathbf{n}}, \xi) - \eta_L^{(f)}(\mu_{v,\mathbf{n}}, \xi)}{\Delta\mu_{v,\mathbf{n}}} \\ &= \lim_{\Delta\mu_{v,\mathbf{n}} \to 0} \frac{1}{L} \frac{\Delta F_L}{\Delta\mu_{v,\mathbf{n}}}. \end{aligned} \tag{4.29}$$

By (4.8), the sample derivative of $\eta_{T_L}^{(f)}$ with respect to $\mu_{v,\mathbf{n}}$ can be computed as

$$\frac{\partial \eta_{T_L}^{(f)}}{\partial \mu_{v,\mathbf{n}}} = \frac{\partial}{\partial \mu_{v,\mathbf{n}}}\{\frac{\eta_L^{(f)}}{\eta_L^{(I)}}\} = \frac{1}{\eta_L^{(I)}}\frac{\partial \eta_L^{(f)}}{\partial \mu_{v,\mathbf{n}}} - \frac{\eta_L^{(f)}}{(\eta_L^{(I)})^2}\frac{\partial \eta_L^{(I)}}{\partial \mu_{v,\mathbf{n}}}. \qquad (4.30)$$

We shall focus on the sample derivative of the customer average $\eta_L^{(f)}$ in (4.6); the sample derivative of the time average $\eta_{T_L}^{(f)}$ can be easily obtained via (4.30).

Let $c_L^{(f)}(\mathbf{n}, v, l)$ be the factor representing the effect of a perturbation Δ of server v at state $X_l = \mathbf{n}$ on F_L. By Corollary 4.1, as L goes to infinity, this factor converges to the sample realization factor; i.e.,

$$\lim_{L\to\infty} c_L^{(f)}(\mathbf{n}, v, l) = c^{(f)}(\mathbf{n}, v, l) \qquad v = 1, 2, \cdots, M, \quad w.p.1.$$

We have the following equation, which is similar to (3.38),

$$\Delta F_L = -\sum_{l=0}^{L-1} S_l \frac{\Delta \mu_{v,\mathbf{n}}}{\mu_{v,\mathbf{n}}} \chi_{\mathbf{n}}(X_l) c_L^{(f)}(\mathbf{n}, v, l),$$

where $S_l[\Delta\mu_{v,\mathbf{n}}/\mu_{v,\mathbf{n}}]$ is the perturbation generated in $[T_l, T_{l+1})$ (see (4.3)). Substituting this equation into (4.29) gives a simple expression for the sample derivative over the interval $[0, T_L]$:

$$\mu_{v,\mathbf{n}} \frac{\partial \eta_L^{(f)}}{\partial \mu_{v,\mathbf{n}}} = -\frac{1}{L}\sum_{l=0}^{L-1} S_l \chi_{\mathbf{n}}(X_l) c_L^{(f)}(\mathbf{n}, v, l). \qquad (4.31)$$

Our next step is to prove something similar to Theorem 3.3 for state-dependent networks. To this end, we need to impose some conditions on $c^{(f)}(\mathbf{n}, i, l)$'s (they are bounded in the state-independent case). Recall that $\mathbf{n}_1 = (N, 0, 0, ..., 0)$, $L_0 = 0$, and $L_k = min\{l > L_{k-1}, X_l = \mathbf{n}_1\}$. Let $L_d \le L < L_{d+1}$. With the same reason as that for (3.49), we obtain

$$\begin{aligned} &\lim_{L\to\infty} \frac{1}{L}\sum_{l=0}^{L-1} S_l \chi_{\mathbf{n}}(X_l) |c_L^{(f)}(\mathbf{n}, v, l) - c^{(f)}(\mathbf{n}, v, l)| \\ = &\lim_{L\to\infty} \frac{1}{L}\sum_{l=L_d+1}^{L-1} S_l \chi_{\mathbf{n}}(X_l) |c_L^{(f)}(\mathbf{n}, v, l) - c^{(f)}(\mathbf{n}, v, l)|. \qquad (4.32) \end{aligned}$$

Let $\bar{c}_d(\mathbf{n}, v) = max\{|c_L^{(f)}(\mathbf{n}, v, l) - c^{(f)}(\mathbf{n}, v, l)| \ : \ L_d + 1 \le l < L < L_{d+1}\}$; then we have

$$\sum_{l=L_d+1}^{L-1} S_l \chi_{\mathbf{n}}(X_l) |c_L^{(f)}(\mathbf{n}, v, l) - c^{(f)}(\mathbf{n}, v, l)|$$

$$\le \bar{c}_d(\mathbf{n}, v) \sum_{l=L_d+1}^{L-1} S_l \chi_{\mathbf{n}}(X_l) \le \bar{c}_d(\mathbf{n}, v)[T_{L_{d+1}} - T_{L_d}]. \qquad (4.33)$$

Condition 4.1 *The integrability condition:*

$$E[|\bar{c}_d(\mathbf{n}, v)(T_{L_{d+1}} - T_{L_d})|] < \infty.$$

Let $\eta^{(f)}$ be the steady-state customer-average performance. By ergodicity, we have

$$\eta^{(f)} = \lim_{L \to \infty} \eta_L^{(f)} = \sum_{\mathbf{n}} \frac{f(\mathbf{n})\pi(\mathbf{n})}{\mu(\mathbf{n})} \qquad w.p.1. \qquad (4.34)$$

For $f = I \equiv 1$, $\eta_L^{(f)} = \eta_L^{(I)} = T_L/L$ is the the average time between two transitions. We have

$$\lim_{L \to \infty} \eta_L^{(I)} := \eta^{(I)} = \frac{1}{\eta} \qquad w.p.1, \qquad (4.35)$$

where η is the system throughput. For notational simplicity, we write $c(\mathbf{n}, v) := c^{(I)}(\mathbf{n}, v)$. For a state-independent network, $c(\mathbf{n}, v)$ is the realization probability. We call

$$\frac{\mu_{v,\mathbf{n}}}{\eta_L^{(I)}} \frac{\partial \eta_L^{(f)}}{\partial \mu_{v,\mathbf{n}}}$$

the *sample normalized derivative* of $\eta_L^{(f)}$ with respect to $\mu_{v,\mathbf{n}}$. We have the following theorem.

Theorem 4.3 *Under the integrability Condition 4.1, the sample normalized derivative of $\eta_L^{(f)}$ with respect to $\mu_{i,\mathbf{n}}$ in a state-dependent network satisfies*

$$\lim_{L \to \infty} \frac{\mu_{v,\mathbf{n}}}{\eta_L^{(I)}} \frac{\partial \eta_L^{(f)}}{\partial \mu_{v,\mathbf{n}}} = -p(\mathbf{n}) c^{(f)}(\mathbf{n}, v) \qquad w.p.1. \qquad (4.36)$$

Proof: By (4.32), (4.33), and Condition 4.1, we can prove

$$\lim_{L\to\infty} \frac{1}{L}\sum_{l=0}^{L-1} S_l\chi_{\mathbf{n}}(X_l)|c_L^{(f)}(\mathbf{n},v,l) - c^{(f)}(\mathbf{n},v,l)| = 0 \qquad w.p.1.$$

From this equation and (4.31), we have

$$\lim_{L\to\infty} \mu_{v,\mathbf{n}} \frac{\partial \eta_L^{(f)}}{\partial \mu_{v,\mathbf{n}}} = -\lim_{L\to\infty} \frac{1}{L}\sum_{l=0}^{L-1} S_l\chi_{\mathbf{n}}(X_l)c^{(f)}(\mathbf{n},v,l) \qquad w.p.1. \quad (4.37)$$

By ergodicity, this leads to

$$\begin{aligned} &\lim_{L\to\infty} \mu_{v,\mathbf{n}} \frac{\partial \eta_L^{(f)}}{\partial \mu_{v,\mathbf{n}}} = E\{S_l\chi_{\mathbf{n}}(X_l)c^{(f)}(\mathbf{n},v,l)\} \\ = \; & E\{S_l c^{(f)}(\mathbf{n},v,l)|X_l = \mathbf{n}\}\pi(\mathbf{n}) \\ = \; & -\frac{\pi(\mathbf{n})}{\mu(\mathbf{n})} c^{(f)}(\mathbf{n},v), \qquad w.p.1. \end{aligned} \quad (4.38)$$

Equation (4.36) follows directly from (4.35), (4.38), and (3.3). □

Using (4.33), we can also prove that under Condition 4.1,

$$\lim_{L\to\infty} \frac{1}{L}\sum_{l=0}^{L-1} E\{S_l\chi_{\mathbf{n}}(X_l)|c_L^{(f)}(\mathbf{n},v,l) - c^{(f)}(\mathbf{n},v,l)|\} = 0;$$

together with (4.31), this leads to

$$\lim_{L\to\infty} \mu_{v,\mathbf{n}} E\{\frac{\partial \eta_L^{(f)}}{\partial \mu_{v,\mathbf{n}}}\} = -\lim_{L\to\infty} \frac{1}{L}\sum_{l=0}^{L-1} E\{S_l\chi_{\mathbf{n}}(X_l)c^{(f)}(\mathbf{n},v,l)\}. \quad (4.39)$$

Equation (4.39) will be used in the next section to derive formulas for the steady-state sensitivity.

Remark 4.1 The integrability Condition 4.1 holds if, for instance, $\bar{c}_d(\mathbf{n},v)$ is bounded. This is true for systems with state-independent service rates. In general, to establish the boundness of $\bar{c}_d(\mathbf{n},v)$ is not easy, and the study of the integrability of $\bar{c}_d(\mathbf{n},v)(T_{L_{d+1}} - T_{L_d})$ is much more involved. The condition is sufficient and far from necessary. We strongly suspect that Theorem 4.3 holds for almost all cases. □

Let

$$\eta_T^{(f)} := \lim_{L\to\infty} \eta_{T_L}^{(f)}$$

be the steady-state time-average performance. By (4.8) and (4.34), we have

$$\eta_T^{(f)} = \frac{\eta^{(f)}}{\eta^{(I)}} = \sum_{\mathbf{n}} p(\mathbf{n}) f(\mathbf{n}). \tag{4.40}$$

Using (4.30) and Theorem 4.3, we can obtain the sample derivative for $\eta_{T_L}^{(f)}$.

Theorem 4.4 *Under Condition 4.1, the sample derivative of $\eta_{T_L}^{(f)}$ with respect to $\mu_{i,\mathbf{n}}$ satisfies*

$$\lim_{L\to\infty} \mu_{v,\mathbf{n}} \frac{\partial \eta_{T_L}^{(f)}}{\partial \mu_{v,\mathbf{n}}} = \eta_T^{(f)} p(\mathbf{n}) c(\mathbf{n}, v) - p(\mathbf{n}) c^{(f)}(\mathbf{n}, v) \qquad w.p.1. \tag{4.41}$$

4.2.2 Sensitivities of Steady-State Performance

We are ready to derive the formulas for the steady-state performance sensitivities. First, we study the sample performance function.

Lemma 4.2 *For any ξ, the sample performance $F_L(\mu_{i,\mathbf{n}}, \xi)$ is a continuous and piecewise linear function of $\mu_{i,\mathbf{n}}$. The slopes of all the segments of $F_L(\mu_{i,\mathbf{n}}, \xi)$ are bounded by $K_L T_L(\mu_{i,\mathbf{n}}, \xi)/\mu_{i,\mathbf{n}}$, where $K_L < \infty$ is a finite number.*

Proof: Recall that $F_L(\mu_{i,\mathbf{n}}, \xi)$ is a linear combination of all $T_l(\mu_{i,\mathbf{n}}, \xi)$'s for $l \le L$. Since $T_l(u_{i,\mathbf{n}}, \xi)$ is a continuous and piecewise linear function of $\mu_{i,\mathbf{n}}$ by Lemma 3.7, the first part of this lemma follows.

For the second part of the lemma, we observe that if $\Delta_{l,max}$ denotes the biggest perturbation of all servers at time T_l, $\kappa_j = \frac{\mu_{j,\mathbf{n}}}{\mu_{j,\mathbf{n}_{k,m}}}$, and

$$\kappa = max\{|1-\kappa_j| + |\kappa_j|, \ for\ any\ j, k, m,\ and\ \mathbf{n}\},$$

then by the propagation rules, $|\Delta_{l+1,max}| \le \kappa |\Delta_{l,max}|$. This implies that the perturbation propagated to T_l because of any Δ generated in $[0, T_l]$ is always bounded by $\kappa^l \Delta := K_l \Delta$. Since the total perturbation generated in $[0, T_l]$ because of $\Delta\mu_{i,\mathbf{n}}$ is less than $T_l(\mu_{i,\mathbf{n}}, \xi)\frac{\Delta\mu_{i,\mathbf{n}}}{\mu_{i,\mathbf{n}}}$, we have

$$|\Delta T_l| \le K_l T_l(\mu_{i,\mathbf{n}}, \xi) |\frac{\Delta\mu_{i,\mathbf{n}}}{\mu_{i,\mathbf{n}}}|.$$

By (4.9), the above equation, and the fact that $\{T_l\}$ is increasing, ΔF_L can be easily bounded by

$$|\Delta F_L| \leq |f| \sum_{l=0}^{L-1} (|\Delta T_{l+1}| + |\Delta T_l|)$$

$$\leq \ |f| \sum_{l=0}^{L-1} (K_l + K_{l+1}) T_L(\mu_{i,\mathbf{n}}, \xi) | \frac{\Delta \mu_{i,\mathbf{n}}}{\mu_{i,\mathbf{n}}}|,$$

with $|f| = \max_{\mathbf{n} \in \Phi} |f(\mathbf{n})| < \infty$. Set $K_L = |f| \sum_{l=0}^{L-1} (K_l + K_{l+1})$. Thus, we have

$$|\frac{\Delta F_L}{\Delta \mu_{i,\mathbf{n}}}| \leq K_L \frac{T_L(\mu_{i,\mathbf{n}}, \xi)}{\mu_{i,\mathbf{n}}}.$$

This completes the proof. □

By Lemma 4.2 and the Lebesgue dominated convergence theorem, we can prove the following Lemma.

Lemma 4.3 *For state-dependent networks,*

$$E\{\frac{\partial}{\partial \mu_{i,\mathbf{n}}} [\eta_L^{(f)}(\mu_{i,\mathbf{n}}, \xi)] | \mathbf{n}_0\} = \frac{\partial}{\partial \mu_{i,\mathbf{n}}} \{E[\eta_L^{(f)}(\mu_{i,\mathbf{n}}, \xi) | \mathbf{n}_0]\}, \qquad (4.42)$$

where $\mathbf{n}_0$ *is the initial state of the sample path.*

Next, suppose that the initial state probability is the steady-state probability $\pi(\mathbf{n})$. The same reasoning as that in Theorem 3.5 leads to the next theorem.

Theorem 4.5 *The sample derivative of* $\eta_L^{(f)}$ *with respect to* $\mu_{v,\mathbf{n}}$ *is an asymptotically unbiased estimate of the derivative of the steady-state performance* $\eta^{(f)}$*; i.e.,*

$$\frac{\partial \eta^{(f)}}{\partial \mu_{v,\mathbf{n}}} = \lim_{L \to \infty} E\{\frac{\partial \eta_L^{(f)}(\mu_{v,\mathbf{n}}, \xi)}{\partial \mu_{v,\mathbf{n}}}\}. \qquad (4.43)$$

Finally, from (4.38) and (4.39), we can easily derive the formula for the normalized derivative of the steady-state performance $\eta^{(f)}$.

Theorem 4.6 *Under the integrability Condition 4.1, the normalized derivative of the steady-state customer-average performance in a state-dependent network takes the following form:*

$$\frac{\mu_{v,\mathbf{n}}}{\eta^{(I)}} \frac{\partial \eta^{(f)}}{\partial \mu_{v,\mathbf{n}}} = -p(\mathbf{n}) c^{(f)}(\mathbf{n}, v) \qquad w.p.1. \qquad (4.44)$$

The derivative of the steady-state time-average performance satisfies

$$\mu_{v,\mathbf{n}} \frac{\partial \eta_T^{(f)}}{\partial \mu_{v,\mathbf{n}}} = \eta_T^{(f)} p(\mathbf{n}) c(\mathbf{n}, v) - p(\mathbf{n}) c^{(f)}(\mathbf{n}, v) \qquad w.p.1. \tag{4.45}$$

The following equations follow directly from (4.17), (4.44), and (4.45).

$$\sum_{i=1}^{M} \frac{\mu_{i,\mathbf{n}}}{\eta^{(I)}} \frac{\partial \eta^{(f)}}{\partial \mu_{i,\mathbf{n}}} = -p(\mathbf{n}) f(\mathbf{n}),$$

$$\sum_{i=1}^{M} \mu_{i,\mathbf{n}} \frac{\partial \eta_T}{\partial \mu_{i,\mathbf{n}}} = \eta^{(f)} p(\mathbf{n}) - p(\mathbf{n}) f(\mathbf{n}),$$

$$\sum_{\mathbf{n}} \sum_{i=1}^{M} \frac{\mu_{i,\mathbf{n}}}{\eta^{(I)}} \frac{\partial \eta^{(f)}}{\partial \mu_{i,\mathbf{n}}} = -\sum_{\mathbf{n}} p(\mathbf{n}) f(\mathbf{n}) = -\eta_T^{(f)},$$

and

$$\sum_{\mathbf{n}} \sum_{i=1}^{M} \mu_{i,\mathbf{n}} \frac{\partial \eta_{T_L}}{\partial \mu_{i,\mathbf{n}}} = \eta_T^{(f)} - \sum_{\mathbf{n}} p(\mathbf{n}) f(\mathbf{n}) = 0.$$

More general cases will be discussed in the next section.

If there is a server i in the network whose service rate depends on only n_i; i.e., $\mu_{i,\mathbf{n}} = \mu_{i,n_i}$, we have

$$\frac{\mu_{i,k}}{\eta^{(I)}} \frac{\partial \eta^{(f)}}{\partial \mu_{i,k}} = -\sum_{n_i=k} p(\mathbf{n}) c^{(f)}(\mathbf{n}, i), \tag{4.46}$$

and

$$\mu_{i,k} \frac{\partial \eta_T^{(f)}}{\partial \mu_{i,k}} = \sum_{n_i=k} p(\mathbf{n}) \{\eta_T^{(f)} c(\mathbf{n}, i) - c^{(f)}(\mathbf{n}, i)\}.$$

If some service rates are zero, then

$$\frac{\mu_{i,k}}{\eta^{(I)}} \frac{\partial \eta^{(f)}}{\partial \mu_{i,k}} = -\sum_{\mathbf{n} \in \Phi_0,\ n_i=k} p(\mathbf{n}) c^{(f)}(\mathbf{n}, i), \tag{4.47}$$

and

$$\mu_{i,k} \frac{\partial \eta_T^{(f)}}{\partial \mu_{i,k}} = \sum_{\mathbf{n} \in \Phi_0,\ n_i=k} p(\mathbf{n}) \{\eta_T^{(f)} c(\mathbf{n}, i) - c^{(f)}(\mathbf{n}, i)\},$$

where Φ_0 is one of the recurrent subsets of the states in which the system is in. If server i in the network has a constant service rate, i.e., $\mu_{i,\mathbf{n}} = \mu_i$, we have

$$\frac{\mu_i}{\eta^{(I)}} \frac{\partial \eta^{(f)}}{\partial \mu_i} = -\sum_{\mathbf{n}} p(\mathbf{n}) c^{(f)}(\mathbf{n}, i),$$

and

$$\mu_i \frac{\partial \eta_T^{(f)}}{\partial \mu_i} = \sum_{\mathbf{n}} p(\mathbf{n}) \{\eta_T^{(f)} c(\mathbf{n}, i) - c^{(f)}(\mathbf{n}, i)\}.$$

Note that in all the above cases, the other servers' service rates may be state dependent. The above equations are special cases of the linear formulas to be derived in the next section.

Example 4.3 Consider the same network as the one in Example 4.1. Solving the flow-balance equation, we obtain the steady-state probabilities as

$$p(1,1) = \frac{1}{11+\mu},$$

$$p(0,2) = \frac{10}{11+\mu},$$

and

$$p(2,0) = \frac{\mu}{11+\mu}.$$

The steady-state throughput η and the quantity $\eta^{(f)}$ can be easily computed as

$$\eta = \sum_{all\ \mathbf{n}} p(\mathbf{n}) \mu(\mathbf{n}) = \frac{2(10+\mu)}{11+\mu} = \frac{1}{\eta^{(I)}},$$

and

$$\eta^{(f)} = \eta^{(I)} \sum_{all\ \mathbf{n}} p(\mathbf{n}) f(\mathbf{n}) = \frac{1+2\mu}{2(10+\mu)}.$$

By directly taking the derivative with respect to μ and using the realization factors calculated in Example 4.1, we can verify Theorem 4.6. In fact, they both give the same value as

$$\frac{\mu}{\eta} \frac{\partial \eta}{\partial \mu} = p(1,1) c[(1,1),2] = \frac{\mu}{(10+\mu)(11+\mu)},$$

and

$$\frac{\mu}{\eta^{(I)}} \frac{\partial \eta^{(f)}}{\partial \mu} = -p(1,1) c^{(f)}((1,1),2) = \frac{19\mu}{(10+\mu)(11+\mu)}.$$

n	$-p(\mathbf{n})c^{(f)}(\mathbf{n},1)$	$-p(\mathbf{n})c^{(f)}(\mathbf{n},2)$	$-p(\mathbf{n})c^{(f)}(\mathbf{n},3)$
(4,0,0)	-0.0168	0.0000	0.0000
(3,1,0)	-0.0395	0.0027	0.0000
(3,0,1)	-0.0702	0.0000	0.0025
(2,2,0)	-0.0406	-0.0059	0.0000
(2,1,1)	-0.0813	-0.0268	0.0005
(2,0,2)	-0.1596	0.0000	0.0263
(1,3,0)	-0.0285	0.0056	0.0000
(1,2,1)	-0.0367	-0.0306	-0.0107
(1,1,2)	-0.1660	-0.0750	0.0325
(1,0,3)	-0.1594	0.0000	0.0504
(0,4,0)	0.0000	0.0000	0.0000
(0,3,1)	0.0000	0.0002	-0.0002
(0,2,2)	0.0000	-0.0157	0.0157
(0,1,3)	0.0000	-0.0338	0.0338
(0,0,4)	0.0000	0.0000	0.0000

Table 4.3: A Numerical Example of Normalized Derivatives

Example 4.4 In this example, we continue the numerical calculation in Example 4.2. Recall that $f(\mathbf{n}) = n_1$. The values of the realization factors are shown in Table 4.2. The values of $-p(\mathbf{n})c^{(f)}(\mathbf{n},i)$ for all n's and $i = 1, 2, 3$ are calculated and listed in Table 4.3.

To verify Theorem 4.6, the normalized derivatives are also obtained by the numerical method based on taking the derivative of the probability flow-balance equations. The values thus obtained are exactly the same as those listed in Table 4.3.

4.3 Linear Formulation of Performance Sensitivities

4.3.1 System Dynamics Revisited

From Section 2.4, the solution to the dynamic equation

$$\dot{x}(t) = A(t)x(t) + B(t)u(t) \tag{4.48}$$

with $x(0) = 0$ is

$$x(t) = \int_0^t \phi(t,s)B(s)u(s)ds, \tag{4.49}$$

where $\phi(t, s)$ is the state transition function of the system. The kth column vector of the matrix $\phi(t, s)$ is the basic solution to (Szidarovszky and Bahill [102])

$$\dot{x}(t) = A(t)x(t), \qquad x(s) = e_k, \tag{4.50}$$

where e_k is the unit vector with the kth component being one. Equations (4.49) and (4.50) imply that the solution to (4.48) is completely determined by the n basic solutions to (4.50), and that the system trajectory for any $B(t)$ and $u(t)$ can be obtained by an integration of a linear combination of these basic solutions.

For a nonlinear system

$$\dot{x} = g(x, \theta(t), t), \tag{4.51}$$

where $\theta(t)$ is a time-varying parameter, we have

$$\Delta\dot{x}(t) = \frac{\partial}{\partial x}\{g(x, \theta(t), t)\}\Delta x(t) + \frac{\partial}{\partial \theta}\{g(x, \theta(t), t)\}\Delta\theta(t).$$

Thus, $\Delta x(t)$ can be determined by the n basic solutions to the following equation:

$$\Delta\dot{x}(t) = \frac{\partial}{\partial x}\{g(x, \theta(t), t)\}\Delta x(t).$$

Consequently, the performance sensitivities of a CVDS, which depends on $\Delta x(t)$, can also be determined by those basic solutions.

The purpose of this section is to establish similar results for the performance sensitivity of queuing networks. Because of the similarity of the realization probability theory and the linearization of a perturbed nonlinear system, we can expect that there exists a set of basic solutions for the sensitivities of performance in queuing networks. We shall show that there is a fundamental matrix, called the *sensitivity matrix*, and that the sensitivity of any performance measure is a linear combination of the entries of the matrix. In other words, the column vectors of the matrix can be viewed as the basic solutions in sensitivity analysis. Each component of the vector corresponds to a sensitivity of a basic performance measure with respect to a service parameter, and the sensitivities of all other performance measures with respect to the same parameter are linear combinations of the corresponding components in all the basic vectors.

4.3.2 The Sensitivity Matrix

The realization factor $c^{(f)}(\mathbf{n}, i)$ measures the final effect of a perturbation of server i at state $\mathbf{n}$ on the performance measure. It can be viewed as the basic element in sensitivity study. The set of equations for realization factors has the matrix form stated in (4.24):

$$\mathbf{A}\mathbf{c}^{(f)} = \mathbf{g}^{(f)}, \tag{4.52}$$

where $\mathbf{c}^{(f)}$ is an $a \times 1$ vector whose components are the realization factors, $\mathbf{g}^{(f)}$ is a $b \times 1$ vector corresponding to the constants in the equations, and $\mathbf{A}$ is an $a \times b$ matrix, with $a = (M+1)K$ and $b = MK$. The order of the equations in the matrix form is not important; thus, the order of the components in $\mathbf{g}^{(f)}$ and $\mathbf{c}^{(f)}$ and that of the row vectors in $\mathbf{A}$ are not important as long as they are consistent with Equations (4.14), (4.17), and (4.18). Hereafter, we shall refer to each component of $\mathbf{g}^{(f)}$ and $\mathbf{c}^{(f)}$ as the $(\mathbf{n}, i)$th component and to each column vector in $\mathbf{A}$ as the $(\mathbf{n}, i)$th vector.

The equations in (4.52) contain $M \times K$ linearly independent ones. Note that the matrix $\mathbf{A}$ does not depend on f. From the fundamental theory of linear equations, there are $M \times K$ basic solutions to the set of equations; each of them is an $(M \times K)$-dimensional vector.

Instead of the pure algebraic method, we shall use a direct method to derive the set of basic vectors. We shall show that for calculating the performance sensitivity, the number of basic vectors can be reduced to K $(M \times K)$-dimensional row vectors, or, equivalently, to $(M \times K)$ K-dimensional column vectors.

As the first step, we define an indicator function for any state $\mathbf{m} \in \Phi$:

$$\delta_{\mathbf{m}}(\mathbf{n}) = \begin{cases} 1 & if\ \mathbf{n} = \mathbf{m}, \\ 0 & if\ \mathbf{n} \neq \mathbf{m}. \end{cases}$$

Then any performance function takes the form

$$f(\mathbf{n}) = \sum_{all\ \mathbf{m}} f(\mathbf{m})\delta_{\mathbf{m}}(\mathbf{n}).$$

Substituting this equation into $F_L^{(f)}$ defined in (4.6), we get

$$\begin{aligned} F_L^{(f)} &= \int_0^{T_L} f(\mathbf{N}(t))dt \\ &= \sum_{all\ \mathbf{m}} \{f(\mathbf{m}) \int_0^{T_L} \delta_{\mathbf{m}}(\mathbf{N}(t))dt\} \\ &= \sum_{all\ \mathbf{m}} \{f(\mathbf{m}) F_L^{(\delta_{\mathbf{m}})}\}. \end{aligned} \tag{4.53}$$

From this, we have

$$\Delta F_L^{(f)} = \sum_{all\ \mathbf{m}} \{f(\mathbf{m}) \Delta F_L^{(\delta_{\mathbf{m}})}\}.$$

Then, from the definition of the realization factor, we obtain

$$c^{(f)}(\mathbf{n}, i) = \sum_{all\ \mathbf{m}} f(\mathbf{m}) c_{\mathbf{m}}(\mathbf{n}, i), \tag{4.54}$$

where, for notational simplicity, we use $c_{\mathbf{m}}(\mathbf{n}, i)$ for the realization factor of $\delta_{\mathbf{m}}(\mathbf{n})$.

From Equation (4.54), all the performance sensitivities can be calculated by using $c_{\mathbf{m}}(\mathbf{n}, i)$.

Example 4.5 The derivative of the steady-state time-average performance in (4.45) equals

$$\mu_{i,\mathbf{n}} \frac{\partial \eta_T^{(f)}}{\partial \mu_{i,\mathbf{n}}} = p(\mathbf{n}) \{ \eta_T^{(f)} c(\mathbf{n}, i) - \sum_{all\ \mathbf{m}} f(\mathbf{m}) c_{\mathbf{m}}(\mathbf{n}, i) \}, \tag{4.55}$$

where $c(\mathbf{n}, i)$ is the realization factor for

$$f \equiv 1 = \sum_{all\ \mathbf{m}} \delta_{\mathbf{m}}(\mathbf{n})$$

and, therefore, equals

$$c(\mathbf{n}, i) = \sum_{all\ \mathbf{m}} c_{\mathbf{m}}(\mathbf{n}, i).$$

We can easily verify Equation (4.55). In fact, from (4.40), the steady-state performance defined by $\delta_{\mathbf{m}}(\mathbf{n})$ is

$$\eta_T^{(\delta \mathbf{m})} = \sum_{all\ \mathbf{n}} p(\mathbf{n}) \delta_{\mathbf{m}}(\mathbf{n}) = p(\mathbf{m}).$$

Thus, (4.45) gives us

$$\mu_{i,\mathbf{n}} \frac{\partial p(\mathbf{m})}{\partial \mu_{i,\mathbf{n}}} = p(\mathbf{n}) p(\mathbf{m}) c(\mathbf{n}, i) - p(\mathbf{n}) c_{\mathbf{m}}(\mathbf{n}, i). \tag{4.56}$$

Substituting (4.40) into (4.55) and using (4.56), we get

$$\begin{aligned} \mu_{i,\mathbf{n}} \frac{\partial \eta_T^{(f)}}{\partial \mu_{i,\mathbf{n}}} &= p(\mathbf{n}) \{ [\sum_{all\ \mathbf{m}} p(\mathbf{m}) f(\mathbf{m})] c(\mathbf{n}, i) - \sum_{all\ \mathbf{m}} f(\mathbf{m}) c_{\mathbf{m}}(\mathbf{n}, i) \} \\ &= \sum_{all\ \mathbf{m}} \{ p(\mathbf{n}) [p(\mathbf{m}) c(\mathbf{n}, i) - c_{\mathbf{m}}(\mathbf{n}, i)] f(\mathbf{m}) \} \\ &= \mu_{i,\mathbf{n}} \{ \sum_{all\ \mathbf{m}} \frac{\partial p(\mathbf{m})}{\partial \mu_{i,\mathbf{n}}} f(\mathbf{m}) \}. \end{aligned} \tag{4.57}$$

This shows that (4.55) coincides with the value given by directly taking the derivative of Equation (4.40).

Equation (4.54) shows that the realization factor for any performance function f, $c^{(f)}(\mathbf{n}, i)$, can be obtained from $c_{\mathbf{m}}(\mathbf{n}, i)$. Therefore, $c_{\mathbf{m}}(\mathbf{n}, i)$, $\mathbf{m}, \mathbf{n} \in \Phi$, $i = 1, 2, ..., M$, form a base for the realization factors. Let $\mathbf{c}^{(f)}$ be an $(M \times K)$-dimensional vector whose $(\mathbf{n}, i)$th component is $c^{(f)}(\mathbf{n}, i)$, $\mathbf{f}$ be an M-dimensional vector whose mth component is $f(\mathbf{m})$, and $\mathbf{c}_{\mathbf{m}}$ (for a fixed $\mathbf{m}$) be an $(M \times K)$-dimensional vector whose $(\mathbf{n}, i)$th component is $c_{\mathbf{m}}(\mathbf{n}, i)$. Then $\mathbf{c}^{(f)}$ is a linear combination of $\mathbf{c}_{\mathbf{m}}$'s. Let $\mathbf{C}$ be a matrix whose mth column is $\mathbf{c}_{\mathbf{m}}$. In a matrix form, (4.45) becomes

$$\mathbf{c}^{(f)} = \mathbf{C}\mathbf{f}. \tag{4.58}$$

$\mathbf{C}$ is an $(M \times K) \times K$ matrix and is called the *realization matrix.*

The basic vectors $\mathbf{c}_{\mathbf{m}}$'s can be obtained from (4.14), (4.17), and (4.18), or the matrix equation (4.52). Note that the constant vector $\mathbf{g}^{(f)}$ in (4.52) corresponding to $\mathbf{c}_{\mathbf{m}}$ is not a unit vector (a vector whose components are all zeros except for one, which is 1). The constant vector $\mathbf{g}^{(f)}$, $f = \delta_{\mathbf{m}}(\mathbf{n})$, can be determined as follows. In (4.17), choose $f(\mathbf{n}) = 1$, if $\mathbf{n} = \mathbf{m}$ and $f(\mathbf{n}) = 0$, otherwise. In (4.18), if $\mathbf{n} = \mathbf{m}$, then the constant term is

$$-\sum_{j=1}^{M} \mu_{k,\mathbf{n}} q_{k,j}[1 - \delta_{\mathbf{m}}(\mathbf{n}_{k,j})] = -\sum_{j \neq k}^{M} \mu_{k,\mathbf{n}} q_{k,j} = -\mu_{k,\mathbf{n}}(1 - q_{k,k}).$$

If $\mathbf{m}$ is a neighboring state of $\mathbf{n}$, e.g., $\mathbf{m} = \mathbf{n}_{k,j}$, then the constant term for the $(\mathbf{n}, k)$th equation is $\mu_{k,\mathbf{n}} q_{k,k}$. In all other cases, the constant terms in (4.18) are zero.

Now, let us study the sensitivity of the performance measures. From (4.44), the sensitivity can be expressed in terms of $p(\mathbf{n})c^{(f)}(\mathbf{n}, i)$. Thus, we define

$$d_{\mathbf{m}}(\mathbf{n}, i) = p(\mathbf{n}) c_{\mathbf{m}}(\mathbf{n}, i)$$

and let $\mathbf{d}_{\mathbf{m}}$ be a vector obtained from $\mathbf{c}_{\mathbf{m}}$ by replacing $c_{\mathbf{m}}(\mathbf{n}, i)$ with $d_{\mathbf{m}}(\mathbf{n}, i)$ and $\mathbf{D}$ be a matrix obtained from $\mathbf{C}$ by replacing $\mathbf{c}_{\mathbf{m}}$ with $\mathbf{d}_{\mathbf{m}}$. $\mathbf{D}$ is called the *sensitivity matrix.*

The Linear Formulation

Let $\mathbf{e}_{\mathbf{n},i}$ be an $(M \times K)$-dimensional vector whose $(\mathbf{n}, i)$th component is one and whose other components are zeros. Then, using (4.54), we can write the normalized sensitivity in (4.44) in a matrix form:

$$\frac{\mu_{i,\mathbf{n}}}{\eta^{(f)}} \frac{\partial \eta^{(f)}}{\partial \mu_{i,\mathbf{n}}} = -p(\mathbf{n}) \sum_{all\ \mathbf{m}} f(\mathbf{m}) c_{\mathbf{m}}(\mathbf{n}, i)$$

$$= -\sum_{all\ \mathbf{m}} f(\mathbf{m}) d_{\mathbf{m}}(\mathbf{n}, i)$$
$$= -\mathbf{e}_{\mathbf{n},i}^T \mathbf{D}\mathbf{f},$$

where the superscript "T" denotes the transpose of a matrix.

Now, let us study the general case. Let $\Sigma = \Phi \times \Gamma = \{(\mathbf{n}, i) : all\ \mathbf{n} \in \Phi,\ and\ i \in \Gamma\}$ be the set of all $(\mathbf{n}, i)$'s, and $\Lambda \subseteq \Sigma$ be a subset of Σ. Suppose that all the service rates $\mu_{i,\mathbf{n}}$ for $(\mathbf{n}, i) \in \Lambda$ depend on the common variable α in the following fashion: $\mu_{i,\mathbf{n}} = \alpha \nu_{i,\mathbf{n}}$, where $\nu_{i,\mathbf{n}}$'s are independent of α. Then the normalized derivative of $\eta^{(f)}$ with respect to α is

$$\begin{aligned}
\frac{\alpha}{\eta^{(I)}} \frac{\partial \eta^{(f)}}{\partial \alpha} &= \frac{\alpha}{\eta^{(I)}} \{ \sum_{(\mathbf{n},i)\in\Lambda} \frac{\partial \eta^{(f)}}{\partial \mu_{i,\mathbf{n}}} \frac{\partial \mu_{i,\mathbf{n}}}{\partial \alpha} \} \\
&= \frac{\alpha}{\eta^{(I)}} \{ \sum_{(\mathbf{n},i)\in\Lambda} \frac{\partial \eta^{(f)}}{\partial \mu_{i,\mathbf{n}}} \nu_{i,\mathbf{n}} \} \\
&= \sum_{(\mathbf{n},i)\in\Lambda} \frac{\mu_{i,\mathbf{n}}}{\eta^{(I)}} \frac{\partial \eta^{(f)}}{\partial \mu_{i,\mathbf{n}}}.
\end{aligned}$$

Therefore, by (4.44), we have

$$\frac{\alpha}{\eta^{(I)}} \frac{\partial \eta^{(f)}}{\partial \alpha} = - \sum_{(\mathbf{n},i)\in\Lambda} p(\mathbf{n}) c^{(f)}(\mathbf{n}, i). \tag{4.59}$$

We can write this equation in a matrix form by using matrix $\mathbf{D}$. Let δ_Λ be the indicator function of the set Λ. That is,

$$\delta_\Lambda(\mathbf{n}, i) = \begin{cases} 1 & if\ (\mathbf{n}, i) \in \Lambda, \\ 0 & otherwise. \end{cases}$$

Let $\mathbf{e}_\Lambda$ be the $(M \times K)$-dimensional vector whose $(\mathbf{n}, i)$th component is $\delta_\Lambda(\mathbf{n}, i)$; then (4.59) takes the form

$$\frac{\alpha}{\eta^{(I)}} \frac{\partial \eta^{(f)}}{\partial \alpha} = -\mathbf{e}_\Lambda^T \mathbf{D}\mathbf{f}. \tag{4.60}$$

Equation (4.60) covers a wide range of sensitivities. Let us consider an example.

Example 4.6 If we choose $\Lambda = \Lambda_i = \{(\mathbf{n}, i) : all\ \mathbf{n},\ with\ i\ fixed\}$, $\nu_{i,\mathbf{n}} \equiv 1$, and replace α by μ_i, we get the formula for Jackson networks:

$$\frac{\mu_i}{\eta^{(I)}} \frac{\partial \eta^{(f)}}{\partial \mu_i} = -\mathbf{e}_{\Lambda_i}^T \mathbf{D}\mathbf{f}.$$

If we choose $\Lambda = \Lambda_{i,k} = \{(\mathbf{n}, i) : n_i = k, \textit{ with } i \textit{ fixed}\}$, $\nu_{i,\mathbf{n}} \equiv 1$, $\alpha = \mu_{i,k}$, we obtain

$$\frac{\mu_{i,k}}{\eta^{(f)}} \frac{\partial \eta^{(f)}}{\partial \mu_{i,k}} = -\mathbf{e}_{\Lambda_{i,k}}^T \mathbf{D}\mathbf{f} = -\sum_{n_i=k} p(\mathbf{n}) c^{(f)}(\mathbf{n}, i).$$

This is the formula for load-dependent networks.

Finally, we choose $\Lambda = \Sigma$. This means that the service rates of all the servers at any state change proportionally to $\Delta\alpha$. Using the above notations, we have

$$\begin{aligned} \frac{\alpha}{\eta^{(f)}} \frac{\partial \eta^{(f)}}{\partial \alpha} &= -\mathbf{e}_{\Sigma}^T \mathbf{D}\mathbf{f} \\ &= -\sum_{all\ \mathbf{n}} \sum_{i=1}^{M} p(\mathbf{n}) c^{(f)}(\mathbf{n}, i) = -\sum_{all\ \mathbf{n}} \{p(\mathbf{n}) \sum_{i=1}^{M} c^{(f)}(\mathbf{n}, i)\} \\ &= -\sum_{all\ \mathbf{n}} p(\mathbf{n}) f(\mathbf{n}) = -\eta_T^{(f)} = -\frac{\eta^{(f)}}{\eta^{(f)}}. \end{aligned}$$

Therefore,

$$\frac{\alpha}{\eta^{(f)}} \frac{\partial \eta^{(f)}}{\partial \alpha} = -1,$$

where α is the coefficient in $\mu_{i,\mathbf{n}} = \alpha \nu_{i,\mathbf{n}}$ for all $(\mathbf{n}, i) \in \Sigma$. The equation shows that the elasticity of any steady-state performance measure $\eta^{(f)}$ with respect to α is -1. This result can be explained intuitively. Since all the service rates change according to $\frac{\Delta \mu_{i,\mathbf{n}}}{\mu_{i,\mathbf{n}}} = \frac{\Delta \alpha}{\alpha}$, the system behaves as if the rate of the clock changes to $(1 + \frac{\Delta\alpha}{\alpha})$ times faster; this results in $\frac{\Delta \eta^{(f)}}{\eta^{(f)}} = -\frac{\Delta\alpha}{\alpha}$.

Gradients

We define the *normalized gradient* of a performance function, ∇f, as a vector whose $(\mathbf{n}, i)$th component is $\frac{\mu_{i,\mathbf{n}}}{\eta^{(f)}} \frac{\partial \eta^{(f)}}{\partial \mu_{i,\mathbf{n}}}$. According to this definition, we have

$$\nabla f = -\mathbf{D}\mathbf{f}.$$

Let $\mathbf{e}$ be an $(M \times K)$-dimensional vector whose $(\mathbf{n}, i)$th component is $e(\mathbf{n}, i)$, with $e(\mathbf{n}, i)$ being real numbers (maybe zero). Then the normalized derivative along the direction of $\mathbf{e}$ is

$$d_{\mathbf{e}} = \sum_{all\ (\mathbf{n},i)} e(\mathbf{n}, i) \frac{\mu_{i,\mathbf{n}}}{\eta^{(f)}} \frac{\partial \eta^{(f)}}{\partial \mu_{i,\mathbf{n}}} = -\mathbf{e}^T \mathbf{D}\mathbf{f}. \tag{4.61}$$

Equation (4.61) is the general form of the performance sensitivity. This equation shows that all the sensitivities of any steady-state performance measure can be expressed in a simple algebraic form by using the sensitivity matrix **D**.

4.3.3 Reducing the Dimension

Since the number of states, K, is usually very large, the sizes of the matrices **C** and **D** are big. In practice, it is necessary to reduce the sizes. This can be done by utilizing the special features of a real problem. For example, since the number of states is large, sometimes it is neither possible nor necessary to specify the performance function for all the states. We have to partition the state space into groups and to assign performance values to each group. Using this feature, we can partition **C** and **D** into submatrices and replace each submatrix by an entry which equals the sum of all the entries in the submatrix. This will dramatically reduce the sizes of **C** and **D**.

Let us first consider how to reduce the number of columns in **C**. As discussed above, the performance function f is usually defined on subsets of Φ. For example, sometimes we are concerned only about whether the number of customers in certain servers exceeds a predetermined number. More precisely, let Ψ_r, $r = 1, 2, ..., R$, be a partition of the state space; i.e.,

$$\cup_{r=1}^{R} \Psi_r = \Phi, \qquad \Psi_r \cap \Psi_s = \emptyset, \qquad 1 \leq r, s \leq R. \tag{4.62}$$

Suppose that the performance function is defined as

$$f(\mathbf{n}) = f(\Psi_r) \quad if\ \mathbf{n} \in \Psi_r. \tag{4.63}$$

Then

$$f(\mathbf{n}) = \sum_{r=1}^{R} f(\Psi_r)\delta_{\Psi_r}(\mathbf{n}),$$

where δ_{Ψ_r} is the indicator function of the set Ψ_r. We have an equation similar to (4.53):

$$F_L^{(f)} = \sum_{r=1}^{R}\{f(\Psi_r)\int_0^{T_L} \delta_{\Psi_r}(\mathbf{N}(t))dt\} = \sum_{r=1}^{R} f(\Psi_r)F_L^{(\delta_{\Psi_r})}.$$

From this, we have

$$c^{(f)}(\mathbf{n}, i) = \sum_{r=1}^{R} f(\Psi_r)c_{\Psi_r}(\mathbf{n}, i). \tag{4.64}$$

We have

$$c_{\Psi_r}(\mathbf{n},i) = \sum_{\mathbf{m}\in\Psi_r} c_{\mathbf{m}}(\mathbf{n},i);$$

this is similar to (4.54). Using (4.64), we can reduce the number of columns in matrix **C** from K to R. The number of columns in matrix **D** can be reduced in the same way. In some cases, R may be only 2. For example, if we are concerned only about the probability that the number of customers in server j is at least k, $p(n_j \geq k)$, then we can choose $\Psi = \{\mathbf{n} : n_j \geq k\}$. Ψ and $\bar{\Psi} = \Phi - \Psi$ form a partition of Φ. We define $f(\Psi) = 1$ and $f(\bar{\Psi}) = 0$. Since $f(\bar{\Psi}) = 0$, $c_{\bar{\Psi}}(\mathbf{n},i)$ is irrelevant. We have

$$c_{\Psi}(\mathbf{n},i) = \sum_{\mathbf{m}\in\Psi} c_{\mathbf{m}}(\mathbf{n},i).$$

Then Equation (4.59) becomes

$$\frac{\alpha}{\eta^{(I)}} \frac{\partial[p(n_j \geq k)\eta^{(I)}]}{\partial\alpha} = - \sum_{(\mathbf{n},i)\in\Lambda} p(\mathbf{n})c_{\Psi}(\mathbf{n},i), \tag{4.65}$$

where α is the common variable for the service rates $\mu_{i,\mathbf{n}}$ with $(\mathbf{n},i)$ in Λ.

The same approach can be applied to reduce the number of rows in **D**. Let $\mathbf{\Lambda} := \{\Lambda_w,\ w = 1,2,...,W\}$ be a partition of Σ; i.e,

$$\cup_{w=1}^{W} \Lambda_w = \Sigma, \quad \Lambda_w \cap \Lambda_u = \emptyset, \qquad 1 \leq u, w \leq W. \tag{4.66}$$

Suppose that

$$e(\mathbf{n},i) = e(\Lambda_w), \quad if\ (\mathbf{n},i) \in \Lambda_w; \tag{4.67}$$

i.e., all the $e(\mathbf{n},i)$'s in Λ_w have the same value of $e(\Lambda_w)$. Then we can define

$$d_{\mathbf{m}}(\Lambda_w) = \sum_{(\mathbf{n},i)\in\Lambda_w} d_{\mathbf{m}}(\mathbf{n},i)$$

and can replace the rows in **D** corresponding to $(\mathbf{n},i) \in \Lambda_w$ with one row vector $\mathbf{d}(\Lambda_w)$ whose **m**th component is $d_{\mathbf{m}}(\Lambda_w)$. The vector **e** in (4.61) becomes a W-dimensional vector $\mathbf{e}_{\mathbf{\Lambda}} = (e(\Lambda_1), ..., e(\Lambda_W))^T$. Note that some $e(\Lambda_w)$'s may be zero.

In summary, if Φ and Σ can be partitioned according to (4.62), (4.63) and (4.66), (4.67), respectively, then the matrix **D** can be partitioned into $W \times R$ submatrices. The (w,r)th submatrix, $1 \leq w \leq W$, $1 \leq r \leq R$, corresponding to the same values of $e(\Lambda_w)$ and $f(\Psi_r)$, can be replaced by

$$d_{\Psi_r}(\Lambda_w) \quad = \quad \sum_{(\mathbf{n},i)\in\Lambda_w} \sum_{\mathbf{m}\in\Psi_r} d_{\mathbf{m}}(\mathbf{n},i)$$

$$= \sum_{(\mathbf{n},i)\in\Lambda_w} \sum_{\mathbf{m}\in\Psi_r} [p(\mathbf{n})c_{\mathbf{m}}(\mathbf{n},i)].$$

Equations (4.60) and (4.61) hold for this new matrix with the understanding that vector $\mathbf{e}$ becomes $\mathbf{e}_{\mathbf{\Lambda}} = (e(\Lambda_1), ..., e(\Lambda_W))^T$ and f is defined on ψ_r. More precisely, suppose that all the service rates $\mu_{i,\mathbf{n}}$ for $(\mathbf{n}, i) \in \Lambda_w$ depend on the common variable α_w: $\mu_{i,\mathbf{n}} = \alpha_w \nu_{i,\mathbf{n}}$, $w = 1, 2, ..., W$. Then the normalized derivative along the direction of $\mathbf{e}_{\mathbf{\Lambda}} = (e(\Lambda_1), ..., e(\Lambda_W))^T$ is

$$
\begin{aligned}
d_{\mathbf{e}_{\mathbf{\Lambda}}} &= \sum_{w=1}^{W} e(\Lambda_w) \frac{\alpha_w}{\eta^{(f)}} \frac{\partial \eta^{(f)}}{\partial \alpha_w} \\
&= -(e(\Lambda_1), ..., e(\Lambda_W)) \begin{pmatrix} d_{\Psi_1}(\Lambda_1) & ... & d_{\Psi_R}(\Lambda_1) \\ ... & ... & ... \\ d_{\Psi_1}(\Lambda_W) & ... & d_{\Psi_R}(\Lambda_W) \end{pmatrix} \begin{pmatrix} f(\Psi_1) \\ \cdot \\ \cdot \\ f(\Psi_R) \end{pmatrix} \\
&= -\mathbf{e}_{\mathbf{\Lambda}}^T \mathbf{D} \mathbf{f}.
\end{aligned} \tag{4.68}
$$

Note that we may have $f(\Psi_{r_0}) = 0$ or $e(\Lambda_{w_0}) = 0$ for some r_0 or w_0. Then the r_0th column or the w_0th row in $\mathbf{D}$ is irrelevant. This further reduces the size of $\mathbf{D}$.

Example 4.7 Suppose that in a closed queuing network, server i is load dependent; i.e., its service rate, μ_{i,n_i}, depends on the number of customers in its own queue, and that the performance of interest is a function of the queue length of server j; i.e., $f(\mathbf{n}) = g(n_j)$. Suppose that we are interested in the sensitivity of $\eta^{(f)}$ with respect to μ_{i,n_i}. We choose

$$\Psi_r = \{\mathbf{n},\ n_j = r\} \qquad 0 \le r \le N,$$

and

$$\Lambda_w = \{\mathbf{n},\ n_i = w\} \qquad 0 \le w \le N.$$

In this example, $\mathbf{D}$ in (4.68) is an $N \times N$ matrix. All the gradients of any function $g(n_j)$ with respect to μ_{i,n_i} are linear combinations of these $N \times N$ basic sensitivities, $d_{\Psi_r}(\Lambda_w)$.

Implementation

To obtain $c_{\Psi_r}(\mathbf{n}, i)$, we only need to solve (4.52) for $\delta_{\Psi_r}(\mathbf{n})$ instead of for $\delta_{\mathbf{m}}(\mathbf{n})$. But to obtain $d_{\Psi_r}(\Lambda_w)$, we still have to solve (4.52) for all the $M \times K$ variables. Therefore, if we do the sensitivity analysis by solving Equation (4.52), the above approach saves computation when applied to reducing the number of columns but does not save any computation for row reduction. However, $d_{\Psi_r}(\Lambda_w)$ can be estimated by simulation. Using simulation, we do not have to estimate all the $d_{\Psi_r}(\mathbf{n}, i)$ for obtaining $d_{\Psi_r}(\Lambda_w)$; therefore, the approach also saves simulation efforts when applied to reducing the number of rows.

The algorithms for estimating performance sensitivities are known as infinitesimal perturbation analysis (IPA) algorithms in literature (Glasserman [57], Ho and Cao [68]). One of the most important and attractive features of the IPA algorithms is that only one sample path of the network is needed for obtaining sensitivity estimates of all the performance measures. The following algorithm (a modified version of the algorithm in Cao and Ma [33]) gives the estimates of $c_\Psi(\mathbf{n}, i)$.

Algorithm.

1. Set $\Delta_j := 0$, for $j = 1, 2, ..., M$, $L := 0$, and $\Delta F := 0$.

2. At each transition time T_l, if $X_{l-1} := X(T_l-) = \mathbf{n}$, then set $\Delta_i := \Delta_i + 1$ and $L := L + 1$.

3. Suppose that at T_l a customer transfers from server k to server m. At this transition time,

   (a) if server m is not idle before T_l, then

      i. Δ_k remains unchanged.

      ii. for all $j \neq k$, set $\Delta_j := (1 - \kappa_j)\Delta_k + \kappa_j \times \Delta_j$, where $\kappa_j = \mu_{j,X_{l-1}}/\mu_{j,T_l}$.

   (b) if server m is idle before T_l, then

      i. Δ_k remains unchanged.

      ii. set $\Delta_m := \Delta_k$ (perturbation propagation through an idle period).

      iii. for all $j \neq k, m$, set $\Delta_j := (1 - \kappa_j)\Delta_k + \kappa_j \times \Delta_j$, where $\kappa_j = \mu_{j,X_{l-1}}/\mu_{j,X_l}$.

   (c) finally,

      i. if $X_{l-1} \notin \Psi$ and $X_l \in \Psi$, then set $\Delta F := \Delta F - \Delta_k$.

      ii. if $X_{l-1} \in \Psi$ and $X_l \notin \Psi$, then set $\Delta F := \Delta F + \Delta_k$.

4. At the end of the simulation, set $c_\Psi(\mathbf{n}, i) := \Delta F/L$.

In the algorithm, Δ_j is the perturbation of server j; $\Delta_i := \Delta_i + 1$ reflects the fact that a new perturbation of size 1 is added to server i at the end of the period in which the system is in state $\mathbf{n}$; L counts the number of perturbations; ΔF records the changes in $F := \int_0^{T_L} \delta_\Psi(\mathcal{N}(t))dt$. Note that at the end of a simulation, some perturbations may not have been realized; however, as $L \to \infty$, the error caused by this fact is negligible.

The algorithm for $d_\Psi(\Lambda)$ can be obtained by replacing steps 1, 2, and 4 in the above algorithm by

1' Set $\Delta_j := 0$, $j = 1, 2, ..., M$, and $\Delta F := 0$.

2' At each transition time T_l, if $(X_{l-1}, i) \in \Lambda$ for an $i \in \Gamma$, then set $\Delta_i := \Delta_i + (T_l - T_{l-1})[\Delta\mu_{i,X_{l-1}}/\mu_{i,X_{l-1}}]$.

4' At the end of the simulation, set $c_\Psi(\Lambda) := \Delta F/T_L$.

The term $(T_l - T_{l-1})[\Delta\mu_{i,X_{l-1}}/\mu_{i,X_{l-1}}]$ in step 2' is the perturbation generated in the period of $(T_l - T_{l-1})$, in which the system is in state $\mathbf{n}$, and $(\mathbf{n}, i) \in \Lambda$; $(T_l - T_{l-1})/T_L$ contributes to estimating the $p(\mathbf{n})$ term in $d_\Psi(\Lambda)$.

It is important to note that the same simulation can be applied to estimating all the $c_\Psi(\mathbf{n}, i)$'s and $d_\Psi(\Lambda)$'s. As discussed above, if the performance function and the sensitivities are such that the spaces Σ and Φ can be properly partitioned, then the number of $c_{\Psi_r}(\Lambda_w)$'s to be estimated is only $W \times R$. All the other performance sensitivities can be obtained by linear algebraic calculations.

Discussion

We have developed a linear algebraic formulation for the performance sensitivities of closed queuing networks with state-dependent service rates. The results are analogous to the fundamental properties of continuous variable systems. These results are derived by emploring the intrinsic dynamic property of queuing networks.

Although the equations derived in this paper provide a new view of the sensitivity analysis of queuing networks, the practical application of these equations is limited by the computation complexity. For practical purposes, the significance of the results in this section is twofold. First, to obtain all the performance sensitivities in the form of (4.6) or (4.7), only one matrix inversion is needed. Second, all the fundamental sensitivities or realization factors can be obtained by one simulation generating one sample path of the queuing network.

4.4 Sensitivity Formulas for Open Networks

In this section, we shall extend the realization theory to open networks. The major difference between open and closed networks is that open networks consist of infinitely many states. Thus, the performance function f may be unbounded. Besides, as we have seen, the main task of deriving sensitivity formulas based on realization factors is to establish the interchangeability of the derivative and the expectation or that of the derivative and the limit when the length of a sample path goes to infinity. With open systems which have infinitely many states, we encounter another difficult task: to prove the interchangeability of the derivative or the expectation and an infinite sum over all the states. Additional conditions for the performance function are needed to guarantee the interchangeability. We shall show that one of the sufficient conditions is the *quasi-Lipschitz condition* introduced in this section.

4.4.1 Realization Factors for Open Networks

We first review the basic concepts. Consider an open network with M single-server nodes, each having a buffer with an infinite size. The service discipline is FCFS. The service requirements of customers at all the servers are exponentially distributed with mean one, and the service rates depend on the system state $\mathbf{n} = (n_1, n_2, ..., n_M)$. The state space $\Phi := \{\mathbf{n}\}$ is countable, but it consists of infinitely many states. Let $\mu_{i,\mathbf{n}}$ be the service rate of server i when the system is in state $\mathbf{n}$. Customers arrive from an outside source, denoted as server 0. The arrival process is a Poisson process with a state-dependent rate $\mu_{0,\mathbf{n}}$. We assume that $0 < w < \mu_{i,\mathbf{n}} < W < \infty$ for all $i = 0, 1, \cdots, M$ and $\mathbf{n} \in \Phi$.

With the source denoted as server 0, we have

$$\mu(\mathbf{n}) = \sum_{i=0}^{M} \epsilon(n_i)\mu_{i,\mathbf{n}},$$

where we assume that $n_0 > 0$ always holds.

After the service completion at server i, a customer will enter server j with probability $q_{i,j}$, $i, j = 1, 2, ..., M$, and will leave the system with probability $q_{i,0}$. We have $\sum_{j=0}^{M} q_{i,j} = 1$. A customer in the arrival process has a probability of $q_{0,i}$ to arrive at server i. $\lambda_{i,\mathbf{n}} := \mu_{0,\mathbf{n}} q_{0,i}$ is the external arrival rate at server i. Set $q_{0,0} = 0$ and we have $\sum_{i=1}^{M} q_{0,i} = 1$. In this setting, the routing probability matrix of the system is an $(M+1) \times (M+1)$ matrix $Q = [q_{i,j}]$, $i, j = 0, 1, 2, ..., M$.

We assume that the routing probabilities satisfy such conditions that a customer entering the network will leave the network with probability one and that every server will be visited by some customers. This excludes, for example, the unstable systems with $M = 1$ and $q_{0,1} = q_{1,1} = 1$, $q_{1,0} = 0$. The condition implies that the open network is irreducible, or, equivalently, the matrix $Q = [q_{i,j}]$, $i, j = 0, 1, 2, ..., M$, is indecomposable (cf. Section 3.1).

The randomness of the system can be determined by a vector $\xi = (\xi_0, \xi_1, \cdots, \xi_M; \zeta_0, \zeta_1, \cdots, \zeta_M; \mathbf{n}_0)$. This is the same as the closed network case discussed in Section 4.2, except ξ_0 and ζ_0 are added to determine the interarrival times and the arrival destinations.

Because the network is assumed to be irreducible, the embedded Markov chain $\mathbf{X}$ is ergodic. Note that the asymptotical stationarity may not hold for some open networks. For example, the embedded Markov chain of an M/M/1 queue is not asymptotically stationary. In fact, if $X_0 = 0$, then X_l is even for all even l's, and odd for all odd l's. However, using the same argument as that in Section 3.1, we can study an equivalent network whose $\mathbf{X}$ is asymptotically stationary and ergodic.

It is clear that the perturbation generation at the servers and the source and the perturbation propagation between servers for open networks are just the same as those for closed networks. The perturbation propagation between the source (i.e., server 0) and servers 1 to M, however, is different. A perturbation in servers 1 to M will never be propagated through an idle period to server 0, and a perturbation of server 0 will be propagated to servers 1 to M; it will never be lost. If we consider server 0 as a server containing infinitely many customers, i.e., a server which never meets any idle period, then the perturbation propagation rules for servers also apply to the source (i.e., server 0). In this sense, *an open network consisting of M servers can be considered as a closed network consisting of $M + 1$ servers with server 0 having infinitely many customers.*

The realization factors are defined in the same way as those for closed networks. Let $\Gamma = \{0, 1, 2, ..., M\}$ be the set of all the servers including the source. Then Lemma 4.1 holds for open networks (with $n_0 > 0$, and $\sum_{i=1}^{M}$ replaced by $\sum_{i=0}^{M}$). For example, we have

$$\textit{if } n_i = 0, \textit{ then } c^{(f)}(\mathbf{n}, i) = 0 \tag{4.69}$$

and

$$\sum_{i=0}^{M} c^{(f)}(\mathbf{n}, i) = f(\mathbf{n}). \tag{4.70}$$

Theorem 4.1 becomes

Theorem 4.7 *For a state-dependent open network and any performance function f, the following equations hold:*

$$
\begin{aligned}
&\{\sum_{i=0}^{M} \epsilon(n_i)\mu_{i,\mathbf{n}}\} c^{(f)}(\mathbf{n},k) \\
= &\sum_{i=0, i\neq k}^{M} \sum_{j=0}^{M} \epsilon(n_i)\mu_{i,\mathbf{n}} q_{i,j} \kappa_k(\mathbf{n},\mathbf{n}_{i,j}) c^{(f)}(\mathbf{n}_{i,j},k) \\
+ &\sum_{j=0}^{M} \mu_{k,\mathbf{n}} q_{k,j} \{ c^{(f)}(\mathbf{n}_{k,j},k) + \sum_{i=0,i\neq k}^{M} [1-\kappa_i(\mathbf{n},\mathbf{n}_{k,j})]\epsilon(n_i) c^{(f)}(\mathbf{n}_{k,j},i) \\
+ &[1-\epsilon(n_j)] c^{(f)}(\mathbf{n}_{k,j},j) + f(\mathbf{n}) - f(\mathbf{n}_{k,j}) \}, \\
&\qquad n_k > 0, \; k \in \Gamma,
\end{aligned} \tag{4.71}
$$

where $\kappa_i(\mathbf{n},\mathbf{n}_{k,j}) = \mu_{i,\mathbf{n}}/\mu_{i,\mathbf{n}_{k,j}}$, $\kappa_k(\mathbf{n},\mathbf{n}_{i,j}) = \mu_{k,\mathbf{n}}/\mu_{k,\mathbf{n}_{i,j}}$, *and* $\epsilon(n_0) \equiv 1$. $\mathbf{n}_{i,j}$ *is a neighboring state of* $\mathbf{n}$*; especially,* $\mathbf{n}_{0,j} = (n_1, ..., n_j+1, ..., n_M)$*, and* $\mathbf{n}_{i,0} = (n_1, ..., n_i - 1, ..., n_M)$.

The following equation comes immediately from Equations (4.69) and (4.70):

$$c^{(f)}[(0,0,...,0),0] = f[(0,0,...,0)]. \tag{4.72}$$

4.4.2 Sensitivity Formulas

Many of the results derived in Section 4.2 hold for open networks without any additional conditions. For example, we have

$$p(\mathbf{n}) = \eta \frac{\pi(\mathbf{n})}{\mu(\mathbf{n})}.$$

An alternative proof of this equation is based on the ergodic property.

$$
\begin{aligned}
p(\mathbf{n}) &= \lim_{L\to\infty} \frac{1}{T_L} \sum_{l=0}^{L-1} S_l \chi_{\mathbf{n}}(X_l) \\
&= \lim_{L\to\infty} \frac{L}{T_L} \{\frac{1}{L}\sum_{l=0}^{L-1} \chi_{\mathbf{n}}(X_l)\} \{\frac{\sum_{l=0}^{L-1} S_l \chi_{\mathbf{n}}(X_l)}{\sum_{l=0}^{L-1} \chi_{\mathbf{n}}(X_l)}\} \\
&= \eta \frac{\pi(\mathbf{n})}{\mu(\mathbf{n})},
\end{aligned}
$$

where

$$\eta = \lim_{L\to\infty} \frac{L}{T_L} = \lim_{L\to\infty} \frac{1}{\eta_L^{(I)}},$$

$$\pi(\mathbf{n}) = \lim_{L\to\infty} \frac{1}{L}\sum_{l=0}^{L-1} \chi_{\mathbf{n}}(X_l),$$

and

$$\frac{1}{\mu(\mathbf{n})} = \lim_{L\to\infty} \frac{\sum_{l=0}^{L-1} S_l \chi_{\mathbf{n}}(X_l)}{\sum_{l=0}^{L-1} \chi_{\mathbf{n}}(X_l)},$$

which is the mean of S_l given $X_l = \mathbf{n}$.

Theorem 4.3 becomes

Theorem 4.8 *Under the integrability Condition 4.1, the sample normalized derivative of $\eta_L^{(f)}$ with respect to $\mu_{i,\mathbf{n}}$ in a state-dependent open network satisfies*

$$\lim_{L\to\infty} \frac{\mu_{v,\mathbf{n}}}{\eta_L^{(I)}} \frac{\partial \eta_L^{(f)}}{\partial \mu_{v,\mathbf{n}}} = -p(\mathbf{n})c^{(f)}(\mathbf{n}, v) \qquad w.p.1. \tag{4.73}$$

In (4.73), $\mu_{0,\mathbf{n}}$ is the mean interarrival rate. Besides Theorem 4.8, we have

$$E\{\frac{\partial}{\partial \mu_{i,\mathbf{n}}}[\eta_L(\mu_{i,\mathbf{n}}, \xi)|\mathbf{n}_0]\} = \frac{\partial}{\partial \mu_{i,\mathbf{n}}}\{E[\eta_L(\mu_{i,\mathbf{n}}, \xi)|\mathbf{n}_0]\} \tag{4.74}$$

for any finite L, with $\mathbf{n}_0$ being the initial state of the sample path.

To study the steady-state performance, we assume that $E[|f|]$ exists. Thus, the sum in

$$\eta^{(f)} = \sum_{\mathbf{n}} \frac{f(\mathbf{n})\pi(\mathbf{n})}{\mu(\mathbf{n})}$$

is well defined in the sense that it does not depend on the order of the summation (Courant and John [44]).

Now, assume that the initial probability distribution of $\mathbf{n}_0$ is the steady-state probability $\pi(\mathbf{n}_0)$ and that $\frac{\partial \pi(\mathbf{n}_0)}{\partial \mu_{i,\mathbf{n}}}$ exists. Following the same argument as for Theorems 3.5 and 4.5, we first have

$$\frac{\partial \eta^{(f)}}{\partial \mu_{v,\mathbf{n}}} = \frac{\partial}{\partial \mu_{v,\mathbf{n}}} \sum_{\mathbf{n}_0 \in \Phi} \{E[\eta_L^{(f)}(\mu_{v,\mathbf{n}}, \xi)|\mathbf{n}_0]\pi(\mathbf{n}_0)\}. \tag{4.75}$$

Since Φ contains infinitely many states, it requires some condition on f to interchange the order of $\sum_{\mathbf{n}_0\in\Phi}$ and $\frac{\partial}{\partial \mu_{v,\mathbf{n}}}$. It will be proved in the next subsection that under the quasi-Lipschitz condition, this interchangeability holds. Thus, under the quasi-Lipschitz condition, we have

$$\begin{aligned}\frac{\partial \eta^{(f)}}{\partial \mu_{v,\mathbf{n}}} &= \sum_{\mathbf{n}_0 \in \Phi} \frac{\partial}{\partial \mu_{v,\mathbf{n}}} \{E[\eta_L^{(f)}(\mu_{v,\mathbf{n}}, \xi)|\mathbf{n}_0]\pi(\mathbf{n}_0)\} \\ &= E\{\frac{\partial}{\partial \mu_{v,\mathbf{n}}}[\eta_L^{(f)}(\mu_{v,\mathbf{n}}, \xi)]\} \\ &+ \sum_{\mathbf{n}_0 \in \Phi} E[\eta_L^{(f)}(\mu_{v,\mathbf{n}}, \xi)|\mathbf{n}_0] \frac{\partial \pi(\mathbf{n}_0)}{\partial \mu_{v,\mathbf{n}}}. \end{aligned} \tag{4.76}$$

Before we derive the main results, we first prove a lemma.

Lemma 4.4 *There exists a positive number $R_1 < \infty$ such that*

$$\left|\frac{\partial \pi(\mathbf{n}_0)}{\partial \mu_{i,\mathbf{n}}}\right| < R_1 \pi(\mathbf{n}_0)$$

holds for all $\mathbf{n}_0$.

Proof: The embedded Markov chain $\mathbf{X}$ is a regenerative process with regenerative points $X_l = \mathbf{n}$ (a fixed state). That is, the state process of $\mathbf{X}$ consists of regenerative periods starting with a state $\mathbf{n}$ and ending with the next $\mathbf{n}$. Let $\mathbf{n}_{k,j}$ be the next state that the system enters after the first state $\mathbf{n}$ in a regenerative period. Then, if we do not count the first state $\mathbf{n}$, each regenerative period becomes a period starting with $\mathbf{n}_{k,j}$ and ending with $\mathbf{n}$. Let $D_{k,j}$ denote such a period. For any pair $\{k, j\}$, $k, j, = 0, 1, ..., M$, all $D_{k,j}$'s are independent and statistically identical. Let $\alpha_{k,j}$ be the expected value of the number of states in $D_{k,j}$, and let $\beta_{k,j}(\mathbf{n}_0)$ be the expected value of the number of states $\mathbf{n}_0$ in $D_{k,j}$. Note that $\beta_{k,j}(\mathbf{n}) = 1$ by definition. The probability that a regenerative period is $D_{k,j}$ is

$$P(\mathbf{n}_{k,j}|\mathbf{n}) = \epsilon(n_k)\frac{\mu_{k,\mathbf{n}} q_{k,j}}{\mu(\mathbf{n})}. \tag{4.77}$$

The expected value of the number of the states in each period is

$$\alpha := \sum_{k,j=0}^{M} P(\mathbf{n}_{k,j}|\mathbf{n})\alpha_{k,j};$$

the expected value of the number of the states $\mathbf{n}_0$ in each period is

$$\beta(\mathbf{n}_0) := \sum_{k,j=0}^{M} P(\mathbf{n}_{k,j}|\mathbf{n})\beta_{k,j}(\mathbf{n}_0).$$

Thus, by the regenerative theory (see, e.g., Crane and Lemoine [48]), we have

$$\pi(\mathbf{n}_0) = \frac{\beta(\mathbf{n}_0)}{\alpha} = \frac{\sum_{k,j=0}^{M} P(\mathbf{n}_{k,j}|\mathbf{n})\beta_{k,j}(\mathbf{n}_0)}{\sum_{k,j=0}^{M} P(\mathbf{n}_{k,j}|\mathbf{n})\alpha_{k,j}}. \tag{4.78}$$

Note that both $\alpha_{k,j}$ and $\beta_{k,j}(\mathbf{n}_0)$ do not depend on $\mu_{i,\mathbf{n}}$ because within $D_{k,j}$ the system does not reach state $\mathbf{n}$. Therefore, from (4.78) we have

$$\begin{aligned}\frac{1}{\pi(\mathbf{n}_0)}\frac{\partial\pi(\mathbf{n}_0)}{\partial\mu_{i,\mathbf{n}}} &= \frac{\sum_{k,j=0}^{M} \frac{\partial P(\mathbf{n}_{k,j}|\mathbf{n})}{\partial\mu_{i,\mathbf{n}}}\beta_{k,j}(\mathbf{n}_0)}{\sum_{k,j=0}^{M} P(\mathbf{n}_{k,j}|\mathbf{n})\beta_{k,j}(\mathbf{n}_0)} \\ &- \frac{\sum_{k,j=0}^{M} \frac{\partial P(\mathbf{n}_{k,j}|\mathbf{n})}{\partial\mu_{i,\mathbf{n}}}\alpha_{k,j}}{\sum_{k,j=0}^{M} P(\mathbf{n}_{k,j}|\mathbf{n})\alpha_{k,j}}.\end{aligned} \tag{4.79}$$

From (4.77), for any $\mathbf{n}$ there are at most $(M+1)\times(M+1)$ different values of $P(\mathbf{n}_{k,j}|\mathbf{n})$. Therefore, there must exist a $K < \infty$ such that $|\frac{\partial P(\mathbf{n}_{k,j}|\mathbf{n})}{\partial\mu_{i,\mathbf{n}}}| < KP(\mathbf{n}_{k,j}|\mathbf{n})$, thus both the first and the second terms on the right-hand side of (4.79) are bounded by K for all $\mathbf{n}_0$. This proves the lemma. □

Now, we are ready to prove the main theorem.

Theorem 4.9 *Under the integrability Condition 4.1 and the quasi-Lipschitz Condition 4.2, for a state-dependent open network, the following equation holds for* $v = 0, 1, ..., M$, $\mathbf{n} \in \Phi$:

$$\frac{\mu_{v,\mathbf{n}}}{\eta^{(I)}}\frac{\partial\eta^{(f)}}{\partial\mu_{v,\mathbf{n}}} = -p(\mathbf{n})c^{(f)}(\mathbf{n}, v) \qquad w.p.1. \tag{4.80}$$

Proof: First, assume that $f(\mathbf{n})$ is bounded. Taking the limit $\lim_{L\to\infty}$ on both sides of (4.76), we can interchange the order of $\lim_{L\to\infty}$ and $\sum_{\mathbf{n}_0\in\Phi}$ and obtain

$$\frac{\partial\eta^{(f)}}{\partial\mu_{v,\mathbf{n}}} = \lim_{L\to\infty} E\{\frac{\partial}{\partial\mu_{v,\mathbf{n}}}[\eta_L^{(f)}(\mu_{i,\mathbf{n}}, \xi)]\}.$$

Then, using a method similar to that of Theorem 4.6, we can prove that (4.80) holds.

Next, suppose $f(\mathbf{n})$ is not bounded. For any $U > 0$ we define a bounded function

$$f_U(\mathbf{n}) = \begin{cases} f(\mathbf{n}) & if\ |f(\mathbf{n})| \le U, \\ U & if\ |f(\mathbf{n})| > U. \end{cases}$$

Let $\eta^{(f_U)}$ and $c^{(f_U)}(\mathbf{n}, i)$, respectively, be the performance and the realization factor of the problem with $f(\mathbf{n})$ replaced by $f_U(\mathbf{n})$. Note that $\eta^{(I)}$ and the steady-state probabilities $p(\mathbf{n})$ and $\pi(\mathbf{n})$ do not depend on U. Since $f_U(\mathbf{n})$ is bounded, we have

$$\frac{\mu_{v,\mathbf{n}}}{\eta^{(I)}}\frac{\partial \eta^{(f_U)}}{\partial \mu_{v,\mathbf{n}}} = -p(\mathbf{n})c^{(f_U)}(\mathbf{n}, v). \tag{4.81}$$

For a fixed $\mathbf{n}$, we choose $U > f(\mathbf{n})$. Thus, for this particular $\mathbf{n}$, $f(\mathbf{n}) = f_U(\mathbf{n})$. We have

$$\begin{aligned} \left|\frac{\partial}{\partial \mu_{v,\mathbf{n}}}[\eta^{(f_U)} - \eta^{(f)}]\right| &= \sum_{\mathbf{m}:f(\mathbf{m})>U} \left\{\frac{f(\mathbf{m}) - U}{\mu(\mathbf{m})}\left|\frac{\partial \pi(\mathbf{m})}{\partial \mu_{v,\mathbf{n}}}\right|\right\} \\ &< \frac{R_1}{H} \sum_{\mathbf{m}:f(\mathbf{m})>U} \{[f(\mathbf{m}) - U]\pi(\mathbf{m})\} \\ &< \frac{R_1}{H} \sum_{\mathbf{m}:f(\mathbf{m})>U} |f(\mathbf{m})|\pi(\mathbf{m}). \end{aligned}$$

In the above, we employed Lemma 4.4 and the property $\mu(\mathbf{m}) > H > 0$, $H = 1/[(M+1)w]$, for all $\mathbf{m}$, where w is the lower bound of $\mu_{i,\mathbf{n}}$. Since $E[|f(\mathbf{n})|]$ exists, we have

$$\lim_{U\to\infty} \frac{\partial}{\partial \mu_{v,\mathbf{n}}}[\eta^{(f_U)} - \eta^{(f)}] = 0.$$

Thus,

$$\lim_{U\to\infty} \frac{\mu_{v,\mathbf{n}}}{\eta^{(I)}}\frac{\partial \eta^{(f_U)}}{\partial \mu_{v,\mathbf{n}}} = \frac{\mu_{v,\mathbf{n}}}{\eta^{(I)}}\frac{\partial \eta^{(f)}}{\partial \mu_{v,\mathbf{n}}}. \tag{4.82}$$

Furthermore, by definition, we have

$$c^{(f)}(\mathbf{n}, v, \xi) = \sum_{l=0}^{L^*-1} f(X_l)\delta_l,$$

where $\delta_l = \Delta S_l/\Delta$. For the system with the performance function $f_U(\mathbf{n})$, we have

$$c^{(f_U)}(\mathbf{n}, v, \xi) = \sum_{l=0}^{L^*-1} f_U(X_l)\delta_l.$$

It is clear that on any sample path ξ, we have

$$\lim_{U\to\infty} c^{(f_U)}(\mathbf{n}, v, \xi) = c^{(f)}(\mathbf{n}, v, \xi).$$

Also, since $|f_U(X_l)| \le |f(X_l)|$, we have $|c^{(f_U)}(\mathbf{n}, v, \xi)| \le \sum_{l=0}^{L^*-1} |f(X_l)\delta_l|$. By the Lebesgue theorem, we have

$$\begin{aligned} \lim_{U\to\infty} c^{(f_U)}(\mathbf{n}, v) &= \lim_{U\to\infty} E\{c^{(f_U)}(\mathbf{n}, v, \xi)\} \\ &= E\{\lim_{U\to\infty} c^{(f_U)}(\mathbf{n}, v, \xi)\} = c^{(f)}(\mathbf{n}, v). \end{aligned} \tag{4.83}$$

Finally, (4.80) follows immediately from (4.81)-(4.83). □

Now let us study the time-average performance $\eta_{T_L}^{(f)}$. First, (4.45) holds for open networks. In (4.45), $c(\mathbf{n}, i)$ is the realization factor for $f \equiv 1$. As discussed in Section 4.4.1, a perturbation at any server will always be lost and a perturbation at the source will always be realized; i.e., $c(\mathbf{n}, i) = 0$, for $i \neq 0$ and $c(\mathbf{n}, 0) = 1$. Thus, (4.45) becomes

$$\mu_{i,\mathbf{n}} \frac{\partial \eta_T^{(f)}}{\partial \mu_{i,\mathbf{n}}} = \begin{cases} -p(\mathbf{n}) c^{(f)}(\mathbf{n}, i), & i \neq 0. \\ p(\mathbf{n})[\eta_T^{(f)} - c^{(f)}(\mathbf{n}, i)], & i = 0. \end{cases}$$

A special case is the load-dependent networks, for which we have

$$\frac{\mu_{v,k}}{\eta^{(I)}} \frac{\partial \eta^{(f)}}{\partial \mu_{v,k}} = -\sum_{n_v = k} p(\mathbf{n}) c^{(f)}(\mathbf{n}, v). \tag{4.84}$$

For open Jackson networks where the service rates are state-independent, we have

$$\frac{\mu_v}{\eta^{(I)}} \frac{\partial \eta^{(f)}}{\partial \mu_v} = -\sum_{all\ \mathbf{n}} p(\mathbf{n}) c^{(f)}(\mathbf{n}, v). \tag{4.85}$$

4.4.3 The Sensitivities of an M/M/1 Queue

As an example, we study the sensitivities of an M/M/1 queue with arrival rate λ and service rate μ. The system state is simply the number of customers in the queue; i.e., $\mathbf{n} = n$. The equations for realization factors are

$$c^{(f)}(0, 1) = 0, \tag{4.86}$$

$$c^{(f)}(n, 0) + c^{(f)}(n, 1) = f(n), \qquad n \ge 0, \tag{4.87}$$

$$c^{(f)}(0, 0) = c^{(f)}(1, 0) + c^{(f)}(1, 1) + f(0) - f(1), \tag{4.88}$$

$$\begin{aligned} (\lambda + \mu) c^{(f)}(n, 0) &= \mu c^{(f)}(n-1, 0) \\ &\quad + \lambda[c^{(f)}(n+1, 0) + f(n) - f(n+1)], \qquad n > 0, \end{aligned} \tag{4.89}$$

and

$$(\lambda+\mu)c^{(f)}(n,1) = \lambda c^{(f)}(n+1,1) \\ +\mu[c^{(f)}(n-1,1)+f(n)-f(n-1)], \qquad n>0. \tag{4.90}$$

Let us first consider the response time $r_L = \frac{1}{L}\int_0^{T_L} n(t)dt$. We have $f(n) = n$. Equations (4.86)-(4.90) become

$$c^{(r)}(0,1) = 0, \tag{4.91}$$

$$c^{(r)}(n,0) + c^{(r)}(n,1) = n, \qquad n \geq 0, \tag{4.92}$$

$$(\lambda+\mu)c^{(r)}(n,0) = \mu c^{(r)}(n-1,0) + \lambda c^{(r)}(n+1,0) - \lambda, \qquad n>0, \tag{4.93}$$

and

$$(\lambda+\mu)c^{(r)}(n,1) = \lambda c^{(r)}(n+1,1) + \mu c^{(r)}(n-1,1) + \mu, \qquad n>0. \tag{4.94}$$

To solve for $c^{(r)}(n,i)$, for $i = 0,1$, we need a boundary condition. Consider a busy period of the M/M/1 queue. Suppose that the server obtains a perturbation at its first customer's service completion time. Then every customer served in the same busy period will have the same perturbation. Thus, $c^{(r)}(1,1)$ equals the average number of customers served in a busy period. We have (Kleinrock [76])

$$c^{(r)}(1,1) = \frac{\mu}{\mu-\lambda} = \frac{1}{1-\rho}, \qquad \rho = \frac{\lambda}{\mu}. \tag{4.95}$$

From (4.91)-(4.95), we obtain

$$c^{(r)}(n,1) = \frac{n}{1-\rho} \tag{4.96}$$

and

$$c^{(r)}(n,0) = -\frac{n\rho}{1-\rho}. \tag{4.97}$$

Because of the structure of the sample path of an M/M/1 queue, we can also find another equation for $c^{(r)}(n,i)$, $i = 0,1$. The equation is derived by using the idea of sub-busy periods (see Section 2.3 or Kleinrock [76]). Suppose that the server obtains a perturbation when the system is in state n. This perturbation will be propagated to the next service completion time in the same busy period. The period from the time that the system enters state n to the first time that the system enters state $n-1$ can be viewed as a sub-busy period of the M/M/1 queue starting from state $n = 1$.

The effect of a perturbation at state n is then equal to the sum of the effect of a perturbation at state 1 plus that of a perturbation at state $n-1$; i.e.,

$$c^{(r)}(n,1) = c^{(r)}(1,1) + c^{(r)}(n-1,1). \tag{4.98}$$

Thus,

$$c^{(r)}(n,1) = nc^{(r)}(1,1). \tag{4.99}$$

Equations (4.91), (4.94), and (4.99) lead to the same solution as (4.96) and (4.97).

Now, the steady-state response time of an M/M/1 queue is (Kleinrock [76]):

$$\eta^{(r)} = \frac{1}{\mu - \lambda}.$$

Also,

$$\eta^{(I)} = \frac{1}{\lambda}.$$

Thus, by taking derivatives, we have

$$\frac{\mu}{\eta^{(I)}}\frac{d\eta^{(r)}}{d\mu} = -\frac{\lambda\mu}{(\mu-\lambda)^2} = -\frac{\rho}{(1-\rho)^2} \tag{4.100}$$

and

$$\frac{\lambda}{\eta^{(I)}}\frac{d\eta^{(r)}}{d\lambda} = \frac{\lambda^2}{(\mu-\lambda)^2} = \frac{\rho^2}{(1-\rho)^2}. \tag{4.101}$$

For an M/M/1 queue, the quasi-Lipschitz condition holds (with $g(n) = n$). Using (4.85), we get

$$\frac{\mu}{\eta^{(I)}}\frac{d\eta^{(r)}}{d\mu} = -\sum_{n=0}^{\infty} p(n)c^{(r)}(n,1)$$

$$= -\sum_{n=0}^{\infty}(1-\rho)\rho^n\{\frac{n}{1-\rho}\} = -\sum_{n=0}^{\infty} n\rho^n = -\frac{\rho}{(1-\rho)^2}$$

and

$$\frac{\lambda}{\eta^{(I)}}\frac{d\eta^{(r)}}{d\lambda} = -\sum_{n=0}^{\infty}(1-\rho)\rho^n\{-\frac{n\rho}{1-\rho}\} = \frac{\rho^2}{(1-\rho)^2},$$

which are the same as (4.100) and (4.101). This verifies the theory of realization factors in the M/M/1 queue case.

As we previously explained, the realization factor contains more information than the parametric derivatives. For example, (4.96) shows that on the average the effect of a service-rate increase on the response time is proportional to the number of customers when the increase occurs. Equation (4.97) shows that the effect of an arrival-rate increase on the response time is only ρ times that of the service-rate increase.

Finally, for the system throughput, we have $\eta = \frac{1}{\eta^{(I)}}$ with $I(n) \equiv 1$ for all $n \geq 0$. From (4.86)-(4.90) and (4.98), we have

$$c^{(I)}(n,0) = 1, \qquad c^{(I)}(n,1) = 0, \qquad n \geq 0.$$

Thus,

$$\frac{\mu}{\eta}\frac{d\eta}{d\mu} = -\frac{\mu}{\eta^{(I)}}\frac{d\eta^{(I)}}{d\mu} = \sum_{n=0}^{\infty} p(n)c^{(I)}(n,1) = 0$$

and

$$\frac{\lambda}{\eta}\frac{d\eta}{d\lambda} = \frac{\lambda}{\eta^{(I)}}\frac{d\eta^{(I)}}{d\lambda} = \sum_{n=0}^{\infty} p(n)c^{(I)}(n,0) = 1, \qquad n \geq 0.$$

These are, of course, intuitively true.

4.4.4 The Quasi-Lipschitz Condition

In this section, we provide a sufficient condition for the interchangeability required in (4.75). For any two states $\mathbf{n}_1$ and $\mathbf{n}_2$, we define a vector $\mathbf{n}_1 - \mathbf{n}_2$, whose kth component is $n_{1,k} - n_{2,k}$, $k = 1, 2, ..., M$. $\mathbf{n}_1 - \mathbf{n}_2$ may not represent a state since its components may be negative integers. Now, we introduce the following condition.

Condition 4.2 *The Quasi-Lipschitz Condition: There is a function g such that $|f(\mathbf{n}_1) - f(\mathbf{n}_2)| \leq g(\mathbf{n}_1 - \mathbf{n}_2)$, for all $\mathbf{n}_1$ and $\mathbf{n}_2$.*

Note that $g(\mathbf{n}_1 - \mathbf{n}_2)$ need not be a metric. Although $g(\mathbf{n} - \mathbf{n}) = 0$ and $g(\mathbf{n}_1 - \mathbf{n}_2) = g(\mathbf{n}_2 - \mathbf{n}_1)$ are desirable, the triangle inequality for a metric is not needed. Hence, we call the condition the *"quasi"-Lipschitz condition*. It is called the *Lipschitz condition* if the function g is a metric (see, e.g., Kreyszig [78]). In words, the quasi-Lipschitz condition requires that the "local growth" of the performance function f around any point is uniformly bounded by the "local growth" of a function g around a fixed point. The condition guarantees that starting from any initial state and for a fixed L, the sample-path performance $\eta_L^{(f)}$ is uniformly bounded.

The quasi-Lipschitz condition is weaker than the Lipschitz condition and is satisfied by most common performance functions, including the metric-type functions. For example, the following functions and those that are bounded by these functions satisfy the quasi-Lipschitz condition:

a. $\alpha_1 n_1 + \alpha_2 n_2 + \cdots + \alpha_M n_M$, $\alpha_i \in (-\infty, \infty)$;

b. $(\alpha_1 n_1^{\beta_1} + \alpha_2 n_2^{\beta_2} + \cdots + \alpha_M n_M^{\beta_M})^{\gamma}$, $\alpha_i > 0$, $1 \geq \beta_i \geq 0$, $1 \geq \gamma \geq 0$;

c. $(\alpha_1 n_1^p + \alpha_2 n_2^p + \cdots + \alpha_M n_M^p)^{1/p}$, $p \geq 1$;

d. $(\frac{\alpha_1 n_1^{p_1}}{\beta_1 + n_1^{p_1}} + \cdots + \frac{\alpha_M n_M^{p_M}}{\beta_M + n_M^{p_M}})^{\gamma}$, $\alpha_i > 0$, $\beta_i > 0$, $p_i > 0$, $\gamma \leq 1$;

e. $(n_1 n_2 \cdots n_M)^{1/M}$.

Theorem 4.10 *Under the quasi-Lipschitz condition, the following interchangeability holds:*

$$\frac{\partial}{\partial \mu_{v,\mathbf{n}}} \sum_{\mathbf{n}_0 \in \Phi} \{E[\eta_L^{(f)}(\mu_{v,\mathbf{n}}, \xi)|\mathbf{n}_0]\pi(\mathbf{n}_0)\}$$
$$= \sum_{\mathbf{n}_0 \in \Phi} \frac{\partial}{\partial \mu_{v,\mathbf{n}}} \{E[\eta_L^{(f)}(\mu_{v,\mathbf{n}}, \xi)|\mathbf{n}_0]\pi(\mathbf{n}_0)\}. \qquad (4.102)$$

Proof: The right-hand side of (4.102) equals

$$\sum_{\mathbf{n}_0 \in \Phi} \{\frac{\partial}{\partial \mu_{v,\mathbf{n}}} E[\eta_L^{(f)}(\mu_{v,\mathbf{n}}, \xi)|\mathbf{n}_0]\pi(\mathbf{n}_0)\}$$
$$= \sum_{\mathbf{n}_0 \in \Phi} \{\frac{\partial}{\partial \mu_{v,\mathbf{n}}} E[\eta_L^{(f)}(\mu_{v,\mathbf{n}}, \xi)|\mathbf{n}_0]\}\pi(\mathbf{n}_0)$$
$$+ \sum_{\mathbf{n}_0 \in \Phi} \{E[\eta_L^{(f)}(\mu_{v,\mathbf{n}}, \xi)|\mathbf{n}_0] \frac{\partial}{\partial \mu_{v,\mathbf{n}}} \pi(\mathbf{n}_0)\}. \qquad (4.103)$$

We have

$$E[\eta_L^{(f)}|\mathbf{n}_0] = \frac{1}{L} \sum_{l=0}^{L-1} E[f(X_l)S_l|\mathbf{n}_0]$$
$$= \frac{1}{L} \sum_{l=0}^{L-1} \sum_{\mathbf{m}} E[f(X_l)S_l|X_l = \mathbf{m}]P(X_l = \mathbf{m}|\mathbf{n}_0), \qquad (4.104)$$

where $P(X_l = \mathbf{m}|\mathbf{n}_0)$ is the probability of $X_l = \mathbf{m}$, given that $X_0 = \mathbf{n}_0$. Next,

$$E[f(X_l)S_l|X_l = \mathbf{m}] = f(\mathbf{m})E[S_l|X_l = \mathbf{m}] = \frac{f(\mathbf{m})}{\mu(\mathbf{m})}. \qquad (4.105)$$

From (4.104) and (4.105),

$$E[\eta_L^{(f)}|\mathbf{n}_0] = \frac{1}{L}\sum_{l=0}^{L-1}\{\sum_{\mathbf{m}}\frac{f(\mathbf{m})}{\mu(\mathbf{m})}P(X_l = \mathbf{m}|\mathbf{n}_0)\}. \qquad (4.106)$$

For any finite L and a given $\mathbf{n}_0$, there are only a finite number of states $\mathbf{m}$ for which $P(X_l = \mathbf{m}|\mathbf{n}_0)$, $0 \le l < L$, may be nonzero. Let $\Phi_{\mathbf{n}_0}$ be the set of those states and $J_{\mathbf{n}_0}$ be the number of those states; $J_{\mathbf{n}_0}$ also depends on L. For a given L, $J_{\mathbf{n}_0}$ is the same for many states $\mathbf{n}_0'$s and takes only a finite number of different values. This can be verified as follows. $J_{\mathbf{n}_0}$ is the same for all states with $n_{0,i} \ge L$ for all $i = 1, 2, ..., M$. Also, for any state $\mathbf{n}_0$, let $\mathbf{n}_0'$ be a state defined by $n_{0,i}' = min\{n_i, L\}$. It is easy to check that $J_{\mathbf{n}_0} = J_{\mathbf{n}_0'}$. Therefore, $J_{\mathbf{n}_0}$ has at most L^M different values. Let $J = max\{J_{\mathbf{n}_0},\ for\ all\ \mathbf{n}_0\}$. Then $J < \infty$.

Let us first examine the first term on the right-hand side of (4.103). Since the sum $\sum_{\mathbf{m}}$ in (4.106) contains only a finite number of nonzero terms, we have

$$\begin{aligned}&\frac{\partial}{\partial\mu_{v,\mathbf{n}}}E[\eta_L^{(f)}|\mathbf{n}_0]\\ &= \frac{1}{L}\sum_{l=0}^{L-1}\{\sum_{\mathbf{m}\in\Phi_{\mathbf{n}_0}} f(\mathbf{m})\frac{\partial[P(X_l = \mathbf{m}|\mathbf{n}_0)/\mu(\mathbf{m})]}{\partial\mu_{v,\mathbf{n}}}\}. \qquad (4.107)\end{aligned}$$

Again, for any finite L and a fixed $\mathbf{n}$, there are only a finite number of nonzero terms in the above sum. We have

$$|\frac{\partial[P(X_l = \mathbf{m}|\mathbf{n}_0)/\mu(\mathbf{m})]}{\partial\mu_{v,\mathbf{n}}}| < K < \infty \qquad for\ all\ \mathbf{m}. \qquad (4.108)$$

Furthermore, with a finite L and a fixed $\mathbf{n}$, for all $\mathbf{n}_0 \in \Phi$ there are only a finite number of $\mathbf{m}$'s for which $\frac{\partial}{\partial\mu_{v,\mathbf{n}}}\{[P(X_l = \mathbf{m}|\mathbf{n}_0)]/\mu(\mathbf{m})\} \ne 0$. Thus, we can choose K big enough so that it does not depend on $\mathbf{n}_0$. From (4.107) and (4.108), we get

$$|\{\frac{\partial}{\partial\mu_{v,\mathbf{n}}}E[\eta_L^{(f)}|\mathbf{n}_0]\}\pi(\mathbf{n}_0)| < K\{\sum_{\mathbf{m}\in\Phi_{\mathbf{n}_0}} |f(\mathbf{m})|\}\pi(\mathbf{n}_0). \qquad (4.109)$$

Under the quasi-Lipschitz condition, for any state $\mathbf{m}$, we have

$$|f(\mathbf{m})| < |f(\mathbf{n}_0)| + g(\mathbf{m} - \mathbf{n}_0).$$

For any fixed L, there are at most J nonzero terms on the right-hand side of (4.109). We have

$$\sum_{\mathbf{m}\in\Phi_{\mathbf{n}_0}} |f(\mathbf{m})| \le J|f(\mathbf{n}_0)| + J max_{\mathbf{m}\in\Phi_{\mathbf{n}_0}}\{g(\mathbf{m}-\mathbf{n}_0)\}.$$

The second term on the right-hand side is finite. Thus,

$$\begin{aligned}&|\{\frac{\partial}{\partial\mu_{v,\mathbf{n}}}E[\eta_L^{(f)}|\mathbf{n}_0]\}\pi(\mathbf{n}_0)|\\ &\quad < \{K_1|f(\mathbf{n}_0)|+K_2\}\pi(\mathbf{n}_0), \qquad K_1,\ K_2<\infty. \qquad (4.110)\end{aligned}$$

From Lemma 4.4, we have

$$|\frac{\partial\pi(\mathbf{n}_0)}{\partial\mu_{v,\mathbf{n}}}| < R_1\pi(\mathbf{n}_0), \qquad R_1<\infty. \qquad (4.111)$$

To indicate the dependency on $\mu_{v,\mathbf{n}}$ explicitly, we write the steady-state probability as $\pi(\mathbf{n}_0,\mu_{v,\mathbf{n}})$. From (4.111), if $\Delta\mu_{v,\mathbf{n}}$ is small enough, then

$$\begin{aligned}\pi(\mathbf{n}_0,\mu_{v,\mathbf{n}}+\Delta\mu_{v,\mathbf{n}}) &< \pi(\mathbf{n}_0,\mu_{v,\mathbf{n}})+R_1\Delta\mu_{v,\mathbf{n}}\pi(\mathbf{n}_0,\mu_{v,\mathbf{n}})\\ &< R_2\pi(\mathbf{n}_0,\mu_{v,\mathbf{n}}), \qquad R_2<\infty.\end{aligned}$$

Since $E\{|f(\mathbf{n})|\}$ exists and $\sum_{\mathbf{n}_0}\pi(\mathbf{n}_0)$ converges, from the above inequality, we conclude that both $\sum_{\mathbf{n}_0}\{|f(\mathbf{n}_0)|\pi(\mathbf{n}_0)\}$ and $\sum_{\mathbf{n}_0}\pi(\mathbf{n}_0)$ converge uniformly in $[\mu_{v,\mathbf{n}},\mu_{v,\mathbf{n}}+\Delta\mu_{v,\mathbf{n}}]$. Therefore, from (4.110), the first term on the right-hand side of (4.103) converges absolutely and uniformly in $[\mu_{v,\mathbf{n}},\mu_{v,\mathbf{n}}+\Delta\mu_{v,\mathbf{n}}]$.

Next, from (4.106), we have

$$|E[\eta_L^{(f)}|\mathbf{n}_0]| < R_0\sum_{\mathbf{m}\in\Phi_{\mathbf{n}_0}}|f(\mathbf{m})|, \qquad (4.112)$$

where R_0 is the bound of $\{P(X_l=\mathbf{m}|\mathbf{n}_0)/\mu(\mathbf{n})\}$ for $\mathbf{m}\in\Phi_{\mathbf{n}_0}$ (e.g., $R_0=\frac{1}{(M+1)w}$). From (4.111), (4.112), and the quasi-Lipschtiz condition, the second term on the right-hand side of (4.103) converges absolutely and uniformly in $[\mu_{v,\mathbf{n}},\mu_{v,\mathbf{n}}+\Delta\mu_{v,\mathbf{n}}]$. This justifies the interchangeability in (4.102). □

Remark 4.2 In the open network studied in this section, all the customers arriving at the network are from the same source. They may enter different servers in the network. In this setting, a perturbation of the arrival time to one server will affect the subsequent arrival times to other servers. This is the case where there is only one class of customers. For networks in which customers come from more than one source, a perturbation of the arrival time from one source cannot affect the arrival times from the other sources; the results in this section may not hold. As an example, consider a system in which two streams of arrivals at servers 1 and 2 come from two different sources. Suppose that an arrival instant obtains a perturbation (e.g., the arrival instant is delayed by Δ); this perturbation will be propagated to all the arrival instants from the same source. (All the arrival instants from that source thereafter will be delayed by Δ.) On the other hand, all the arrival instants from the other source will remain unchanged. As a result, the perturbation will never be realized or lost by the system. Therefore, the theory developed so far does not hold for the perturbation in one of the arrival streams. In other words, the theorems developed in this section cannot be applied to the sensitivity of the parameter changes in the distribution functions of the interarrival times if the customers come from more than one source (multiclass open networks). However, the theorems in this section can still be applied to the sensitivities of the performance measures with respect to service distribution parameters, even if the system has more than one stream of arrival customers. More discussion of multiclass networks will be in Chapter 6.

and sense only affect the successive arrival times to other extrema. This is the case where there is only one channel of reaction. For networks in which molecules come from more than one source, a perturbation of the arrival rates from one source can also affect the arrival times from the other sources. The double minimum then may not hold. As an example, consider a system in which [illegible] with [illegible] 1 and 2 [illegible] different [illegible]. Suppose that [illegible] a perturbation [illegible] the [illegible] this perturbation will be propagated to [illegible] arrival [illegible] from the same [illegible] from [illegible]. On the other hand, all the arrival [illegible] from the [illegible]. As a result, [illegible] the system. [illegible] the theory [illegible] does not hold for [illegible] the [illegible]. In other words, the [illegible] to the [illegible] of the parameter changes in the [illegible] the [illegible] if the [illegible] from [illegible] the [illegible] to the [illegible] the [illegible] if the [illegible].

Chapter 5

Systems with General Distributions

In this chapter, we shall develop the realization theory for queuing networks, closed or open, in which the service and the interarrival time distributions are nonexponential. The state of such a network consists of two parts: the discrete part, indicating the number of customers in each server, and the continuous part, indicating the remaining or the elapsed service times of the customers in each server. The state process of such a network is usually not a Markov process with a countable state space. We shall derive a set of linear differential equations for the realization probabilities and the realization factors, define a perturbation generation function, and prove that the normalized derivative of a performance equals the negative expected value of the product of the generation function and the realization factor. The main reference for this chapter is Cao [16].

5.1 Closed Networks

5.1.1 The Network and Its Steady-State Probabilities

We consider a closed queuing network consisting of M single-server nodes and N single-class customers. The routing probabilities are $q_{i,j}$, $i, j = 1, 2, ..., M$. Each server has a buffer with an infinite size. Let $F_i(s)$ be the cumulative distribution function of the service time of server i, $i = 1, 2, \cdots, M$. We assume that the following condition holds.

Condition 5.1 *The distribution functions $F_i(s)$, $i = 1, 2, \cdots, M$, are absolutely continuous.*

Under this condition, for each i there exists a probability density function $f_i(s)$, which satisfies $F_i(s) = \int_0^s f_i(t)dt$ and $f_i(s) < \infty$ for all $0 \leq s < \infty$. Thus, $F_i(s)$ does not contain atoms. This is necessary for the analysis of this chapter.

Given the elapsed service time r, the probability density function of the remaining service time s is

$$h_i(s) = \frac{f_i(s+r)}{1 - F_i(r)}.$$

Let $\mathbf{x} = (\mathbf{n}, \mathbf{r})$ be the state of the system, where $\mathbf{n} = (n_1, n_2, ..., n_M)$ is the discrete part of the state with n_i being the number of customers in server i, and $\mathbf{r} = (r_1, r_2, ..., r_M)$ is the continuous part of the state with r_i being the elapsed service time of the customer being served at server i. We asuume that $r_i = 0$ if $n_i = 0$. $\mathbf{n}$ is also called the *discrete state* of the system. The set of the discrete states is called the *discrete state space* and is denoted as Φ. The state process is denoted as $\mathcal{X}(t) = (\mathcal{N}(t), \mathcal{R}(t))$. The process is considered right-continuous. We denote by R_+^M the quadrant with $r_i \geq 0$, $i = 1, 2, ..., M$, in the M-dimensional real space R^M. The state space of the process is then $\Phi \times R_+^M$. For convenience, we employ the following notations:

$$\mathbf{r} - \Delta\mathbf{t} = (r_1 - \Delta t, r_2 - \Delta t, ..., r_M - \Delta t)$$

and

$$\mathbf{r}_{-i} = (r_1, r_2, ..., r_i = 0, ..., r_M),$$

$$\mathbf{r} + \tau_i = (r_1, r_2, ..., r_i + \tau_i, ..., r_M).$$

Using these notations, we can write

$$\mathbf{r}_{-i} + \tau_i = (r_1, r_2, ..., \tau_i, ..., r_M),$$

$$\mathbf{r}_{-(i+j)} = \{\mathbf{r}_{-i}\}_{-j} = (r_1, ..., r_i = 0, ..., r_j = 0, ..., r_M), \qquad i \neq j.$$

We also define $\mathbf{r} > \mathbf{r}'$ ($\mathbf{r} \geq \mathbf{r}'$) as $r_i > r_i'$ ($r_i \geq r_i'$) for all i.

If the distribution functions $F_i(s)$, $i = 1, 2, ..., M$, are not exponential, then the state process $\mathcal{X}(t)$ is not a Markov process in a countable state space. Let $(\Omega, \mathcal{F}, P)$ be the probability space on which the stochastic process $\mathcal{X}(t)$ is defined, $P(\mathbf{x}, t) = P(\mathbf{n}, \mathbf{r}, t) = Pr\{\mathcal{N}(t) = \mathbf{n}, \mathcal{R}(t) \leq \mathbf{r} = (r_1, r_2, ..., r_M)\}$ be the probability distribution function of the state $\mathbf{x}$ at time t, and $p(\mathbf{x},t) = p(\mathbf{n},\mathbf{r},t) = \frac{d}{d\mathbf{r}}P(\mathbf{n},\mathbf{r},t)$ be the probability density function of the state process. As was seen in previous chapters, ergodicity of the state space is crucial to our study; thus, we assume that the following condition holds.

Condition 5.2 *The state process $\mathcal{X}(t)$ is ergodic and asymptotically stationary with a unique invariant measure $p(\mathbf{x})$.*

$p(\mathbf{x}) = p(\mathbf{n},\mathbf{r})$ is also called the steady-state probability density of the system state. Condition 5.2 implies that the steady-state probability density $p(\mathbf{x})$ exists and $p(\mathbf{x},t) \to p(\mathbf{x})$ as $t \to \infty$ and that for any measurable function $f: \ \Phi \times R_+^M \to R$ with $E[|f(\mathbf{x})|] < \infty$,

$$\lim_{t\to\infty} \frac{1}{t}\int_0^t f[\mathbf{x}(t)]dt = E[f(\mathbf{x})]$$

holds with probability one for any given initial state (see Section 2.2).

The ergodicity of a stochastic process per se is a very complex issue and is beyond the scope of this book. Interested readers may refer to, for example, Sigman [97] and [98], which present sufficient conditions for the ergodicity of closed and open queuing networks, Borovkov [6], [7], and [8], which discuss the ergodicity and related subjects, and Whitt [106], which proves the existence of the steady-state distribution in the context of a generalized semi-Markov process. Here, we just point out that we require that the network is indecomposable, or the routing probability matrix $Q = [q_{i,j}]$ is irreducible. That is, any customer in any server should have a positive probability of reaching every other server in the network, either directly or by going through some other servers. This is necessary for the system to reach every possible state. The irreducibility of Q is also sufficient for the ergodicity if all the service distributions are exponential (cf. Chapter 2).

In the remainder of this section, we derive the equations for the probability density functions. Clearly, we have

$$p(\mathbf{n},\mathbf{r}) = 0, \qquad \textit{if } r_i > 0 \textit{ and } n_i = 0 \textit{ for some } i. \qquad (5.1)$$

Next, if $r_i > 0$ for all $n_i > 0$, then the only possible case that at time $t+\Delta t$ the system is at state $(\mathbf{n},\mathbf{r})$ is that the system was at state $(\mathbf{n},\mathbf{r}-\Delta \mathbf{t})$ at time t and there is no customer transition in period $[t, t+\Delta t]$. Note that the probability that there is a customer transition from server i in $[t, t+\Delta t)$ given that the elapsed time is r_i is $g_i(r_i)\Delta t$, where

$$g_i(r_i) = h_i(s=0) = \frac{f_i(r_i)}{1 - F_i(r_i)}$$

is the *hazard rate* (see Section 2.1). Therefore,

$$p(\mathbf{n},\mathbf{r},t+\Delta t) = p(\mathbf{n},\mathbf{r}-\Delta\mathbf{t},t)\{1-\sum_{i=1}^{M}\epsilon(n_i)g_i(r_i)\Delta t\}$$

$$= \{p(\mathbf{n},\mathbf{r},t) - \sum_{i=1}^{M}\frac{\partial p(\mathbf{n},\mathbf{r},t)}{\partial r_i}\Delta t\}\{1-\sum_{i=1}^{M}\epsilon(n_i)g_i(r_i)\Delta t\}.$$

From this, we obtain a partial differential equation for $p(\mathbf{n},\mathbf{r},t)$:

$$\frac{\partial p(\mathbf{n},\mathbf{r},t)}{\partial t} = -\sum_{i=1}^{M}\frac{\partial p(\mathbf{n},\mathbf{r},t)}{\partial r_i} - p(\mathbf{n},\mathbf{r},t)\{\sum_{i=1}^{M}\epsilon(n_i)g_i(r_i)\}. \tag{5.2}$$

Letting $t\to\infty$, we get the differential equation for the steady-state probability distribution:

$$\sum_{i=1}^{M}\frac{\partial p(\mathbf{n},\mathbf{r})}{\partial r_i} = -p(\mathbf{n},\mathbf{r})\{\sum_{i=1}^{M}\epsilon(n_i)g_i(r_i)\}, \qquad r_i>0, \quad \textit{for all } n_i>0. \tag{5.3}$$

Now consider the case in which there is one server, say server i, for which $n_i>0$ but $r_i=0$. Since $r_i=0$, server i must start a new service at time t. There are two possible situations in which this may happen: server i just completed its service to a customer and started its service to the next customer, or server i was idle and just received a customer from another server j. However, in the second situation, we should have $r_i=r_j=0$ after the customer transition. Therefore, the first situation is the sole case if we have only one $r_i=0$. This explains the following equation:

$$\begin{aligned}p(\mathbf{n},\mathbf{r},t) &= \sum_{k\neq i}^{M} q_{i,k}\epsilon(n_k-1)\int_0^\infty p(\mathbf{n}_{k,i},\mathbf{r}_{-i}+\tau_i,t)g_i(\tau_i)d\tau_i \\ &+ q_{i,i}\int_0^\infty p(\mathbf{n},\mathbf{r}_{-i}+\tau_i,t)g_i(\tau_i)d\tau_i, \\ &\qquad \textit{for } n_i>0, \quad r_i=0.\end{aligned} \tag{5.4}$$

Letting $t\to\infty$, we get the equation for steady-state probabilities:

$$\begin{aligned}p(\mathbf{n},\mathbf{r}) &= \sum_{k\neq i}^{M} q_{i,k}\epsilon(n_k-1)\int_0^\infty p(\mathbf{n}_{k,i},\mathbf{r}_{-i}+\tau_i)g_i(\tau_i)d\tau_i \\ &+ q_{i,i}\int_0^\infty p(\mathbf{n},\mathbf{r}_{-i}+\tau_i)g_i(\tau_i)d\tau_i, \\ &\qquad \textit{for } n_i>0 \quad r_i=0.\end{aligned} \tag{5.5}$$

Next, consider the case in which at time t the system is in a state $\mathbf{n}$ with $r_i = r_j = 0$, $i \neq j$, $n_j = 1$ and $n_i \neq 1$. This implies that an idle period of server j is just terminated by a customer from server i. Thus, we have

$$p(\mathbf{n},\mathbf{r}) = q_{i,j} \int_0^\infty p(\mathbf{n}_{j,i}, \mathbf{r}_{-(i+j)} + \tau_i) g_i(\tau_i) d\tau_i, \quad for \ r_i = r_j = 0,\ n_j = 1,\ and\ n_i \neq 1. \tag{5.6}$$

Similarly, if $r_i = r_j = 0$, $n_j = n_i = 1$, we have

$$\begin{aligned} p(\mathbf{n},\mathbf{r}) &= q_{i,j} \int_0^\infty p(\mathbf{n}_{j,i}, \mathbf{r}_{-(i+j)} + \tau_i) g_i(\tau_i) d\tau_i \\ &+ q_{j,i} \int_0^\infty \pi(\mathbf{n}_{i,j}, \mathbf{r}_{-(i+j)} + \tau_j) g_j(\tau_j) d\tau_j, \\ &\qquad for \ r_i = r_j = 0,\ n_j = n_i = 1. \end{aligned} \tag{5.7}$$

Finally, for all other cases, $p(\mathbf{n},\mathbf{r}) = 0$ (the probability of more than two elapsed times being zero is zero). The steady-state probability density function satisfies the normalizing condition:

$$\sum_{all\ \mathbf{n}} \int_{R_+^M} p(\mathbf{n},\mathbf{r}) d\mathbf{r} = 1. \tag{5.8}$$

Equations (5.1),(5.3), and (5.5)-(5.8) specify the steady-state probabilities $p(\mathbf{n},\mathbf{r})$.

Let $\pi(\mathbf{n},\mathbf{r})$ be the steady-state probability density seen by an observer at the transition instants. $\pi(\mathbf{n},\mathbf{r})$ corresponds to the steady-state probability $\pi(\mathbf{n})$ of the embedded Markov chain in networks with exponential service time distributions. The normalizing condition for $\pi(\mathbf{n},\mathbf{r})$ is

$$\sum_{all\ \mathbf{n}} \int_{R_+^M} p(\mathbf{n},\mathbf{r}) d\mathbf{r} = 1. \tag{5.9}$$

5.1.2 Differential Equations for Realization Factors

Since the service rates in the network considered here do not depend on the system states, the perturbation propagation rules are the same as those in Section 3.2.

Denote by $(\mathbf{x}, V, t) = (\mathbf{n}, \mathbf{r}, V, t)$ the perturbation in set V at time t when the system state is $\mathcal{X}(t) = (\mathbf{n},\mathbf{r})$. If there is no need to specify the time, we write the perturbation as $(\mathbf{x}, V) = (\mathbf{n}, \mathbf{r}, V)$. If $V = \{i\}$, we simply write the perturbation as $(\mathbf{x}, i) = (\mathbf{n}, \mathbf{r}, i)$. Since the service rates are assumed to be state independent in this chapter, any perturbation will be either realized or lost with probability one; this will happen, e.g., when the system reaches the discrete state $(N, 0, \cdots, 0)$.

Let f be a mapping from the discrete state space Φ to $R = (-\infty, \infty)$. The customer-average and time-average sample performance measures, $\eta_L^{(f)}$ and $\eta_{T_L}^{(f)}$, are defined the same as (4.6) and (4.7). For example,

$$\eta_L^{(f)} = \frac{F_L}{L},$$

with

$$F_L = \int_0^{T_L} f(\mathcal{N}(t))dt.$$

Let $N_l := \mathcal{N}(T_l)$; then

$$F_L = \sum_{l=0}^{L-1} f(N_l)S_l, \quad S_l = T_{l+1} - T_l, \quad l = 1, 2, \cdots.$$

The realization probability and the realization factor of a perturbation with respect to f are defined in the same way as in Chapters 3 and 4. That is, the realization factor $c^{(f)}(\mathbf{x}, V)$ is defined as the expected value of $\Delta F_L/\Delta$, where $L > L^*$ and at T_{L^*} the perturbation has already been realized or lost.

The proof of the next lemma is the same as that of Lemma 4.1.

Lemma 5.1

$$If\ n_i = 0,\ then\ c^{(f)}(\mathbf{n}, \mathbf{r}, i) = 0; \tag{5.10}$$

$$c^{(f)}(\mathbf{n}, \mathbf{r}, \Gamma) = f(\mathbf{n}); \tag{5.11}$$

If $V_1 \cap V_2 = \emptyset$ *and* $V_1 \cup V_2 = V_3$, *then*

$$c^{(f)}(\mathbf{n}, \mathbf{r}, V_1) + c^{(f)}(\mathbf{n}, \mathbf{r}, V_2) = c^{(f)}(\mathbf{n}, \mathbf{r}, V_3); \tag{5.12}$$

and

$$\sum_{i=1}^{M} c^{(f)}(\mathbf{n}, \mathbf{r}, i) = f(\mathbf{n}). \tag{5.13}$$

In what follows, we shall derive a differential equation for the realization factors. By the superposition property of propagation, the realization factor of a perturbation at server k at time t, $(\mathbf{x}, k, t)$, consists of two parts. The first part equals the sum of the realization factors of the perturbations at time $t+\Delta t$ that are resultants of the propagation of the perturbation $(\mathbf{x}, k, t)$

from time t to time $t+\Delta t$, weighted by the probabilities that the system will be in the corresponding states at time $t+\Delta t$; the second part is the direct effect of the perturbation at time t on $\Delta F_L/\Delta$. Note that a perturbation Δ of server k delays the transition from server k to server j; this increases the sojourn time that the system stays in state $\mathbf{n}$ by Δ, resulting in an increase of $f(\mathbf{n})\Delta$ in ΔF_L, and decreases the sojourn time that the system stays in state $\mathbf{n}_{k,j}$ by Δ, resulting in a decrease of $f(\mathbf{n}_{k,j})\Delta$ in ΔF_L. Thus, the direct effort of Δ of server k at t on $\Delta F_L/\Delta$ is $f(\mathbf{n}) - f(\mathbf{n}_{k,j})$.

To help in understanding the derivation, we explicitly denote the time t in the realization factors and write it as $c^{(f)}(\mathbf{x}, V, t)$. Thus, from the above discussion, we have

$$\begin{aligned}
c^{(f)}(\mathbf{n},\mathbf{r},k,t) &= c^{(f)}(\mathbf{n},\mathbf{r}+\Delta\mathbf{t},k,t+\Delta t)[1-\sum_{i=1}^{M}\epsilon(n_i)g_i(r_i)\Delta t] \\
&+ \sum_{i=1}^{M}\sum_{j=1}^{M}\epsilon(n_i)\epsilon(n_j)g_i(r_i)\Delta r_i q_{i,j}c^{(f)}(\mathbf{n}_{i,j},\mathbf{r}_{-i}+\Delta\mathbf{t},k,t+\Delta t) \\
&+ \sum_{i=1}^{M}\sum_{j=1}^{M}\epsilon(n_i)[1-\epsilon(n_j)]g_i(r_i)\Delta r_i q_{i,j}c^{(f)}(\mathbf{n}_{i,j},\mathbf{r}_{-i-j}+\Delta\mathbf{t},k,t+\Delta t) \\
&+ \sum_{j=1}^{M}[1-\epsilon(n_j)]g_k(r_k)\Delta r_k q_{k,j}c^{(f)}(\mathbf{n}_{k,j},\mathbf{r}_{-k-j}+\Delta\mathbf{t},j,t+\Delta t) \\
&+ \sum_{j=1}^{M}\{g_k(r_k)\Delta r_k q_{k,j}[f(\mathbf{n})-f(\mathbf{n}_{k,j})]\}+o(\Delta t). \qquad (5.14)
\end{aligned}$$

In the above equation, $\Delta r_i = \Delta t$, $i = 1, 2, ..., M$, and $o(\Delta t)$ represents the high order terms of Δt. $g_i(r_i)$ is the hazard rate of $F_i(s_i)$. The probability of one customer transition from server i in $[t, t+\Delta t)$ given that the elapsed time is r_i is $g_i(r_i)\Delta t$. Thus, the first term on the right-hand side of (5.14) represents the event in which no transition occurs in $[t, t+\Delta t)$. The second term represents the case in which a customer transfers from server i to server j and server j is not idle before the transition; in this case, there is no propagation and $r_i = 0$ after the transition. The third term describes the situation in which an idle period of server j is terminated by server i in $[t, t+\Delta t)$; in this case, $r_i = r_j = 0$ after the transition. The fourth term represents the perturbation propagation effect; if server k terminates an idle period of server j, the perturbation will be propagated to server j. Finally, the term containing $f(\mathbf{n}) - f(\mathbf{n}_{k,j})$ represents the direct effect of the perturbation at time t.

Substituting the equation

$$
\begin{aligned}
& c^{(f)}(\mathbf{n}, \mathbf{r}+\Delta \mathbf{t}, k, t+\Delta t) \\
= \quad & c^{(f)}(\mathbf{n}, \mathbf{r}, k, t+\Delta t)+\sum_{i=1}^{M} \frac{\partial c(\mathbf{n}, \mathbf{r}, k, t+\Delta t)}{\partial r_i} \Delta t
\end{aligned}
$$

into (5.14) and letting $\Delta t \rightarrow 0$, we obtain

$$
\begin{aligned}
\frac{\partial c^{(f)}(\mathbf{n}, \mathbf{r}, k, t)}{\partial t} = \quad & c^{(f)}(\mathbf{n}, \mathbf{r}, k, t) \sum_{i=1}^{M} \epsilon(n_i) g_i(r_i)-\sum_{i=1}^{M} \frac{\partial c^{(f)}(\mathbf{n}, \mathbf{r}, k, t)}{\partial r_i} \\
- \quad & \sum_{i=1}^{M} \sum_{j=1}^{M} \epsilon(n_i) \epsilon(n_j) g_i(r_i) q_{i,j} c^{(f)}(\mathbf{n}_{i,j}, \mathbf{r}_{-i}, k, t) \\
- \quad & \sum_{i=1}^{M} \sum_{j=1}^{M} \epsilon(n_i)[1-\epsilon(n_j)] g_i(r_i) q_{i,j} c^{(f)}(\mathbf{n}_{i,j}, \mathbf{r}_{-i-j}, k, t) \\
- \quad & \sum_{j=1}^{M} g_k(r_k) q_{k,j}\{[1-\epsilon(n_j)] c^{(f)}(\mathbf{n}_{k,j}, \mathbf{r}_{-k-j}, j, t) \\
& \qquad +f(\mathbf{n})-f(\mathbf{n}_{k,j})\}.
\end{aligned}
$$

Since the realization probability depends only on the system state $(\mathbf{n}, \mathbf{r})$ and is otherwise independent of t, $\frac{\partial c^{(f)}(\mathbf{n},\mathbf{r},k,t)}{\partial t}=0$ and $c^{(f)}(\mathbf{n}, \mathbf{r}, k, t)= c^{(f)}(\mathbf{n}, \mathbf{r}, k)$. Finally, we have the following theorem.

Theorem 5.1 *For a closed queuing network with general service-time distributions, the realization factor satisfies*

$$
\begin{aligned}
\sum_{i=1}^{M} \frac{\partial c^{(f)}(\mathbf{n}, \mathbf{r}, k)}{\partial r_i} = \quad & c^{(f)}(\mathbf{n}, \mathbf{r}, k) \sum_{i=1}^{M} \epsilon(n_i) g_i(r_i) \\
- \quad & \sum_{i=1}^{M} \sum_{j=1}^{M} \epsilon(n_i) \epsilon(n_j) g_i(r_i) q_{i,j} c^{(f)}(\mathbf{n}_{i,j}, \mathbf{r}_{-i}, k) \\
- \quad & \sum_{i=1}^{M} \sum_{j=1}^{M} \epsilon(n_i)[1-\epsilon(n_j)] g_i(r_i) q_{i,j} c^{(f)}(\mathbf{n}_{i,j}, \mathbf{r}_{-i-j}, k) \\
- \quad & \sum_{j=1}^{M} g_k(r_k) q_{k,j}\{[1-\epsilon(n_j)] c^{(f)}(\mathbf{n}_{k,j}, \mathbf{r}_{-k-j}, j) \\
& \qquad +f(\mathbf{n})-f(\mathbf{n}_{k,j})\}. \qquad (5.15)
\end{aligned}
$$

For $f(\mathbf{n}) = I(\mathbf{n}) \equiv 1$, $\eta_L^{(f)} = \eta_L^{(I)} = T_L/L = 1/\eta_L$, with η_L being the sample throughput. In this case, Equations (5.10)-(5.13) and (5.15) become the equations for realization probabilities $c(\mathbf{n}, \mathbf{r}, i) := c^{(I)}(\mathbf{n}, \mathbf{r}, i)$.

For networks with exponential service distributions, the left-hand side of (5.15) is zero, and the differential Equation (5.15) reduces to (4.20).

5.1.3 The Perturbation Generation Function

Recall that the randomness of a closed queuing network can be represented by a vector $\xi = (\xi_1, \cdots, \xi_M; \zeta_1, \cdots, \zeta_M; \mathcal{X}(0))$, where $\xi_i = (\xi_{i,1}, \xi_{i,2}, \cdots)$, and $\xi_{i,k}$, $i = 1, 2, \cdots, M$, $k = 1, 2, \cdots$, are random variables uniformly distributed on $[0, 1)$.

Let $F_i(s, \theta)$ be the distribution function of the service times of the customers at server i. In this expression, we explicitly denote the dependency on a parameter θ. The service times of server i can be obtained by using the inverse function of the distribution function, which is defined as (see Section 2.1)

$$F_i^{-1}(\xi, \theta) = sup\{s : \ F_i(s, \theta) \leq \xi\}.$$

With the uniform random variable $\xi_{i,k}$, we can determine the service time of the kth customer at server i as

$$s_{i,k} = F_i^{-1}(\xi_{i,k}, \theta), \tag{5.16}$$

which has the distribution function $F_i(s, \theta)$.

In sensitivity analysis, we assume that a service distribution parameter θ changes to $\theta + \Delta\theta$. The change $\Delta\theta$ will induce changes in $s_{i,k}$ and the service completion times $t_{i,k}$, $i = 1, 2, \cdots, M$, $k = 1, 2, \cdots$. From (5.16), the service time of a customer is

$$s = F^{-1}(\xi, \theta).$$

For ease of notation, we dropped the subscript in the expression. For the system with a parameter $\theta + \Delta\theta$, the service time of the customer is

$$s' = F^{-1}(\xi, \theta + \Delta\theta).$$

Let $s' = s + \Delta s$. Then we have

$$\Delta s = F^{-1}(\xi, \theta + \Delta\theta) - F^{-1}(\xi, \theta) = \frac{\partial F^{-1}(\xi, \theta)}{\partial \theta}|_{\xi = F(s,\theta)} \Delta\theta. \tag{5.17}$$

Δs is the *perturbation generated* during the service period because of $\Delta\theta$. The same random variable ξ is used for both s and s'.

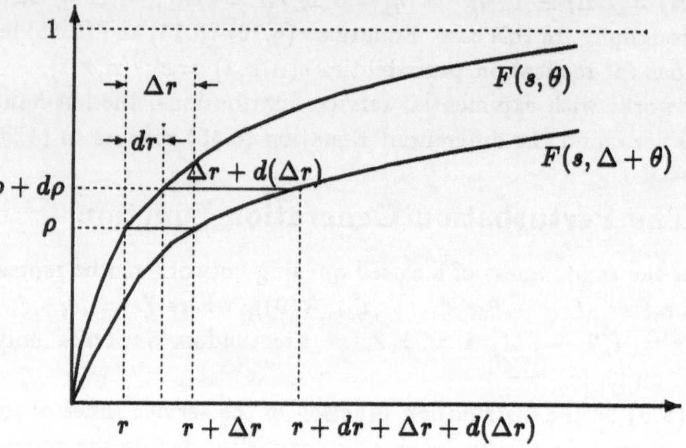

Figure 5.1: The Perturbation Generation Function

An important conceptual extension for networks with general service time distributions is that the perturbation generated in the entire service period can be "spread" throughout the service period. Let r, $0 \le r \le s$ be the elapsed service time and let $\rho = F(r,\theta)$. As r increases from 0 to s, ρ increases from 0 to $\xi = F(s,\theta)$. Now, what is the perturbation generated in the small interval $[r, r+dr)$? Again, if θ changes to $\theta + \Delta\theta$, the elapsed time r changes to $r + \Delta r$ (corresponding to the same ρ) with

$$\Delta r = \frac{\partial F^{-1}(\rho,\theta)}{\partial\theta}|_{\rho=F(r,\theta)}\Delta\theta.$$

Δr can be viewed as the perturbation generated in the elapsed service period $[0,r)$ because of $\Delta\theta$. Therefore, the perturbation generated in the service interval $[r, r+dr)$ is

$$\begin{aligned} d(\Delta r) &= \frac{\partial}{\partial r}\left\{\frac{\partial F^{-1}(\rho,\theta)}{\partial\theta}|_{\rho=F(r,\theta)}\right\}\Delta\theta dr \\ &= G(r,\theta)\Delta\theta dr, \end{aligned} \tag{5.18}$$

where

$$G(r,\theta) = \frac{\partial}{\partial r}\left\{\frac{\partial F^{-1}(\rho,\theta)}{\partial\theta}|_{\rho=F(r,\theta)}\right\} \tag{5.19}$$

is called the *perturbation generation function*. Figure 5.1 illustrates the meaning of dr, Δr, $\Delta r + d(\Delta r)$.

From (5.19), we have

$$\Delta s = \int_0^s d(\Delta r) = \{\int_0^s G(r,\theta)dr\}\Delta\theta.$$

That is, the perturbation generated in the entire service period, Δs, equals the integral of the perturbation generated in $[r, r + dr)$.

For the exponential service distribution $F(s,\theta) = 1 - exp(-s\theta)$ with $\theta = \mu$ being the mean service rate, we have

$$r = F^{-1}(\rho,\theta) = -\frac{1}{\theta}ln(1-\rho),$$

$$G(r,\theta) = \frac{\partial}{\partial r}\left\{\frac{\partial F^{-1}(\rho,\theta)}{\partial\theta}|_{\rho=F(r,\theta)}\right\} = \frac{\partial}{\partial r}\{-\frac{r}{\theta}\} = -\frac{1}{\theta}.$$

Therefore, for exponential distributions, the perturbation generated in any service interval $[r, r + dr)$ is $-\frac{\Delta\theta}{\theta}dr$, which is proportional to dr. This linear property holds in many other situations. In fact, if the distribution function takes the form $F(\frac{s}{\theta})$ (θ is a scale parameter), then $r = \theta F^{-1}(\rho)$ and $G(r,\theta) = \frac{\partial}{\partial r}(\frac{r}{\theta}) = \frac{1}{\theta}$. In this case, the perturbation generated in $[r, r + dr)$ is $\frac{\Delta\theta}{\theta}dr$, proportional to dr. For another example, let θ be the service rate of a server. In this case, the service time takes the form $s = \frac{F^{-1}(\xi)}{\theta}$. $F^{-1}(\xi)$ may be considered the service requirement of a customer on the server. Simple calculation gives $G(r,\theta) = -\frac{1}{\theta}$.

For further study, we need the following assumption.

Condition 5.3 *There is a positive number $K > \infty$ such that in a neighborhood of θ, $[\theta_a, \theta_b]$, $|G(r,\theta)| < K$ holds for all $r \geq 0$.*

This condition guarantees that for a finite $\Delta\theta$, the perturbation generated in any finite interval $[r, r + dr)$ is finite. Condition 5.3 holds for many functions. For example, it holds for the cases discussed above in which the perturbation generated is proportional to dr. Condition 5.3 is almost the same as, but somewhat stronger than, the following condition:

$$|\frac{\partial F^{-1}(\xi,\theta)}{\partial\theta}| < KF^{-1}(\xi,\theta). \tag{5.20}$$

In fact, if Condition 5.3 holds, so does Equation (5.20). Equation (5.20) assures that the perturbation generated in the entire service period $[0, s)$, $s = F^{-1}(\xi,\theta)$, is bounded by $Ks\Delta\theta$; it has the same effect as Condition 5.3 except for the perturbation generated in an interval starting in the middle of a service period. In other words, Condition 5.3 assures that the perturbation generated in interval $[t_0, t_{i,1}]$ is also bounded by $K(t_{i,1}-t_0)\Delta\theta$, even if $r_i(t_0) > 0$.

5.1.4 Sample Derivatives

The Similarity Between the Nominal and the Perturbed Paths

For systems with general distributions, (3.29) holds; i.e.,

$$\Delta_{i,k} = \begin{cases} \Delta_{i,k-1} + \Delta s_{i,k} & if\ n_i(t_{i,k-1}+) \neq 0, \\ \Delta_{j,h} + \Delta s_{i,k} & if\ n_i(t_{i,k-1}+) = 0. \end{cases} \tag{5.21}$$

Equation (3.30) becomes

$$\Delta s_{i,k} = \begin{cases} 0 & if\ i \neq v, \\ \frac{\partial F^{-1}(\xi,\theta)}{\partial \theta}\Big|_{\xi=F(s,\theta)} \Delta\theta & if\ i = v. \end{cases} \tag{5.22}$$

Under Condition 5.3, (5.20) holds; thus, from (5.22), the perturbation generated in a service interval $[0, s]$ is less than $Ks\Delta\theta$. Therefore, for any sample path ξ with a finite length T_L, we can always choose $\Delta\theta(\xi)$ small enough so that $|\Delta_{i,k}| < S_{L,min}(\xi)$, and the nominal and the perturbed paths are similar. Therefore, (5.21) holds for any sample path with a finite L. The equation shows that the superposition holds for perturbation propagation (cf. Chapter 3).

The Sample Derivatives

Suppose that θ is a parameter of the service distribution function of server v. By (5.18), in any interval $[t, t+dt)$, there is a perturbation $G_v[r_v(t), \theta]\Delta\theta dt$ generated at server v. This perturbation will be accumulated at the end of the service period of the customer being served at t; then the perturbation will be propagated to other servers and will have a final effect on the performance function F_L. Let Δ denote the perturbation generated in $[t, t+\Delta t)$. The effect of this Δ on F_L is denoted as $c_L^{(f)}[\mathcal{N}(t), \mathcal{R}(t), v, t]\Delta$. That is,

$$c_L^{(f)}[\mathcal{N}(t), \mathcal{R}(t), v, t] = \frac{1}{\Delta}\{\sum_{t<T_l\leq T_L} f(N_l)\Delta S_l\}.$$

Let

$$H := max\{f(\mathbf{n}) : \mathbf{n} \in \Phi\}.$$

Suppose $t \geq T_{L_0}$. Since $|\Delta S_l/\Delta| = 1$ or 0, we have

$$|c_L^{(f)}[\mathcal{N}(t), \mathcal{R}(t), v, t]| \leq H(L - L_0). \tag{5.23}$$

Next, let

$$c^{(f)}[\mathcal{N}(t), \mathcal{R}(t), v, t] = \lim_{L\to\infty} c_L^{(f)}[\mathcal{N}(t), \mathcal{R}(t), v, t].$$

Under Condition 5.2, the limit on the right-hand side exists with probability one. Both $c_L^{(f)}[\mathcal{N}(t), \mathcal{R}(t), v, t]$ and $c^{(f)}[\mathcal{N}(t), \mathcal{R}(t), v, t]$ depend on the sample path (ξ, θ). By the definition of the realization probability, we have

$$E[c^{(f)}\{\mathcal{N}(t), \mathcal{R}(t), v, t]\} = c^{(f)}(\mathbf{n}, \mathbf{r}, v),$$

$$\textit{for } \mathcal{N}(t) = \mathbf{n} \textit{ and } \mathcal{R}(t) = \mathbf{r}.$$

By the superposition property of the propagation, ΔF_L equals the sum of all the perturbations propagated to F_L from the perturbations generated in $[0, T_L]$. Thus,

$$\Delta F_L = \int_0^{T_L} \epsilon[n_v(t)]\{G_v[r_v(t), \theta]\Delta\theta\}c_L^{(f)}[\mathcal{N}(t), \mathcal{R}(t), v, t]dt.$$

Letting $c_L^{(f)}[\mathcal{N}(t), \mathcal{R}(t), v, t] = 0$ for $r_v(t) = 0$, we have

$$\Delta F_L = \int_0^{T_L} \{G_v[r_v(t), \theta]\Delta\theta\}c_L^{(f)}[\mathcal{N}(t), \mathcal{R}(t), v, t]dt.$$

From this, we get

$$\frac{\partial F_L}{\partial \theta} = \int_0^{T_L} G_v[r_v(t), \theta]c_L^{(f)}[\mathcal{N}(t), \mathcal{R}(t), v, t]dt. \tag{5.24}$$

Under Condition 5.1, $f_i(s) < \infty$, the probability that a service time equals any particular value is zero. Therefore, for any θ, the probability that two service completions occur simultaneously is zero. By the same argument as that in Section 3.4, we have the following result.

Lemma 5.2 *The sample function $\eta_L^{(f)}(\theta, \xi)$ is, with probability one, a continuous and piecewise differentiable function of θ.*

In other words, the probability that the sample function $\eta_L^{(f)}(\xi, \theta)$ is nondifferentiable at any θ (i.e., the two one-sided derivatives are not equal) is zero.

To study the convergence of the sample derivative, we first prove a lemma.

Lemma 5.3

$$\lim_{L\to\infty} \frac{1}{T_L} \int_0^{T_L} G_v[r_v(t), \theta]c_L^{(f)}[\mathcal{N}(t), \mathcal{R}(t), v, t]dt$$

$$= E\{G_v(r_v, \theta)c^{(f)}(\mathbf{n}, \mathbf{r}, v)\}. \tag{5.25}$$

Proof: The Lemma is similar to Theorem 3.3. We shall sketch the proof below. By ergodicity of $\mathcal{X}(t)$ (Condition 5.2), Equation (5.25) is equivalent to

$$\lim_{L\to\infty} \frac{1}{T_L} \int_0^{T_L} G_v[r_v(t),\theta]\{c_L^{(f)}[\mathcal{N}(t),\mathcal{R}(t),v,t] - c^{(f)}[\mathcal{N}(t),\mathcal{R}(t),v,t]\}dt = 0, \quad w.p.1. \tag{5.26}$$

Let $\mathbf{n}_1 = (N,0,0,...,0)$, $L_0 = 0$, and $L_k = min\{l > L_{k-1}, \mathcal{N}(T_l) = \mathbf{n}_1\}$, $k = 1,2,\cdots$. Let d be the integer such that $L_d \leq L < L_{d+1}$. Then $c_L^{(f)}[\mathcal{N}(t),\mathcal{R}(t),v,t] = c^{(f)}[\mathcal{N}(t),\mathcal{R}(t),v,t]$ for all $t \leq T_{L_d}$. Thus,

$$\begin{aligned}
&\lim_{L\to\infty} \frac{1}{T_L} \left| \int_0^{T_L} G_v[r_v(t),\theta]\{c_L^{(f)}[\mathcal{N}(t),\mathcal{R}(t),v,t] - c^{(f)}[\mathcal{N}(t),\mathcal{R}(t),v,t]\}dt \right| \\
&\leq \lim_{L\to\infty} \frac{1}{T_L} \int_{T_{L_d}}^{T_L} |G_v[r_v(t),\theta]| \times |\{c_L^{(f)}[\mathcal{N}(t),\mathcal{R}(t),v,t] - c^{(f)}[\mathcal{N}(t),\mathcal{R}(t),v,t]\}|dt.
\end{aligned}$$

Applying (5.23) to both $c_L^{(f)}[\mathcal{N}(t),\mathcal{R}(t),v,t]$ and $c^{(f)}[\mathcal{N}(t),\mathcal{R}(t),v,t]$ (the perturbation at t is realized at L_{d+1}) and using Condition 5.3, we have

$$\begin{aligned}
&\lim_{L\to\infty} \frac{1}{T_L} \left| \int_0^{T_L} G_v[r_v(t),\theta]\{c_L^{(f)}[\mathcal{N}(t),\mathcal{R}(t),v,t] - c^{(f)}[\mathcal{N}(t),\mathcal{R}(t),v,t]\}dt \right| \\
&< \lim_{L\to\infty} 2KH \frac{(T_{L_{d+1}} - T_{L_d})(L_{d+1} - L_d)}{T_{L_d}}.
\end{aligned} \tag{5.27}$$

It is easy to prove that the limit on the right-hand side is zero. Thus, (5.26) follows. □

Using this lemma and the Lebesgue dominated theorem, we can prove

$$\lim_{L\to\infty} \frac{1}{T_L} E\{\int_0^{T_L} G_v[r_v(t),\theta]c_L^{(f)}[\mathcal{N}(t),\mathcal{R}(t),v,t]dt\} = E\{G_v(r_v,\theta)c^{(f)}(\mathbf{n},\mathbf{r},v)\}. \tag{5.28}$$

The right-hand side is the steady-state mean value. It is important to note that on the left-hand side, the initial state may have an arbitrary distribution.

The next theorem follows immediately from (5.24) and Lemma 5.3.

Theorem 5.2 *Under Conditions 5.1, 5.2, and 5.3, the normalized sample derivative satisfies*

$$\lim_{L\to\infty} \frac{1}{\eta_L} \frac{\partial \eta_L^{(f)}(\xi,\theta)}{\partial \theta} = E\{G_v(r_v,\theta)c^{(f)}(\mathbf{n},\mathbf{r},v)\}$$

$$= \sum_{all\ \mathbf{n}} \int_{R_+^M} G_v(r_v,\theta)c^{(f)}(\mathbf{n},\mathbf{r},v)p(\mathbf{n},\mathbf{r})d\mathbf{r} \quad w.p.1. \qquad (5.29)$$

For exponential distributions, we have $\theta = \mu_v$, $G_v(r_v,\theta) = -(1/\theta)$, and $c^{(f)}(\mathbf{n},\mathbf{r},v) = c^{(f)}(\mathbf{n},v)$. Thus,

$$\int_{R_+^M} G_v(r_v,\theta)c^{(f)}(\mathbf{n},\mathbf{r},v)p(\mathbf{n},\mathbf{r})d\mathbf{r}$$

$$= -\frac{1}{\theta}\int_{R_+^M} c^{(f)}(\mathbf{n},v)p(\mathbf{n},\mathbf{r})d\mathbf{r} = -\frac{1}{\theta}p(\mathbf{n})c^{(f)}(\mathbf{n},v),$$

where

$$p(\mathbf{n}) = \int_{R_+^M} p(\mathbf{n},\mathbf{r})d\mathbf{r} \qquad (5.30)$$

is the steady-state probability of $\mathbf{n}$. Equation (5.29) becomes

$$\lim_{L\to\infty} \frac{\mu_v}{\eta_L^{(f)}} \frac{\partial \eta_L^{(f)}}{\partial \mu_v} = -\sum_{all\ \mathbf{n}} p(\mathbf{n})c^{(f)}(\mathbf{n},v).$$

This is the equation derived in Chapter 3.

5.1.5 Steady-State Sensitivity Formulas

The Steady-State Performance

First, let us define the steady-state throughput as follows:

$$\eta = \sum_{i=1}^{M}\{\frac{1}{\bar{s}_i}\sum_{all\ \mathbf{n}} \epsilon(n_i)p(\mathbf{n})\},$$

where $\bar{s}_i = \int s dF_i(s)$ is the mean service time of server i, and $p(\mathbf{n})$ is the steady-state probability of $\mathbf{n}$, defined in (5.30). Note that $\sum_{all\ \mathbf{n}} \epsilon(n_i)p(\mathbf{n}) = p(n_i > 0)$ is the steady-state probability of server i being busy. Let

$$\mu(\mathbf{n}) = \sum_{i=1}^{M} \frac{\epsilon(n_i)}{\bar{s}_i} = \sum_{i=1}^{M} \epsilon(n_i)\mu_i,$$

where $\mu_i = 1/\bar{s}_i$ is the average service rate of server i. $\mu(\mathbf{n})$ is the average service rate of the system when its discrete state is $\mathbf{n}$. We have

$$\eta = \sum_{all\ \mathbf{n}} p(\mathbf{n})\mu(\mathbf{n}).$$

This equation has the same form as (3.4), which is for Jackson networks.

Now, let L_i be the number of service completions of server i in $[0, T_L]$. Then $L = \sum_{i=1}^{M} L_i$. The length of the busy periods of server i in $[0, T_L]$ is $\int_0^{T_L} \epsilon[n_i(t)]dt$. By the ergodicity of the state process $\mathcal{X}(t)$ and the law of large numbers, we have

$$\begin{aligned}
\lim_{L\to\infty} \frac{L_i}{T_L} &= \frac{\lim_{L\to\infty} \frac{1}{T_L}\int_0^{T_L} \epsilon[n_i(t)]dt}{\lim_{L\to\infty} \frac{1}{L_i}\int_0^{T_L} \epsilon[n_i(t)]dt} \\
&= \frac{p(n_i > 0)}{\bar{s}_i} = \sum_{all\ \mathbf{n}} \epsilon(n_i)p(\mathbf{n})\mu_i, \qquad w.p.1.
\end{aligned}$$

Therefore,

$$\begin{aligned}
\lim_{L\to\infty} \eta_L &= \lim_{L\to\infty} \frac{L}{T_L} \\
&= \sum_{i=1}^{M} \sum_{all\ \mathbf{n}} \epsilon(n_i)p(\mathbf{n})\mu_i = \eta, \qquad w.p.1. \qquad (5.31)
\end{aligned}$$

This implies that the sample throughput converges to the steady-state throughput as L goes to infinity.

For steady-state performance $\eta_T^{(f)}$ and $\eta^{(f)}$, we have

$$\begin{aligned}
\eta_T^{(f)} &:= \lim_{L\to\infty} \eta_{T_L}^{(f)} = \lim_{L\to\infty} \frac{1}{T_L}\int_0^{T_L} f[\mathcal{N}(t)]dt \\
&= \sum_{all\ \mathbf{n}} f(\mathbf{n})p(\mathbf{n}), \qquad w.p.1, \qquad (5.32)
\end{aligned}$$

and

$$\begin{aligned}
\eta^{(f)} &:= \lim_{L\to\infty} \eta_L^{(f)} \\
&= \lim_{L\to\infty} \frac{1}{L}\int_0^{T_L} f[\mathcal{N}(t)]dt = \frac{\eta_T^{(f)}}{\eta}, \qquad w.p.1. \qquad (5.33)
\end{aligned}$$

This is the same as (4.40).

The Unbiasedness

Let $(\mathbf{n}_O, \mathbf{r}_O)$ be the initial state of the sample path in $[0, T_L]$.

Lemma 5.4 *Under Conditions 5.1 and 5.3, we have*

$$E\{\frac{\partial}{\partial\theta}[\eta_L^{(f)}(\theta,\xi)]|\mathbf{n}_O, \mathbf{r}_O\} = \frac{\partial}{\partial\theta}E\{\eta_L^{(f)}(\theta,\xi)|\mathbf{n}_O, \mathbf{r}_O\}. \qquad (5.34)$$

Proof: We have seen that under Condition 5.1, $T_l(\xi, \theta)$ is continuous and piecewise differentiable with probability one. So is $F_L(\theta, \xi)$, since it is a linear conbination of $T_l(\xi, \theta)$, $l = 1, 2, \cdots, L$. From (5.23), (5.24), and Condition 5.3, we have

$$|\frac{\partial F_L}{\partial\theta}| \leq KHLT_L. \qquad (5.35)$$

This holds for all the derivatives (one-sided and two-sided) in interval $[\theta, \theta + \Delta\theta]$. Besides, for any ξ, $T_L(\theta, \xi)$ is monotonic with respect to θ. The monotonicity follows directly from the perturbation propagation rules. (For more general cases, see Shanthikumar and Yao [96].) Thus, we have

$$\frac{\eta_L^{(f)}(\theta + h, \xi) - \eta_L^{(f)}(\theta, \xi)}{h} \leq KHT_L(\theta + \Delta\theta, \xi)$$

$$\textit{for all } h \in [\theta, \theta + \Delta\theta].$$

Note that for any finite L, $E[T_L(\theta + \Delta\theta, \xi)|\mathbf{n}_O, \mathbf{r}_O] \leq \infty$. The theorem follows directly from Lebesgue's dominated convergence theorem. □

The Sensitivity Formulas

The method for deriving the sensitivity formulas is essentially the same as that in Chapters 3 and 4, except that it requires some additional conditions to guarantee the interchangeability regarding the infinite integration $\int_{R_+^M}$.

We assume that the initial state probability is the steady-state probability $\pi(\mathbf{n}_O, \mathbf{r}_O)$ and let

$$\eta_L^{(f)}(\theta) = E[\eta_L^{(f)}(\theta, \xi)].$$

We first study the convergence property of $\frac{\partial \eta_L^{(f)}(\theta)}{\partial\theta}$ as $L \to \infty$. We have

$$\begin{aligned}\frac{\partial \eta_L^{(f)}(\theta)}{\partial\theta} &= \frac{\partial}{\partial\theta}\{\sum_{all\ \mathbf{n}_O}\int_{R_+^M} E[\eta_L^{(f)}(\theta,\xi)|\mathbf{n}_O, \mathbf{r}_O]\pi(\mathbf{n}_O, \mathbf{r}_O)d\mathbf{r}_O\} \\ &= \sum_{all\ \mathbf{n}_O}\frac{\partial}{\partial\theta}\{\int_{R_+^M} E[\eta_L^{(f)}(\theta,\xi)|\mathbf{n}_O, \mathbf{r}_O]\pi(\mathbf{n}_O, \mathbf{r}_O)d\mathbf{r}_O\}. \qquad (5.36)\end{aligned}$$

Let $R_r^M = [0,r)^M$ be a hypercube in R_+^M and $R_{r-}^M = R_+^M - R_r^M$ be the complementary set of R_r^M in R_+^M.

Condition 5.4 *There exist two real numbers $r^* > 0$ and $K^* > 0$ such that $|\frac{\partial \pi(\mathbf{n},\mathbf{r})}{\partial \theta}| < K^* \pi(\mathbf{n},\mathbf{r})$ for all $\mathbf{r} \in R_{r^*-}^M$, and the integration in (5.9) uniformly converges in a neighborhood of θ.*

The condition is similar to Lemma 4.4. Later, we shall prove in Lemma 5.5 that under Conditions 5.3 and 5.4, the two operators $\frac{\partial}{\partial \theta}$ and $\int_{R_+^M}$ in (5.36) are interchangeable. Interchanging the order of $\frac{\partial}{\partial \theta}$ and $\int_{R_+^M}$ and applying Lemma 5.4 leads to

$$\begin{aligned}\frac{\partial \eta_L^{(f)}(\theta)}{\partial \theta} &= E\{\frac{\partial}{\partial \theta}[\eta_L^{(f)}(\xi,\theta)]\} \\ &+ \sum_{all\ \mathbf{n}_0} \int_{R_+^M} E[\eta_L^{(f)}(\xi,\theta)|\mathbf{n}_0,\mathbf{r}_0] \frac{\partial}{\partial \theta}[\pi(\mathbf{n}_0,\mathbf{r}_0)] d\mathbf{r}_0. \quad (5.37)\end{aligned}$$

Using the same argument as that in Lemma 5.5, we can prove that $lim_{L\to\infty}$ and $\int_{R_+^M}$ are interchangeable. Thus,

$$\begin{aligned}&\lim_{L\to\infty} \sum_{all\ \mathbf{n}_0} \int_{R_+^M} E[\eta_L^{(f)}(\xi,\theta)|\mathbf{n}_0,\mathbf{r}_0] \frac{\partial}{\partial \theta}[\pi(\mathbf{n}_0,\mathbf{r}_0)] d\mathbf{r}_0 \\ &= \sum_{all\ \mathbf{n}_0} \int_{R_+^M} \lim_{L\to\infty} E[\eta_L^{(f)}(\xi,\theta)|\mathbf{n}_0,\mathbf{r}_0] \frac{\partial}{\partial \theta}[\pi(\mathbf{n}_0,\mathbf{r}_0)] d\mathbf{r}_0 \\ &= \lim_{L\to\infty} E[\eta_L^{(f)}(\xi,\theta)|\mathbf{n}_0,\mathbf{r}_0] \sum_{all\ \mathbf{n}_0} \int_{R_+^M} \frac{\partial}{\partial \theta}[\pi(\mathbf{n}_0,\mathbf{r}_0)] d\mathbf{r}_0. \quad (5.38)\end{aligned}$$

The last equation is due to the fact that $\lim_{L\to\infty} E[\eta_L^{(f)}(\xi,\theta)|\mathbf{n}_0,\mathbf{r}_0]$ does not depend on the initial state $(\mathbf{n}_0,\mathbf{r}_0)$. Furthermore, under Condition 5.4, we have

$$\sum_{all\ \mathbf{n}_0} \int_{R_+^M} \frac{\partial}{\partial \theta}[\pi(\mathbf{n}_0,\mathbf{r}_0)] d\mathbf{r}_0 = \frac{\partial}{\partial \theta}\{\sum_{all\ \mathbf{n}_0} \int_{R_+^M} [\pi(\mathbf{n}_0,\mathbf{r}_0)] d\mathbf{r}_0\} = 0.$$

Thus,

$$\lim_{L\to\infty} \sum_{all\ \mathbf{n}_0} \int_{R_+^M} E[\eta_L^{(f)}(\xi,\theta)|\mathbf{n}_0,\mathbf{r}_0] \frac{\partial}{\partial \theta}[\pi(\mathbf{n}_0,\mathbf{r}_0)] d\mathbf{r}_0 = 0.$$

Letting $L \to \infty$ on both sides of (5.37), we get

$$\lim_{L\to\infty} \frac{\partial \eta_L^{(f)}(\theta)}{\partial \theta} = \lim_{L\to\infty} E\{\frac{\partial}{\partial \theta}[\eta_L^{(f)}(\xi,\theta)]\}, \qquad w.p.1. \qquad (5.39)$$

Next, consider

$$\eta_L^{(f)}(\theta) = E[\eta_L^{(f)}(\theta,\xi)] = \frac{1}{L}E\{\sum_{l=0}^{L-1} f(N_l)S_l\}. \tag{5.40}$$

Since the initial probability is the steady-state probability, $E\{f(N_l)S_l\}$ are equal for all $l = 1, 2, \cdots, L-1$. Therefore,

$$\eta_L^{(f)}(\theta) = E[f(N_l)S_l].$$

Thus,

$$\lim_{L\to\infty} \eta_L^{(f)}(\theta) = E\{f(N_l)S_l\} = \lim_{L\to\infty} \frac{1}{L}\sum_{l=0}^{L-1} f(N_l)S_l = \eta^{(f)}.$$

Also, we have

$$\lim_{L\to\infty} \frac{\partial \eta_L^{(f)}}{\partial\theta} = \frac{\partial \eta^{(f)}}{\partial\theta}.$$

Then (5.39) becomes

$$\frac{\partial \eta^{(f)}}{\partial\theta} = \lim_{L\to\infty} E\{\frac{\partial}{\partial\theta}[\eta_L^{(f)}(\xi,\theta)]\}. \tag{5.41}$$

Equation (5.41) shows that the sample derivative is an asymptotically unbiased estimate of the derivative of the steady-state performance.

Now, we are ready to prove the main theorem.

Theorem 5.3 *Under Conditions 5.1 - 5.4, the normalized derivative of the steady-state performance equals the expected value of the product of the perturbation generation function and the realization factor. That is,*

$$\begin{aligned}\frac{1}{\eta^{(I)}}\frac{\partial \eta^{(f)}}{\partial\theta} &= E[G_v(r_v,\theta)c^{(f)}(\mathbf{n},\mathbf{r},v)] \\ &= \sum_{all\ \mathbf{n}} \int_{R_+^M} G_v(r_v,\theta)c^{(f)}(\mathbf{n},\mathbf{r},v)p(\mathbf{n},\mathbf{r})d\mathbf{r}. \end{aligned}\tag{5.42}$$

Proof: We have

$$E\{\frac{\partial}{\partial\theta}[\eta_L^{(f)}(\theta,\xi)]\} = \frac{1}{L}E\{\frac{\partial F_L}{\partial\theta}\}.$$

From (5.24) and (5.28), we have

$$\lim_{L\to\infty} E\{\frac{\partial}{\partial\theta}\eta_L^{(f)}(\theta,\xi)\} = \eta^{(I)}E[G_v(r_v,\theta)c^{(f)}(\mathbf{n},\mathbf{r},v)].$$

The theorem follows from this equation and (5.41). □

For the time-average performance, from (4.40), we have

$$\frac{\partial \eta_T^{(f)}}{\partial \theta} = \lim_{L\to\infty} \frac{\partial \eta_{T_L}^{(f)}(\xi,\theta)}{\partial \theta}$$
$$= -\eta_T^{(f)} E[G_v(r_v,\theta)c(\mathbf{n},\mathbf{r},v)] + E[G_v(r_v,\theta)c^{(f)}(\mathbf{n},\mathbf{r},v)]. \quad (5.43)$$

For networks with exponential distributions, (5.42) and (5.43) become

$$\frac{\mu_v}{\eta^{(f)}}\frac{\partial \eta^{(f)}}{\partial \mu_v} = -\sum_{all\ \mathbf{n}} p(\mathbf{n})c^{(f)}(\mathbf{n},v)$$

and

$$\mu_v \frac{\partial \eta^{(f)}}{\partial \mu_v} = \eta_T^{(f)}\{\sum_{all\ \mathbf{n}} p(\mathbf{n})c(\mathbf{n},v)\} - \sum_{all\ \mathbf{n}} p(\mathbf{n})c^{(f)}(\mathbf{n},v).$$

These are the same as the results derived in Chapters 3 and 4 (cf. (4.44) and (4.45)).

A Technical Lemma

Lemma 5.5 *Under Conditions 5.3 and 5.4, we have*

$$\frac{\partial}{\partial\theta}\{\int_{R_+^M} E[\eta_L^{(f)}(\theta,\xi)|\mathbf{n}_0,\mathbf{r}_0]\pi(\mathbf{n}_0,\mathbf{r}_0)d\mathbf{r}_0\}$$
$$= \int_{R_+^M} \frac{\partial}{\partial\theta}\{E[\eta_L^{(f)}(\theta,\xi)|\mathbf{n}_0,\mathbf{r}_0]\pi(\mathbf{n}_0,\mathbf{r}_0)\}d\mathbf{r}_0. \quad (5.44)$$

Proof: The right-hand side of (5.44) equals

$$\int_{R_+^M} \frac{\partial}{\partial\theta}\{E[\eta_L^{(f)}(\theta,\xi)|\mathbf{n}_0,\mathbf{r}_0]\}\pi(\mathbf{n}_0,\mathbf{r}_0)d\mathbf{r}_0$$
$$+ \int_{R_+^M} \{E[\eta_L^{(f)}(\theta,\xi)|\mathbf{n}_0,\mathbf{r}_0]\frac{\partial}{\partial\theta}\pi(\mathbf{n}_0,\mathbf{r}_0)\}d\mathbf{r}_0. \quad (5.45)$$

For any fixed L, $E[\eta_L^{(f)}(\theta,\xi)|\mathbf{n}_0,\mathbf{r}_0]$ is bounded. Thus, under Condition 5.4, the second term in (5.45) converges uniformly in a neighborhood of θ.

By (5.35) and Lemma 5.4, the first term in (5.45) is bounded by

$$KH\int_{R_+^M} E\{[T_L(\theta,\xi)]|\mathbf{n}_0,\mathbf{r}_0\}\pi(\mathbf{n}_0,\mathbf{r}_0)d\mathbf{r}_0.$$

Again, since for any fixed L $E\{[T_L(\theta,\xi)]|\mathbf{n}_0,\mathbf{r}_0\}$ is bounded, from Condition 5.4, the first term in (5.45) converges uniformly in a neighborhood of θ.

In sum, the term on the right-hand side of (5.44) converges uniformly in a neighborhood of θ. Therefore, the interchangeability holds. □

Remark 5.1 Condition 5.4 certainly is not a necessary condition for Theorem 5.3. A better condition should be specified in terms of the service distribution functions $F_i(s)$, $i = 1, 2, ..., M$. A possible conjecture is that Condition 5.4 holds if Condition 5.3 does; but to verify this conjecture involves detailed studies of the differential equations for the steady-state probabilities, and would deviate from the main objective of this book. However, it is worthwhile noting that Condition 5.4 holds if the service distributions are exponential or have a finite support (i.e., there exists a large real number s^* such that $F_i(s) = 1$, $i = 1, 2, ..., M$, for all $s > s^*$). s^* can be chosen as large as we like; therefore, at least we can approximate any network to any degree of accuracy by a network for which Condition 5.4 holds.

5.2 Open Networks

The discussion in this section is similar to that in Section 4.4. In an open network, customers arrive at the network from an outside source, denoted as server 0. The service times of server 0 are customers' interarrival times to the network. Let $F_0(s) = \int_0^s f_0(t)dt$ be the distribution function of the interarrival time. The routing matrix for an open network is an $(M + 1) \times (M + 1)$ matrix $Q = [q_{i,j}]_{i,j=0}^M$, with $q_{0,i}$ denoting the probability of an arrival customer entering server i and $q_{i,0}$ denoting the probability of a departure customer from server i leaving the network. We assume that Q is indecomposable, so that any customer entering the network will eventually leave the network with probability one, and that every server will be visited by some customers. We also assume that the state process $\mathcal{X}(t)$ is ergodic.

There are infinitely many states in an open network. We assume that the performance function f is finite; i.e., $|f(\mathbf{n})| < \infty$ for all $\mathbf{n} \in \Phi$.

As discussed in Section 4.4, the propagation rules among servers 1 to M are the same as those for closed networks. The perturbation propagation between the source (i.e., server 0) and the servers is different. A perturbation in servers 1 to M will never be propagated to server 0, and a perturbation of server 0 will be propagated to servers 1 to M with probability one and will never be lost. Thus, we have

$$c^{(I)}(\mathbf{n},\mathbf{r},0)=1 \tag{5.46}$$

and

$$c^{(I)}(\mathbf{n},\mathbf{r},i)=0, \qquad i=1,2,\cdots,M. \tag{5.47}$$

An open network consisting of M servers can be considered as a closed network consisting of $M+1$ servers with server 0 having infinitely many customers. Thus, in an open network, $\epsilon(n_0)>0$ always holds. Therefore, if we replace $\sum_{i=1}^{M}$ by $\sum_{i=0}^{M}$, then (5.10)-(5.13) and (5.15) hold for open networks. Furthermore, by the ergodicity assumption, Theorem 5.2 holds. Also, for any finite L and a given initial state $(\mathbf{n}_0,\mathbf{r}_0)$, only a finite number of discrete states can be reached in L transitions. Thus Lemma 5.4 holds.

For steady-state sensitivity, we have to overcome some additional technical problems due to the infinite state space. First of all, the sum over all states $\mathbf{n}$ may depend on the order of the items in the summation $\sum_{all\ \mathbf{n}}$. The sum does not depend on the order if it converges absolutely. Next, there is an issue of the interchangeability of the $\sum_{all\ \mathbf{n}}$ and $\lim_{L\to\infty}$ (or $\frac{\partial}{\partial\theta}$) in the proof of Theorem 5.3. To find a reasonable sufficient condition on the performance function $f(\mathbf{n})$ for the interchangeability is very technically involved. Of course, one sufficient condition is $|f|<H<\infty$; in this case, we can prove that the uniform convergence property required for the interchangeability holds. Another possible sufficient condition is the quasi-Lipschitz condition introduced in Section 4.4. We shall leave it to the readers to work out the details.

In particular, the theorems hold for the system throughput. Letting $\eta^{(f)}=\eta=\frac{1}{\eta^{(I)}}$ in (5.42), we get

$$\frac{1}{\eta}\frac{\partial\eta}{\partial\theta}=-E[G_v(r_v,\theta)c^{(I)}(\mathbf{n},\mathbf{r},v)].$$

From (5.46) and (5.47), we have

$$\frac{1}{\eta}\frac{\partial\eta}{\partial\theta}=\begin{cases}0, & if\ v\neq 0,\\ -E[G_0(r_0,\theta)], & if\ v=0.\end{cases}$$

This result is intuitively reasonable. In an open network, the system throughput equals the input rate and is independent of the service time distributions. This explains that $\frac{\partial\eta}{\partial\theta}=0$ for any service distribution parameter θ. If θ is the interarrival rate (or some other scale parameters), then $G_v(r_v,\theta)=-\frac{1}{\theta}$ and $\frac{\theta}{\eta}\frac{\partial\eta}{\partial\theta}=1$. That is, the elasticity of the system throughput with respect to the input rate is 1. This is obviously correct. This result shows that the most interesting applications of the theorems to open networks are for general performance measures instead of the system throughput.

In the above setting, all the customers arrive at the network from the same source; hence, they have the same interarrival distributions. This is the case of only one class of customers. For networks in which customers arriving at different servers have different interarrival service distributions, we have to assume that the customers are from different sources. Strictly speaking, this kind of network is no longer a single-class network. Customers from different sources belong to different classes. We shall discuss the perturbation realization of multiclass networks in Chapter 6.

Chapter 6

Multiclass Networks and Networks with Blocking

In the previous chapters, we developed the sensitivity analysis by employing the systems' dynamic features. The concept of realization plays an important role. Because of the strong conncention among the servers and the customers in an irreducible network, a perturbation will eventually be realized by the network; i.e., every server in the network will have the same size of perturbation (cf. Theorem 3.1 and Corollary 4.1). Based on this property, the realization probability and the realization factor are defined. One can prove, by virtue of ergodicity, that the sample normalized derivative converges to the expected value of the the product of the perturbation generation function and the realization factor (cf. Theorems 3.4, 4.3, 4.8, and 5.2).

To study the sensitivity of the steady-state performance, one has to establish the unbiasedness of the sample derivatives (cf. Lemmas 3.8, 4.3, and 5.4), or equivalently, to prove the interchangeability of the two operators E and $\frac{\partial}{\partial \theta}$. We have shown that the unbiasedness holds for state-dependent networks and networks with general distributions.

In this chapter, we devote to multiclass networks and networks with finite buffer capacities. These two types of networks have some common features. First, the connection among the servers in these two types of networks are in a sense stronger than that in single-class networks with infinite buffer capacities. In a multiclass network, a perturbation of one server may be propagated to another server even if the customers in any of the two servers cannot reach the other; the propagation can be achieved through a third server which the customers in both servers can visit. A necessary and sufficient condition is given for mulitclass networks to be irreducible. (A multiclass network in which any perturbation may be propagated to all servers is called an irreducible network.) In networks with finite capacities, a server may block any other server when its buffer is full, and a perturbation can be propagated through a blocking period. Thus, perturbations have more chance to be propagated. On the other hand, there may not exist a state such as $(N, 0, \cdots, 0)$, which is assumed in the infinite buffer capacity case. Thus, the proof of the irrducibility is more technically involved.

Next, the interchangeability of "E" and $\frac{\partial}{\partial \theta}$ holds only for certain configurations. Thus, while the realization factors can be used to calculate the limiting values of the sample derivatives, these values may not equal the derivatives of the steady-state performance. We give a necessary condition for the interchangeability in multiclass networks and a sufficient condition for that in networks with finite buffer capacities.

The multiclass networks and the networks with blocking are discussed in Sections 6.1 and 6.2, respectively. The interchangeability issue is discussed in Section 6.3. Section 6.4 offers an summary of the results. The study in this chapter will help to gain more insight on the dynamic nature of queuing networks. Readers are refered to Cao [24], [25], [26], [27], [30], Glasserman [58], and Heidelberger et al. [66] for more details.

6.1 Multiclass Networks

In Section 6.1.1, we provide a necessary and sufficient condition for a multiclass network to be irreducible; in such a network, a perturbation will be realized by the network with probability one. In Section 6.1.2, we derive equations for realization factors. In Section 6.1.3, we show that for many multiclass networks the unbiasedness of the sample derivative does not hold. We derive a necessary condition for the unbiasedness and give some examples which show that one can still use the realization factors to calculate the sensitivity of the steady-state performance when the unbiasedness holds.

6.1.1 Irreducible Multiclass Networks

Consider a closed queuing network consisting of M single-server nodes and K classes of customers. The service discipline of each server is first come first served. The service time of class k customers at server i is randomly distributed with a cumulative probability distribution $F_{i,k}(s)$ whose mean is $\bar{s}_{i,k}$, $0 < \bar{s}_{i,k} < \infty$, $1 \leq i \leq M$, $1 \leq k \leq K$. Let $n_{i,k}$ be the number of class k customers in server i, $n_i = \sum_{k=1}^{K} n_{i,k}$, and $N = \sum_{i=1}^{M} n_i$. We assume that each server has a buffer with a size bigger than N ("infinite" buffer size). A customer of class k in server i is referred to as (i,k). After the service completion at server i, a customer (i,k) has a probability $q_{i,k;j,h}$ of going to server j and changing to class h.

If we further assume that all $F_{i,k}(s)$ are exponential, then the system state can be expressed as $\mathbf{x} = (x_1, x_2, \cdots, x_M)$, where x_i is the state of server i, $i = 1, 2, \cdots, M$. $x_i = (k_{i,1}, k_{i,2}, \cdots, k_{i,n_i})$; $k_{i,j}$ is the class of the jth customer in server i. If $n_i = 0$, then set $x_i = 0$. If the distributions are not exponential, then the state defined above is the discrete part of the system state. We shall see that the exponential assumption is not crucial for the results in this subsection; we just need it for the convenience of statement.

As in the single-class case, we use $(\mathbf{x}, V)$ to denote the perturbation state where the system state is $\mathbf{x}$ and the servers in set $V \subseteq \Gamma$ have a perturbation. The evolution of $V(t)$ depends on the sample path and can be determined by the perturbation propagation rules. The realization of a perturbation is defined in Defintion 3.1: The perturbation $(\mathbf{x}, V)$ is said to be realized (or lost) if eventually every server in the network gets (or losses) the perturbation.

We now define the concept of irreducibility of a multiclass network. The irreducibility reflects the interaction among the servers in the network. In the single-class case, the irreducibility is defined by the indecomposibility of its routing matrix Q; if the routing matrix is indecomposable, then a customer in any server can reach every other server and a perturbation of any server can affect every other server; and the opposite is true. In a multiclass network, the situation is different. Even if the routing matrix is decomposable, it is still possible that a perturbation in any server can affect any other server in the system. Besides, we shall see that some special systems may not be decomposed into independent subnetworks.

Definition 6.1 *A multiclass closed queuing network is called irreducible if the perturbation in any perturbation state* $(\mathbf{x}, V)$ *will, with probability one, be either realized or lost.*

It is interesting to note that there is some similarity between the irreducibility of a queuing network and the controllability (Bryson and Ho [12]) of continuous variable dynamic systems. Roughly speaking, if a continuous variable system described in (1.1) is controllable, then one can control the system to any feasible state by choosing properly the control variable $u(t)$; in other words, every component of $x(t)$ is affected by $u(t)$. There may be some systems in which $x(t)$ can be decomposed into two or more parts, each of them can only be controlled by a part of $u(t)$. For example, $x(t) = (x_1(t), x_2(t))$ and $u(t) = (u_1(t), u_2(t))$, where $x_i(t)$ can be controlled by $u_i(t)$, $i = 1, 2$, and $u_i(t)$ cannot affect $x_j(t)$, $j \neq i$.

As in the single-class case, the interaction among servers is strongly related to the system structure. Let Ψ be the set of all possible (i, k)s. The number of elements in Ψ is $R \leq K \times M$. The inequality is possible since customers of some classes may not be able to visit some servers. Let $Q = [q_{i,k;j,h}]$ be the routing matrix. Q is an $R \times R$ Markov matrix. The decomposability of Q is defined in the same way as that in Section 3.1. That is, a Markov matrx is said to be decomposable if, by reordering the indices of servers and the classes of customers, Q can be written as:

$$Q = \begin{bmatrix} Q_{1,1} & O \\ Q_{2,1} & Q_{2,2} \end{bmatrix}, \tag{6.1}$$

where $Q_{1,1}$ is an $r \times r$ matrix, $0 < r < R$. O is a matrix whose elements are all zeros. Otherwise, Q is said to be indecomposable.

Lemma 6.1 *If Q is indecomposable, then any customer (i, k) has a positive probability of going to server i' and changing to class k' in less than R transitions.*

Proof: Consider two subsets $A_1 = \{(i, k)\}$ and $A_2 = \Psi - \{(i, k)\}$. By the indecomposability, there must be an $(i_1, k_1) \in A_2$ such that $q_{i,k;i_1,k_1} > 0$. (Otherwise by choosing $q_{i,k;i,k}$ as the up-left entry, Q can be written in the form of (6.1) with $r = 1$.) Next, consider two subsets $A_1 = \{(i, k), (i_1, i_1)\}$ and $A_2 = \Psi - A_1$. For the same reason, there must be an $(i_2, k_2) \in A_2$ such that either $q_{i,k;i_2,k_2} > 0$ or $q_{i_1,k_1;i_2,k_2} > 0$. In either case, (i, k) has a positive probability of reaching (i_2, k_2) in two steps at most, or (i, k) has a positive probability of reaching any element in $\{(i, k), (i_1, k_1), (i_2, k_2)\}$ in two steps at most. The lemma can be proved by repeating the above procedure $R - 1$ times. □

In (6.1), $Q_{1,1}$ and $Q_{2,2}$ may be further decomposable. Let the final decomposed form of Q be

$$Q = \begin{bmatrix} Q_{1,1} & O & O & \cdots & O \\ Q_{2,1} & Q_{2,2} & O & \cdots & O \\ \cdots & \cdots & \cdots & \cdots & \cdots \\ Q_{m,1} & Q_{m,2} & Q_{m,3} & \cdots & Q_{m,m} \end{bmatrix}, \tag{6.2}$$

where $Q_{l,l}$, $1 \leq l \leq m$ is an indecomposable $r_l \times r_l$ matrix. In a system whose transition matrix has the form in (6.2), all (i, k)s can be decomposed into m chains C_l, $l = 1, 2, \cdots, m$, each corresponding to one $Q_{l,l}$. By Lemma 6.1, any element (i,k) of chain C_l has a positive probability of becoming any other element (i',k') of the same chain in less than r_l transitions. Note that if $Q_{l,l'} = 0$, $l \neq l'$, then an element of C_l can never become an element of $C_{l'}$. Now suppose that $Q_{l,l'} = 0$ and $Q_{l',l} \neq 0$, $l' > l$. Let $(i,k) \in C_{l'}$ and $(j,h) \in C_l$ such that $q_{i,k;j,h} > 0$. Since $Q_{l',l'}$ is indecomposable, every element of $C_{l'}$ has a positive probability of becoming (i,k), which in turn has a positive probability of becoming (j,h) of C_l. However, since $Q_{l,l'} = 0$, once a customer enters C_l it can never go back to $C_{l'}$ again. Therefore, there will be no customer in chain C_l in equilibrium. Thus, for static properties, we can assume $Q_{l,l'} = 0$ for all $l \neq l'$, and consider the following form

$$Q = \begin{bmatrix} Q_{1,1} & O & O & \cdots & O \\ O & Q_{2,2} & O & \cdots & O \\ \cdots & \cdots & \cdots & \cdots & \cdots \\ O & O & O & \cdots & Q_{m,m} \end{bmatrix}. \qquad (6.3)$$

This form implies that a customer in chain C_l can not enter chain $C_{l'}$, $l \neq l'$. Thus, a customer belongs to only one chain.

A major difference between multiclass and single-class networks is that, in a multiclass network, interaction between any two servers in the system may still exist even if the transition matrix Q takes the decomposable form (6.3). The details are discussed later.

Let $I_l = \{i : (i,k) \in C_l\}$, $1 \leq l \leq m$, be the set of all servers that the customers of chain C_l may visit, and let $|I_l|$ be the number of the elements in I_l. I_l is a subset of $\Gamma = \{1, 2, \cdots, M\}$.

Lemma 6.2 *Let $t_{i,g}$ be the service completion time of the gth customer in queue i. For any $\mathbf{x}$ and $V \subseteq \Gamma$ such that $i \notin V$, the probability that $min_{j \in V}\{t_{j,1}\} > t_{i,g}$, $1 \leq g \leq n_i$, is positive.*

In words, the probability that all customers in set V remain in their respective servers before the gth customer in server i, $i \notin V$, leaves server i is positive. If, in addition, $|I_l| > 1$ for all $1 \leq l \leq m$, then all customers in server i have a positive probability of going to some other servers. Server i may become idle after the service completion of its last customer. When $g = n_i$, this statement implies that the probability that server i becomes idle while all customers in set V remain in their respective servers is positive.

Obviously, the lemma holds for exponential distributions. This is simply due to the fact that for any two exponentially distributed random variables

s_1 and s_2, with means $\bar{s}_1 = 1/\mu_1$ and $\bar{s}_2 = 1/\mu_2$, respectively, the probability of $s_1 > s_2$ is positive. In fact, we have

$$P(s_1 > s_2) = \int_0^\infty \mu_2 e^{-\mu_2 s_2}\{\int_{s_2}^\infty \mu_1 e^{-\mu_1 s_1} ds_1\} ds_2 = \frac{\mu_2}{\mu_1 + \mu_2} \tag{6.4}$$

The main result is stated in the following theorem.

Theorem 6.1 *A multiclass queuing network with exponential service distributions is irreducible if and only if its routing matrix satisfies the following conditions.*

1. *There is at most one server, denoted as server i, such that $I_l = \{i\}$, for some chains C_l, $1 \le l \le m$.*

2. *$I_1 \cup I_2 \cup \cdots \cup I_m = \Gamma$, and*

3. *For any two non-null sets of integers $\alpha_1, \alpha_2, \cdots \alpha_u$ and $\beta_1, \beta_2, \cdots, \beta_v$, such that $\{\alpha_1, \alpha_2, \cdots, \alpha_u,\} \cup \{\beta_1, \beta_2, \cdots, \beta_v\} = \{1, 2, \cdots, m\}$ and $\{\alpha_1, \alpha_2, \cdots, \alpha_u,\} \cap \{\beta_1, \beta_2, \cdots, \beta_v\} = \emptyset$, we have*

$$\{I_{\alpha_1} \cup I_{\alpha_2} \cup \cdots \cup I_{\alpha_u}\} \cap \{I_{\beta_1} \cup I_{\beta_2} \cup \cdots \cup I_{\beta_v}\} \neq \emptyset.$$

The theorem indicates that the irreducibility of a multiclass queuing network does not require its routing matrix Q to be indecomposable. This is because customers of different chains can interact on each other when they present to the same server. For an example, let us consider a multiclass closed network in which customers do not change their classes. The routing probability from server i to server j for class k is denoted as $p_{i,j;k}$. The routing matrix Q of such a system can be decomposed into form (6.3) with $m = K$, $Q_{l,l} = [q_{i,j;l}]$, $1 \le l \le K$, and $Q_{l,l'} = O$ for all $l \neq l'$. The simplest case of such a system is shown in Figure 6.1 where $M = 3$, $m = 2$, and $q_{1,1;1} = q_{2,2;2} = q_{3,3;1} = q_{3,3;2} = 0$, $q_{1,3;1} = q_{3,1;1} = q_{2,3;2} = q_{3,2;2} = 1$. For this system, we have

$$Q_{1,1} = \begin{bmatrix} 1 & 0 \\ 0 & 1 \end{bmatrix}$$

and

$$Q_{2,2} = \begin{bmatrix} 1 & 0 \\ 0 & 1 \end{bmatrix}.$$

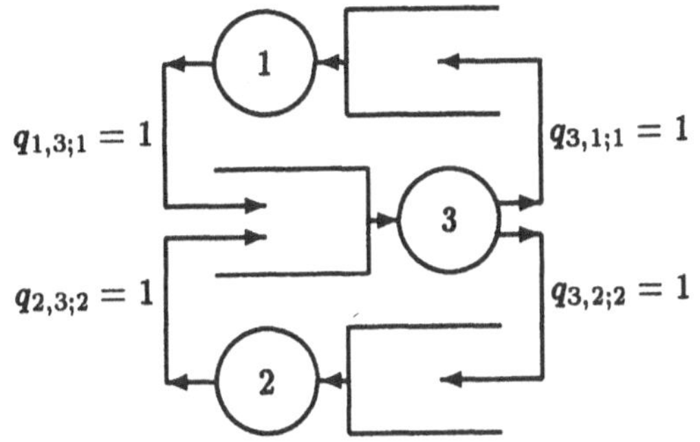

Figure 6.1: An Irreducible Multiclass Network

By the theorem, the system is irreducible, since $I_1 = \{1,3\}$, $I_2 = \{2,3\}$ and $I_1 \cap I_2 = \{3\}$. Customers of Chain C_1 and C_2 interact on each other when they are both in server 2.

To prove the theorem, we need one lemma.

Lemma 6.3 *In a perturbed multiclass queuing network with exponential service distributions, if Conditions 2 and 3 in Theorem 6.1 are satisfied and $|I_l| > 1$ for all $1 \leq l \leq m$, and $(\mathbf{x}, V)$ is a state with $n_i > 0$ for all $i \in V$, then the perturbation in $(\mathbf{x}, V)$ has a positive probability of being realized in a finite number of customer transitions.*

Proof: The proof of the lemma consists of the following three steps:

i) Suppose that the system is in state $(\mathbf{x}, V)$. By Condition 2 of the theorem, there must be a set, denoted as I_1, such that $I_1 \cap V \neq \emptyset$. We can prove that the probability that all customers of chain C_1 in the system move into V in a finite number of transitions while all customers in V remain there is positive. Let $V' = \Gamma - V$. If $I_1 \subset V$, then the statement is trivial. Thus we assume that $I_1 \cap V' \neq \emptyset$. Consider any customer of C_1 which is in server $j \in I_1 \cap V'$. Let this customer be the lth customer of server j. By the indecomposability of the matrix $Q_{1,1}$ and Lemma 6.1, this customer has a positive probability of going through a sequence of servers $j \to j_1 \to j_2 \cdots \to j_\alpha \to i$ to reach a server $i \in I_1 \cap V$. We can assume that servers j to j_α are not in V. Denote the corresponding sequence of servers and classes as $(j,k) \to (j_1,k_1) \to (j_2,k_2) \cdots \to (j_\alpha,k_\alpha) \to (i,k')$. By

Lemma 6.2, the probability that all customers in set V remain there while customer (j, k) leaves server j in a finite number of transitions is positive. Since $q_{j,k;j_1,k_1} > 0$, the probability that this customer moves to server j_1 as a customer (j_1, k_1) is positive. By the same argument, we observe that the probability that all customers in set V remain there while the customer (j_1, k_1) moves to server j_2 in a finite number of transitions is positive, and so on. Finally, the probability that all customers in set V remain there while this customer moves to server $i \in V$ in a finite number of transitions is positive. Applying the above procedure to all other customers of C_1 in V', we conclude that the probability that all customers in set V remain there and, meanwhile, all customers of C_1 in set $I_1 \cap V'$ move into set $I_1 \cap V$ is positive. Note that in the whole process there is no customer transition from set V to other servers. By the propagation rules, after all customers of C_1 move into set V, we still have $V(t) = V$. Thus, the probability of the system going from state $(\mathbf{x}, V)$ to a state in which all customers of C_1 are in set V is positive.

ii) Suppose that after the system takes the process described in step i, the system state is $(\mathbf{x}', V)$. By Condition 3 of the theorem, there must be another set, denoted as I_2, such that $I_1 \cap I_2 \neq \emptyset$. We shall prove that the probability that all customers in chains C_1 and C_2 move, in a finite number of transitions, into the servers that have the perturbation is positive. If $I_2 \cap V \neq \emptyset$, then, following the same procedure used in step i, we can prove that all customers of C_2 will move into set V with a positive probability. Thus the above statement holds. Now we assume that $I_2 \cap V = \emptyset$. We first choose a customer of C_1 in any server $j \in I_1 \cap V$. By the indecomposability, this customer has a positive probability of going through a sequence of servers $j \rightarrow j_1 \rightarrow j_2 \rightarrow \cdots \rightarrow j_\alpha \rightarrow i$ to reach a server $i \in I_1 \cap I_2$. We can assume that in all servers j_1 to j_α there is no customers of C_1. (If any server in this sequence contains a customer of C_1, we simply choose that server as server j.) We can also assume that the chosen customer of C_1 in server j is the first customer in the server. (Otherwise, we apply Lemma 6.2 to server j to conclude that, with a positive probability, the customer will become the first customer while all customers in all other servers remain there.) Since $|I_l| > 1$ for all $1 \leq l \leq m$, by Lemma 6.2, the probability that server j_1 becomes idle in a finite number of transitions while other customers remain in their respective servers is positive. After this, we let the customer of C_1 enter server j_1. Since server j_1 is idle, by the perturbation propagation rule, server j_1 will acquire the perturbation after this transition. Successively using the same procedure for servers j_2, j_3, $\cdots$, j_α, and i, we can move the chosen customer of C_1 and the perturbation it carries to server i. Therefore, the probability that server i obtains the perturbation in a finite number of transitions is positive. Now we can use set $\{i\}$, $i \in I_1 \cap I_2$, as set V in step i. Following the same procedure used there, we can prove that the probability that all customers of C_2 move to server i in a finite number of transitions and obtain the perturbation while all customers of C_1 remain in their respective servers is positive.

iii) Next, if $m > 2$, then there must be a set I_3 such that either $I_3 \cap I_1 \neq \emptyset$ or $I_3 \cap I_2 \neq \emptyset$. By the same technique as in step ii, the probability that all customers of C_1, C_2, and C_3 move, in a finite number of transitions, into servers that possess the perturbation is positive. Repeating this procedure, we can show that the probability that all customers move into servers that have the perturbation in $(\mathbf{x}, V)$ (in other words, the perturbation is realized) in a finite number of transitions is positive. This completes the proof of Lemma 6.3. □

The Proof of the Theorem

1. Part 1: The "if" part of the theorem.

First we assume that $|I_l| > 1$ for all $1 \leq l \leq m$. Let $(\mathbf{x}(t), V(t))$ be the perturbation state at time t. Obviously, $\mathbf{x}(t)$ and $(\mathbf{x}(t), V(t))$ are Markov processes. Let $\mathbf{x}_i$ and $(\mathbf{x}_i, V_i)$, $i = 1, 2, \cdots$, be the embedded Markov chains of these two processes, respectively. For any state $\mathbf{x}$, we define a function $\chi_i(\mathbf{x}) = 1$, if $\mathbf{x}_i = \mathbf{x}$; and $\chi_i(\mathbf{x}) = 0$, otherwise. We also define two similar functions for processes V_i and $(\mathbf{x}_i, V_i)$. Note that if on a sample path at some time t_0, $V(t_0) = \Gamma$ (or $\emptyset$), then by the perturbation propagation rules, $V(t) = \Gamma$ (or $\emptyset$), for all $t > t_0$. From this, we only need to prove that $V(t)$ will reach either Γ or $\emptyset$ with probability one. Now we assume that the opposite is true, that is, with a non-zero probability there exist some sample paths on which $V(t)$ never reaches either Γ or $\emptyset$. Since the number of sets V is finite, there must exist a set $V \neq \Gamma$, and $\emptyset$, such that on these paths the frequency that V_i visits V, $\lim_{n\to\infty} \frac{\sum_{i=1}^{n} \chi_i(V)}{n}$, is positive. Since the number of system states is finite, there must exist at least one state, denoted as $\mathbf{x}$, such that $\lim_{n\to\infty} \frac{\sum_{i=1}^{n} \chi_i(\mathbf{x}, V)}{n}$ is positive.

Now we consider one of these paths. Let $(X_{i_1} V_{i_1})$, $(X_{i_2} V_{i_2})$, $\cdots$ be the sequence on the path such that $(X_{i_k} V_{i_k}) = (\mathbf{x}, V)$, for all k, $k = 1, 2, \cdots$. In Lemma 6.3, we proved that the perturbation in $(\mathbf{x}, V)$ has a positive probability of being realized (or lost) in a finite number of customer transitions; or equivalently, $(\mathbf{x}, V)$ has a positive probability of being followed by a sequence of a finite number of customer transitions which leads the perturbation set V to either Γ, or $\emptyset$. We denote this sequence as $\mathcal{S}$ and the number of transitions in the sequence as D. Next, we choose a subsequence (X_{j_1}, V_{j_1}), (X_{j_2}, V_{j_2}) $\cdots$ from the sequence (X_{i_1}, V_{i_1}), (X_{i_2}, V_{i_2}), $\cdots$ such that $j_k - j_{k-1} > D$ for all k. For each k we define a random variable R_k, $R_k = 1$ if (X_{j_k}, V_{j_k}) is followed by sequence $\mathcal{S}$; $R_k = 0$ otherwise. From the construction of (X_{j_k}, V_{j_k}) and the Markov property of $\{X(t), V(t)\}$, R_k, $k = 1, 2, \cdots$ are independent, and $P(R_k = 1)$ is the probability that the perturbation in $(\mathbf{x}, V)$ is realized through the sequence $\mathcal{S}$. Therefore, from Lemma 6.3, $P(R_k = 1) > 0$. By the law of large numbers,

$$\lim_{L\to\infty} \frac{\sum_{k=1}^{L} R_k}{L} = P(R_k = 1) > 0.$$

Therefore, there must be some k such that $R_k = 1$. Thus, after this (X_{j_k}, V_{j_k}), (X_i, V_i) must visit Γ or $\emptyset$. This contradicts to our original assumption that $V(t)$ never visits Γ or $\emptyset$ with a non-zero probability. The first part of the theorem is thus proved when $|I_l| > 1$ for all l.

Next, we assume that for some chains C_{j_l}, $1 \le j_l \le m$, $I_{j_l} = \{i\}$. Then the customers in chains C_{j_l} feed back to server i immediately after they leave it. Server i never becomes idle. Any perturbation in any other server can never be propagated to server i. Furthermore, whenever a customer of other chains enters server i, the customer loses its perturbation. Because of Conditions 2 and 3, the customers of all chains in the system interact directly or indirectly with server i. Through this interaction, all customers will lose their perturbations with probability one. (i.e., $V(t)$ will reach $\emptyset$ with probability one.) This statement can be proved by rigorously following the same procedure as in steps i-iii of the proof of Lemma 6.3. On the other hand, if server i is perturbed, then this perturbation will be propagated to all other servers, either directly or indirectly, with probability one. In summary, the perturbation in server i will be realized with probability one while the perturbation in any other servers will be lost with probability one. The "if" part of the theorem holds in this case.

2. Part 2: The "only if" part of the theorem.

Condition 2 implies that customers may visit all servers in the system. Obvously, servers which are not in $I_1 \cup \cdots \cup I_m$ can be removed from the system. We can consider only the subsytem consisting of $I_1 \cup \cdots \cup I_m$.

If Condition 3 does not hold, then customers in $\{I_{\alpha_1} \cup I_{\alpha_2} \cup \cdots \cup I_{\alpha_u}\}$ can never visit servers in $\{I_{\beta_1} \cup I_{\beta_2} \cup \cdots \cup I_{\beta_v}\}$ and vice versa. In this case a perturbation in $\{I_{\alpha_1} \cup I_{\alpha_2} \cup \cdots \cup I_{\alpha_u}\}$ can never be propagated to servers in $\{I_{\beta_1} \cup I_{\beta_2} \cup \cdots \cup I_{\beta_v}\}$ and vice versa. The system can be decomposed

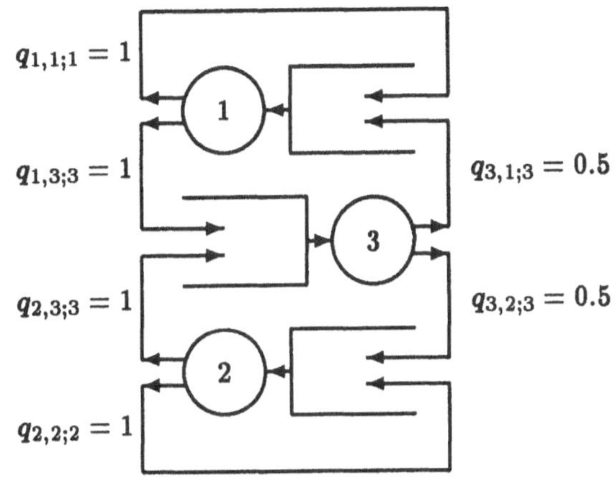

Figure 6.2: Perturbations Are Neither Realized Nor Lost

into two subsystems each consisting of servers in $\{I_{\alpha_1} \cup I_{\alpha_2} \cup \cdots \cup I_{\alpha_u}\}$ and in $\{I_{\beta_1} \cup I_{\beta_2} \cup \cdots \cup I_{\beta_v}\}$ respectively.

As to Condition 1, let us assume that $I_k = \{i\}$, $I_h = \{j\}$. This means that servers i and j never meet idle periods. A perturbation generated at server i can never be propogated to server j and vice versa. Therefore, it can never be realized or lost by the whole system. This completes the proof of the theorem. □

Figure 6.2 shows a system consisting of three servers and three classes of customers. Each class contains one customer, which does not change its class. It is readily verified that $I_1 = \{1\}$, $I_2 = \{2\}$, and $I_3 = \{1, 2, 3\}$. Conditions 2 and 3 in the theorem are satisfied. Suppose that server 1 obtains a perturbation. Then server 1 will always have that perturbation. On the other hand, this perturbation will never be propagated to server 2. Server 3, however, obtains the perturbation after the class 3 customer comes from server 1 and loses the perturbation after the class 3 customer comes from server 2. Therefore, $\lim_{t\to\infty} V(t)$ does not exist. The system can not be decomposed into indecomposable subsystems.

From the proof of the theorem and the above discussion, we have

Corollary 6.1 *If $|I_l| > 1$ for all l, then a multiclass queuing network with exponential service distributions is either irreducible or can be decomposed into some irreducible subnetworks.*

From the corollary, if $|I_l| > 1$ for all l, then $\lim_{t\to\infty} V(t)$ exists with probability one for any initial state $(\mathbf{x}(0), V(0))$.

As we can see, the assumption about the exponential distribution is not essential to the results of this subsection. In fact, what is needed for proving Theorem 6.1 is Lemma 6.2, which holds under a very mild condition. To see this, we use the hazard rate function. For any absolutely continuous distribution function $F(t)$, with $f(t) = \frac{d}{dt}F(t)$, the hazard rate is

$$g(t) := \frac{f(t)}{1 - F(t)} \qquad t \geq 0. \tag{6.5}$$

An important observation is that the hazard rate is invariant with respect to the initial point. To state this precisely, let $t_0 > 0$ be any time instant. The conditional probability distribution of t given $t > t_0$ is

$$F(t|t_0) = \frac{F(t) - F(t_0)}{1 - F(t_0)} \qquad t \geq t_0.$$

The corresponding probability density function is

$$f(t|t_0) = F'(t|t_0) = \frac{F'(t)}{1 - F(t_0)} \qquad t \geq t_0.$$

Therefore, the "conditional" hazard rate at t_0 is

$$g(t|t_0) := \frac{f(t|t_0)}{1 - F(t|t_0)} = g(t) \qquad t \geq 0. \tag{6.6}$$

Finally, for any two distributions with hazard rates g_1 and g_2, (6.4) becomes

$$P(s_1 > s_2) = \int_0^{\infty} g_2 e^{-g_2 s_2} \{ \int_{s_2}^{\infty} g_1 e^{-g_1 s_1} ds_1 \} ds_2. \tag{6.7}$$

By (6.6), (6.7) holds for any given elapsed times of s_1 and s_2. From (6.7), Lemma 6.2 holds if the hazard rates of all the service time distributions have an upper and a lower bound:

$$0 < \epsilon \leq g_{i,k}(t) \leq H < \infty \qquad 0 \leq t < \infty, \tag{6.8}$$

where $g_{i,k}(t)$ is the hazard rate of the service time of class k customers at server i.

It is clear that (6.8) holds for Coxian distributions and the phase-type distributions.

In summary, we have defined the irreducibility for multiclass queuing networks. The irreducibility in fact describes the static characteristic of a perturbed system. A necessary and sufficient condition for a multiclass network to be irreducible is given. It is also shown that in some special networks in which $|I_l| = 1$ for at least two chains of customers (i.e., there are at least two servers which never meet idle periods), a perturbation may be neither realized nor lost as $t \to \infty$.

6.1.2 The Sensitivity Analysis

The Limiting Values of Sample Sensitivities

In an irreducible multiclass network, we can define the realization probability of a perturbation and the realization factor for a performance function in the same way as for the single-class case. We can develop the equations for realization probabilities and realization factors and use them to calculate the limiting values of the sample performance sensitvities. The proofs of these results are similar to those in Chapters 3 and 4, and hence are omitted.

We assume that the service time distributions are exponential. The realization factor of the perturbation $(\mathbf{x}, V)$ for performance function f is denoted as $c^{(f)}(\mathbf{x}, V)$.

Lemma 6.4

$$If\ n_i = 0,\ then\ c^{(f)}(\mathbf{x}, i) = 0: \tag{6.9}$$

$$c^{(f)}(\mathbf{x}, \Gamma) = f(\mathbf{x}); \tag{6.10}$$

If $V_1 \cap V_2 = \emptyset$ *and* $V_1 \cup V_2 = V_3$, *then*

$$c^{(f)}(\mathbf{x}, V_1) + c^{(f)}(\mathbf{x}, V_2) = c^{(f)}(\mathbf{x}, V_3); \tag{6.11}$$

and

$$\sum_{i=1}^{M} c^{(f)}(\mathbf{x}, i) = f(\mathbf{x}). \tag{6.12}$$

Let $\mathbf{x}_{i,j}$ be a neighboring state of $\mathbf{x}$, i.e., the state that follows $\mathbf{x}$ after a customer transition from server i to server j. Let $q(\mathbf{x}; \mathbf{x}_{i,j})$ be the probability that the next state is $\mathbf{x}_{i,j}$ given that the present state is $\mathbf{x}$. For example, consider the network shown in Figure 6.1. Suppose that each class consists of two customers and the service rates of these two classes of customers at all three servers are equal. For $\mathbf{x} = (1, 2, (2, 1))$, we have

$$q[\mathbf{x}; (0, 2, (2, 1, 1))] = \frac{1}{3},$$

$$q[\mathbf{x};(1,0,(2,1,2))]=\frac{1}{3},$$

$$q[\mathbf{x};((1,2),2,1)]=\frac{1}{3},$$

and $q[\mathbf{x};\mathbf{x}']=0$ for $\mathbf{x}$ not equal the above three states.

Theorem 6.2 *For realization factors in a multiclass network, the following equations hold:*

$$c^{(f)}(\mathbf{x},k)=\sum_{i=1}^{M}\sum_{j=1}^{M}\epsilon(x_i)q(\mathbf{x},\mathbf{x}_{i,j})c^{(f)}(\mathbf{x}_{i,j},i)$$

$$+\sum_{j=1}^{M}q(\mathbf{x},\mathbf{x}_{k,j})\{[1-\epsilon(x_j)]c^{(f)}(\mathbf{x}_{k,j},j)+f(\mathbf{x})-f(\mathbf{x}_{k,j})\}$$

$$n_k>0,\ k\in\Gamma, \tag{6.13}$$

where $\epsilon(x_i)=0$ *if* $x_i=0$ *and* $\epsilon(x_i)=1$ *if* $x_i\neq 0$.

The equation is self-explanatory and its proof is omitted.

Let $p(\mathbf{x})$ be the steady-state probability of $\mathbf{x}$. Let Φ be the state space and $\Phi_{i;k}\subset\Phi$ be a set of states such that $\mathbf{x}\in\Phi_{i;k}$ means that a class k customer is the first customer (i.e., the customer being served) in server i. Let $\mu_{i;k}$ be the service rate of server i for the class k customers. (It should be distinguished from $\mu_{i,k}$, which is the service rate of server i containing k customers in a single-class network.) Using the same approach as in Chapters 3 and 4, we can prove the following theorem.

Theorem 6.3 *In an irreducible multiclass network, the sample normalized derivative of* $\eta_L^{(f)}$ *with respect to* $\mu_{v;k}$ *converges to the following value:*

$$\lim_{L\to\infty}E\{\frac{\mu_{v;k}}{\eta_L^{(f)}}\frac{\partial\eta_L^{(f)}}{\partial\mu_{v;k}}\}=\sum_{\mathbf{x}\in\Phi_{v;k}}p(\mathbf{x})c^{(f)}(\mathbf{x},v). \tag{6.14}$$

If at server v, $\mu_{v;k}=\mu_v$ for all classes of customers, then

$$\lim_{L\to\infty}E\{\frac{\mu_{v}}{\eta_L^{(f)}}\frac{\partial\eta_L^{(f)}}{\partial\mu_{v}}\}=\sum_{all\ \mathbf{x}}p(\mathbf{x})c^{(f)}(\mathbf{x},v). \tag{6.15}$$

Let $\eta_{j,L}=L_j/T_L$ be the sample throughput of server j for all classes of customers, with L_j being the number of service completions of server j in

$[0, T_L]$. For $\eta_{j,L}$, if $\mu_{v;k} \equiv \mu_v$, we have

$$\begin{aligned}
&\lim_{L\to\infty} E\{\frac{\mu_v}{\eta_{j,L}} \frac{\partial \eta_{j,L}}{\partial \mu_v}\} \\
= &- \lim_{L\to\infty} E\{\frac{\mu_v}{\eta_{j,L}^{(I)}} \frac{\partial \eta_{j,L}^{(I)}}{\partial \mu_v}\} \\
= &- \sum_{all\ \mathbf{x}} p(\mathbf{x}) c(\mathbf{x}, v), \qquad (6.16)
\end{aligned}$$

where $c(\mathbf{x}, v) = c^{(I)}(\mathbf{x}, v)$ is the realization probability. If $\mu_{v;k}$ are different, then

$$\begin{aligned}
&\lim_{L\to\infty} E\{\frac{\mu_{v;k}}{\eta_{j,L}} \frac{\partial \eta_{j,L}}{\partial \mu_{v;k}}\} \\
= &- \lim_{L\to\infty} E\{\frac{\mu_{v;k}}{\eta_{j,L}^{(I)}} \frac{\partial \eta_{j,L}^{(I)}}{\partial \mu_{v;k}}\} \\
= &- \sum_{all\ \mathbf{x}\in\Phi_{v;k}} p(\mathbf{x}) c(\mathbf{x}, v),
\end{aligned}$$

Let $\eta_{j,h;L} = L_{j,h}/T_L$ be the sample throughput of server j for class h customers, with $L_{j,h}$ being the number of the class h customers' service completions at server j in $[0, T_L]$. For $\eta_{j,h;L}$, we have

$$\begin{aligned}
&\lim_{L\to\infty} E\{\frac{\mu_{v;k}}{\eta_{j,h;L}} \frac{\partial \eta_{j,h;L}}{\partial \mu_{v;k}}\} \\
= &- \lim_{L\to\infty} E\{\frac{\mu_{v;k}}{\eta_{j,h;L}^{(I)}} \frac{\partial \eta_{j,h;L}^{(I)}}{\partial \mu_{v;k}}\} \\
= &- \sum_{all\ \mathbf{x}\in\Psi_{v,k}} p(\mathbf{x}) c(\mathbf{x}, v), \qquad (6.17)
\end{aligned}$$

Equation (6.17) shows that the limiting value for the sample derivatives for $\eta_{j,h;L}$ are the same for all j and h. This is true for the derivative of the steady-state throughput for single-class networks (see Chapter 3). We shall discuss the derivatives of the steady-state performance for multiclass case in Section 6.3.

Example 6.1 Consider the network shown in Figure 6.1. The system consists of 3 servers and 2 classes of customers. Each class consists of only one customer. The class 1 customer (called customer 1 hereafter) goes from

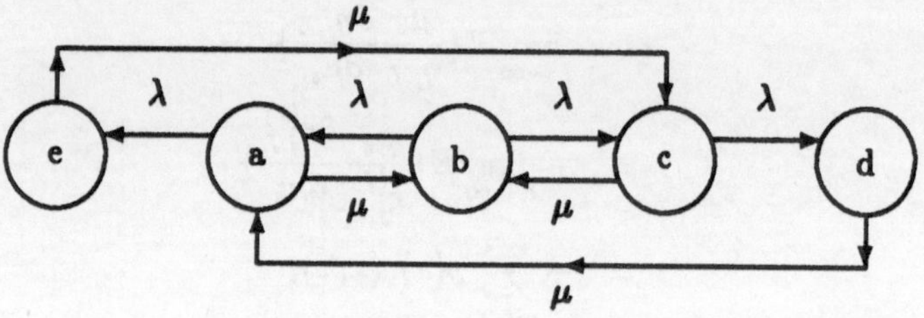

Figure 6.3: The State-Transition Diagram

server 3 to server 1 and back, while the class 2 customer (customer 2) goes from server 3 to server 2 and back. The service rate of server 3 is μ for both classes, and those of servers 1 and 2 are both λ. The state of server 1 can be denoted as 1 or 0, depending on whether customer 1 is in the server or not. Similarly, the state of server 2 is 2 or 0. The state of server 3 is denoted as (i, j), $i, j = 0, 1, 2$, where i (j) is the class of the first (second) customer in the queue, and 0 means that there is no customer in that position. There are a total of 5 states of the network, which can be denoted as $a = (1, 0, (2, 0))$, $b = (1, 2, (0, 0))$, $c = (0, 2, (1, 0))$, $d = (0, 0, (1, 2))$, and $e = (0, 0, (2, 1))$. The state-transition diagram is shown in Figure 6.3. We have

$$\mu p(d) + \lambda p(b) = (\lambda + \mu)p(a),$$

$$\mu p(a) + \mu p(c) = 2\lambda p(b),$$

$$\lambda p(b) + \mu p(e) = (\lambda + \mu)p(c),$$

$$\lambda p(c) = \mu p(d),$$

$$\lambda p(a) = \mu p(e).$$

From the above equations and $\sum_{all\ \mathbf{x}} p(\mathbf{x}) = 1$, we get

$$p(a) = p(c) = \frac{1}{2 + \frac{\mu}{\lambda} + 2\frac{\lambda}{\mu}},$$

$$p(b) = \frac{\frac{\mu}{\lambda}}{2 + \frac{\mu}{\lambda} + 2\frac{\lambda}{\mu}},$$

$$p(d) = p(e) = \frac{\frac{\lambda}{\mu}}{2 + \frac{\mu}{\lambda} + 2\frac{\lambda}{\mu}}.$$

Moreover, from (6.13), letting $f \equiv 1$, we have

$$
\begin{aligned}
(\lambda+\mu)c(a,1) &= \mu c(b,1),\\
(\lambda+\mu)c(a,3) &= \lambda c(e,3)+\mu c(b,2),\\
2c(b,1) &= c(c,3)+c(a,1),\\
2c(b,2) &= c(c,2)+c(a,3),\\
(\lambda+\mu)c(c,2) &= \mu c(b,1),\\
(\lambda+\mu)c(c,3) &= \lambda c(d,3)+\mu c(b,1).
\end{aligned}
$$

By (6.9) and (6.12), we have

$$
\begin{aligned}
c(d,3) = c(e,3) &= 1,\\
c(a,1)+c(a,3) &= 1,\\
c(b,1)+c(b,2) &= 1,\\
c(c,2)+c(c,3) &= 1.
\end{aligned}
$$

Solving the above equations, we get

$$
\begin{aligned}
c(b,1) = c(b,2) &= \frac{1}{2},\\
c(d,3) = c(e,3) &= 1,\\
c(a,1) = c(c,2) &= \frac{\mu}{2(\lambda+\mu)},\\
c(a,3) = c(c,3) &= \frac{\lambda+\frac{1}{2}\mu}{\lambda+\mu}.
\end{aligned}
$$

Finally, we have

$$
\begin{aligned}
&\sum_{\text{all } \mathbf{x}} p(\mathbf{x})c(\mathbf{x},3)\\
= \;& p(a)c(a,3)+p(c)c(c,3)+p(d)c(d,3)+p(e)c(e,3)\\
= \;& \frac{1+4\frac{\lambda}{\mu}+2\frac{\lambda^2}{\mu^2}}{2+\frac{\mu}{\lambda}+2\frac{\lambda}{\mu}}\,\frac{\mu}{\lambda+\mu}.
\end{aligned}
$$

On the other hand, the expected throughputs of server 1 and server 2 are

$$
\begin{aligned}
\eta_1 &= \lambda[p(a)+p(b)]\\
&= \frac{\lambda+\mu}{2+\frac{\mu}{\lambda}+2\frac{\lambda}{\mu}}
\end{aligned}
$$

and

$$\begin{aligned}\eta_2 &= \lambda[p(c)+p(b)] \\ &= \frac{\lambda+\mu}{2+\frac{\mu}{\lambda}+2\frac{\lambda}{\mu}}.\end{aligned}$$

The expected throughputs of customer 1 and customer 2 at server 3 are

$$\begin{aligned}\eta_{3,1} &= \mu[p(c)+p(d)] \\ &= \frac{\lambda+\mu}{2+\frac{\mu}{\lambda}+2\frac{\lambda}{\mu}}\end{aligned}$$

and

$$\begin{aligned}\eta_{3,2} &= \mu[p(a)+p(e)] \\ &= \frac{\lambda+\mu}{2+\frac{\mu}{\lambda}+2\frac{\lambda}{\mu}};\end{aligned}$$

and that of server 3 for both customers is

$$\begin{aligned}\eta_3 &= \mu[p(a)+p(c)+p(d)+p(e)] \\ &= \frac{2(\lambda+\mu)}{2+\frac{\mu}{\lambda}+2\frac{\lambda}{\mu}}.\end{aligned}$$

Taking the derivative with respect to μ of both sides of the above equations, we can verify

$$\begin{aligned}\frac{\mu}{\eta_{3,2}}\frac{\partial\eta_{3,2}}{\partial\mu} &= \frac{\mu}{\eta_{3,1}}\frac{\partial\eta_{3,1}}{\partial\mu} \\ = \frac{\mu}{\eta_3}\frac{\partial\eta_3}{\partial\mu} &= \frac{\mu}{\eta_2}\frac{\partial\eta_2}{\partial\mu} \\ = \frac{\mu}{\eta_1}\frac{\partial\eta_1}{\partial\mu} &= \sum_{all\ \mathbf{x}} p(\mathbf{x})c(\mathbf{x},3).\end{aligned}$$

By Theorem 6.3, this is equivalent to

$$\begin{aligned}&\lim_{L\to\infty} E\{\frac{\mu}{\eta_{3,h;L}}\frac{\partial\eta_{3,h;L}}{\partial\mu}\} \\ &= \lim_{L\to\infty} E\{\frac{\mu}{\eta_{i,L}}\frac{\partial\eta_{i,L}}{\partial\mu}\} \\ &= \frac{\mu}{\eta_{3,h}}\frac{\partial\eta_{3,h}}{\partial\mu} = \frac{\mu}{\eta_i}\frac{\partial\eta_i}{\partial\mu},\end{aligned} \tag{6.18}$$

with $i=1,2$, $h=1,2$.

Equation (6.18) verifies that for the system shown in Figure 6.1, the sample elasticities of the throughputs converge to the elasticities of the steady-state throughputs. We shall see in Section 6.3 that this is not true in general for multiclass networks. A necessary condition will be given there for this fact.

6.2 Networks With Blocking

In this section, we study single-class networks in which the buffer capacities are finite. In such networks, perturbations can be propagated through blocking periods. Thus, the propagation rules have to be modified. In addition, for some finite-capacity networks, there may not exist a state such as $(N, 0, \cdots, 0)$; thus, the proof of perturbation realization is more difficult. Finally, like the multiclass networks, the unbiasedness of the sample derivative does not also hold; thus, the limiting values of the sample normalized derivatives, calculated by realization probabilities or realization factors, may not equal the normalized derivatives of the steady-state performance. These issues will be addressed in this section and the next.

6.2.1 Perturbation Propagation Through Blocking Periods

The Blocking

Consider a closed queuing network consisting of M single-server nodes and N single-class customers. Each server has a buffer. The size of the buffer of server i is B_i. The service times of customers at server i have an exponential distribution with mean $\bar{s}_i = 1/\mu_i$, $i = 1, 2, \cdots, M$. The service discipline is first come first served. At the completion of the service at server i, a customer attempts to go to server j with probability $q_{i,j}$, $i, j = 1, 2, \cdots, M$. If at that time there are B_j customers in server j (including the customer being served), then the customer has to stay in server i until a space in buffer j is released to accommodate it. In this case, we say that server i (or the customer being held in server i) is *blocked* by server j. During the blocked period server i is forced down; i.e., it provides no service to any customer.

If a server blocks more than one server simultaneously, then the blocked customers will enter the server according to their order of being blocked; i.e., the customer who was blocked first will enter the server first. If there is a "blocked chain" in the network, e.g., server 1 blocks server 2 which, in turn, blocks server 3, then this blocked chain is released by the customer in the first server (server 1) of this chain. At the time when a customer leaves the first server (server 1), the customer waiting in each server of the chain enters simultaneously the server which blocks it. A server blocked by another server can be considered as a blocked chain containing only two servers.

Several blocked chains may form a "tree." For example, server 1 blocks server 2 which in turn blocks servers 3 and 4. If server 4 is blocked by server 2 eariler than server 3, then at the time when the customer in server 1 departs, the blocked branch consisting of servers 1, 2 and 4 is released; the blocked branch consisting of servers 2 and 3 still remains.

Sometimes there may even exist a "blocked loop." For instance, if server i_1 was blocked by server i_2, server i_2 was blocked by server i_3, $\cdots$, server i_{k-1} was blocked by server i_k, and at some time later, server i_k becomes blocked by server i_1, then a deadlock loop occurs. In this case we assume that the customers in each server in this loop will enter the server which blocks this customer as soon as the loop appears, regardless of the fact that some other servers may be blocked earlier than the servers in this loop. We say that the blocked loop is released by server i_k, which is the last server to be blocked in the loop. As an example, we consider the following scenario: At time t_1 server 3 is blocked by server 1; at time $t_2 > t_1$, server 2 is blocked by server 1; at time $t_3 > t_2$, server 1 completes the service to its customer which, in turn, attempts to go to server 2, and meanwhile buffer 2 is full. In this case, a dead lock occurs. To break the dead lock, we assume that at time t_3+ the customer in server 1 enters server 2 and the customer in server 2 enters server 1, despite the fact that server 3 was blocked earlier than server 2 by server 1. We say that the blocked loop is released by server 1.

Propagation Via Blocking

Let $t_{i,k}$ be the service completion time of the kth customer at server i. Because of blocking, the recursive equation for $t_{i,k}$ is somewhat more complicated than (3.9), which is for systems with infinite buffer capacities. Figure 6.4 illustrated the situation in a blocking period.

In Figure 6.4, the buffer size of server j is 2. At time t_1, a customer completes its service at server i and attempts to go to server j, whose buffer is full. The customer has to wait until t_2, at which a customer leaves server j. The period from t_1 to t_2 is called a *blocked period.* Suppose that somehow the service completion time of server j, t_2, is delayed by Δ. Then

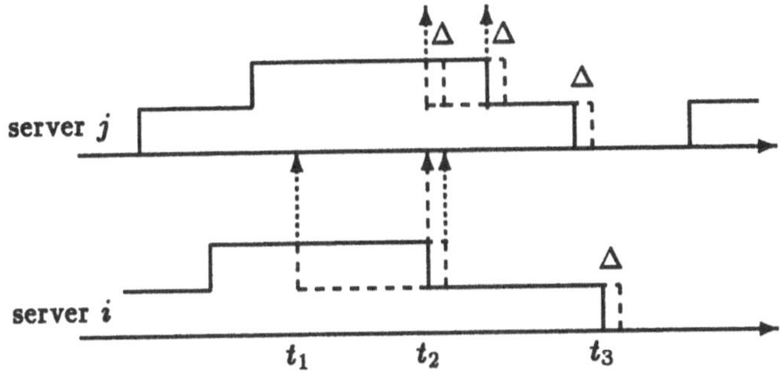

Figure 6.4: Perturbation Propagation Via a Blocking Period

the transition time of the customer from server i to server j will also be delayed by Δ. Thus, the service starting time of the next customer of server i will be delayed by Δ. That is, the perturbation Δ will be propagated from server j to server i, which is blocked by server j.

In Figure 6.4, we say that a blocked period of server i is terminated by a customer in server j. In a blocked chain, all the blocked periods of all the servers in the chain are terminated at the same time by the customer at the head of the chain. In a blocked loop, all the blocked periods of all the servers in the loop are terminated by the customer who completes the loop.

With these terminology, the recursive equation for the service completion times (corresponding to (3.9)) in a network with finite capacity can be written as

$$t_{i,k} = \begin{cases} t_{i,k-1} + s_{i,k} & \textit{if } n_i(t_{i,k-1}+) \neq 0 \textit{ and server } i \\ & \textit{is not blocked;} \\ t_{j,h} + s_{i,k} & \textit{if the idle or the blocked period is} \\ & \textit{terminated by the hth customer} \\ & \textit{of server } j. \end{cases} \tag{6.19}$$

In Figure 6.4, the service completion time of server i, t_3, equals that of server j, t_2 (instead of that of server i, t_1), plus the service time.

By (6.19), the perturbation propagation rules for networks with finite buffer capacities can be stated as follows.

Perturbation propagation rules for networks with blocking:

1. A server will keep its perturbation (real or null) until it meets an idle

or a blocked period.

2. If a customer from server i terminates an idle or a blocked period of server j, then after this idle or blocked period server j will have the same perturbation (real or null) as server i. In this case, we say that the perturbation of server i is propagated to server j.

6.2.2 Perturbation Realization

The system state is denoted as $\mathbf{x} = (\mathbf{n}, \mathbf{b})$, where $\mathbf{n} = (n_1, n_2, \cdots, n_M)$ denotes the numbers of customers in servers, and $\mathbf{b}$ is a record that denotes the blocking status of the system. One possible form of $\mathbf{b}$ is $\mathbf{b}=(b_1, b_2, \cdots, b_M)$, where b_i is an $M-1$ dimensional vector of integers; the lth component $b_{i,l} = j$ means that there are at least l servers blocked by server i, and server j is the lth one according to their order of being blocked. The state space, which contains only a finite number of states, is denoted by Φ.

The concepts of perturbation realization and realization probabilities are defined in the same way as the networks with infinite buffer capacities. However, the proofs of the theorems are entirely different. This is because, for example, there may be no state such as $(N, 0, \cdots, 0)$ in a network with blocking. Also, if $\sum_{j \neq i} b_j < N$, server i never meets any idle period. Thus, in a network with finite buffers, it is possible that every server does not have any idle period. In this case, perturbation is propogated through blocked periods. Therefore, the proofs should be based on only blocked periods.

In a network with blocking, there are three different types of transitions: i) A customer completes its service at server i and enters server j; ii) A customer completes its service at server i and is blocked by server j; and iii) A blocked chain or loop is released because of a customer transition from server i to server j. We define a *feasible transition* for state $\mathbf{x}$ as a transition whose occurrence is permitted by the state $\mathbf{x}$ and the routing probability matrix Q of the network. For example, in a three server Jackson network if $\mathbf{x} = (3, 1, 0)$, and $q_{2,3} \neq 0$, $q_{3,1} \neq 0$, then a customer transition from server 2 to server 3 is feasible for this state; but a transition from server 3 to server 1 is not feasible for the same state. A *sequence* of transitions is said to be *feasible* for state $\mathbf{x}$, if any transition in the sequence is a feasible transition for the state resulted from the initial state $\mathbf{x}$ by performing all preceding transitions in the sequence. For example, in the above three node network for state $\mathbf{x} = (3, 1, 0)$, a sequence consisting of a transition from server 2 to server 3 followed by a transition from server 3 to server 1 is a feasible sequence; while a sequence consisting of a transition from server 2 to server 3 followed by another transition from server 2 to server 3 is not feasible since after the first transition there is no customer in server 2.

The proof of the main result, Theorem 6.4, is based on several lemmas.

Lemma 6.5 *If at time t the system is in state $\mathbf{x}$, then the probability that any feasible sequence for state $\mathbf{x}$ consisting of a finite number of transitions occurs right after t is positive.*

Proof: The lemma follows directly from Lemma 6.2. □

Using this lemma, we can prove:

Lemma 6.6 *The Markov process $\mathcal{X}(t)$ of a queuing network with finite buffer capacities and an indecomposable routing matrix is irreducible.*

Proof: It suffices to prove that all states can be reached from each other. First we assume that $\mathbf{x}_1$ is a state in which no server is blocked and prove that $\mathbf{x}_1$ can reach any other state of the network. The proof consists of three steps:

i) Suppose that $\mathbf{x}_2$ is a neighboring state of $\mathbf{x}_1$ and that in state $\mathbf{x}_2$ there is no blocking. (i.e., there is a pair of indices (i, j) such that $n_{1,i} > 0$, $n_{2,i} = n_{1,i} - 1$, $n_{2,j} = n_{1,j} + 1$, and $n_{2,k} = n_{1,k}$, for all $k \neq i, j$.) Since the routing matrix is indecomposable, there exists a path $i - i_1 - i_2 - \cdots - i_l - j$ such that $q_{i,i_1} q_{i_1,i_2} \cdots q_{i_l,j} > 0$. Thus, if $n_{1,i_k} > 0$ for $k = 1, 2, \cdots, l$, then the sequence consisting of a transition from server i_l to server j, followed by a transition from server i_{l-1} to server i_l, $\cdots$, followed by a transition from server i to server i_1 is a feasible sequence of transitions for state $\mathbf{x}_1$. If in the path there is a server, say server i_l, which is idle, (i.e, $n_{i_l} = 0$,) then in the above sequence the transition from server i_l to server j followed by the transition from server i_{l-1} to server i_l has to be replaced by the transition from server i_{l-1} to server i_l followed by the transition from server i_l to server j. After this sequence of transitions occurs, the system reaches $\mathbf{x}_2$ from $\mathbf{x}_1$. By Lemma 6.5, state $\mathbf{x}_1$ has a positive probability of reaching state $\mathbf{x}_2$.

ii) Suppose that $\mathbf{x}_2$ is any state in which no server is blocked. In this case, by the indecomposability of the network, we can find a sequence of states $\mathbf{x}_1$, $\mathbf{x}_{i_1}$, $\cdots$, $\mathbf{x}_{i_l}$, $\mathbf{x}_2$, such that $\mathbf{x}_{i_k}$, and $\mathbf{x}_{i_{k+1}}$, $k = 0, 1, \cdots, l$, with $\mathbf{x}_{i_0} = \mathbf{x}_1$ and $\mathbf{x}_{i_{l+1}} = \mathbf{x}_2$, are a pair of neighboring states and there is no blocking in $\mathbf{x}_{i_{k+1}}$. Thus, $\mathbf{x}_1$ has a positive probability of reaching $\mathbf{x}_2$ through $\mathbf{x}_{i_1}$, $\mathbf{x}_{i_2}$, $\cdots$, $\mathbf{x}_{i_l}$.

iii) Suppose that $\mathbf{x}_2$ is any state of the network. Let $\mathbf{x}_3$ be the state in which the number of customers in every server is the same as $\mathbf{x}_2$ but there is no blocking in $\mathbf{x}_3$. As proved in ii, $\mathbf{x}_1$ has a positive probability of reaching $\mathbf{x}_3$. Apparently, there exists a feasible sequence of transitions which forms blockings existing in $\mathbf{x}_2$ and leads $\mathbf{x}_3$ to $\mathbf{x}_2$. Thus, $\mathbf{x}_1$ can reach $\mathbf{x}_2$ through $\mathbf{x}_3$.

Finally, we assume that $\mathbf{x}_1$ is any state. It is easy to see that there is a feasible sequence of transitions which releases all the blockings in $\mathbf{x}_1$ and leads $\mathbf{x}_1$ to a state $\mathbf{x}_4$ in which no server blocks others. Therefore, $\mathbf{x}_1$ has a positive probability of reaching any other state through $\mathbf{x}_4$. The proof of the lemma is completed. □

With the above two lemmas, we can prove the following lemma.

Lemma 6.7 *In a closed queuing network with finite buffer capacities, an indecomposable routing matrix, and exponential service time distributions, a perturbation in any set V and state $\mathbf{x}$ has a positive probability of being realized or being lost.*

Proof: Based on Lemma 6.5, it suffices to prove that for any state $\mathbf{x}$ there exists a feasible sequence of a finite number of transitions such that by propagating the perturbation in set V along this sequence of transitions the perturbation set will finally reach Γ or $\emptyset$; or the perturbed state $(\mathbf{x}, V)$ will reach a perturbed state that has a positive probability of being realized or lost.

First we consider the case where there is a server i such that $B_i \geq N$ (i.e., server i has a buffer with an infinite capacity). The proof of the lemma in this case consists of three steps:

a.i) Suppose $n_k = 0$ for all $k \in V$. In this case a customer transition from any server $j \in \Gamma - V$ to any server $k \in V$ will cancel the perturbation of server k. Since the network is indecomposable, customers in servers in $\Gamma - V$ may enter servers in V in a finite number of transitions. Therefore, any feasible sequence will eventually lead the perturbation set to $\emptyset$. The probability of the perturbation in $(\mathbf{x}, V)$ being lost is one.

a.ii) Suppose $n_k = 0$ for all $k \in \Gamma - V$. In this case any customer transition from server $i \in V$ to server $k \in \Gamma - V$ terminates an idle period, hence propagates the perturbation to server k. The probability of the perturbation in $(\mathbf{x}, V)$ being realized is one.

a.iii) Suppose that there is a server i such that $B_i \geq N$. Since the routing matrix is indecomposable, it is feasible for every customer in other servers to move into server i. Thus, there is a feasible sequence of transitions that moves all customers in other servers into server i while keeping the customers who have entered server i staying there. After performing this sequence of transitions, the system state becomes $n_i = N$, $n_j = 0$ for all $j \neq i$. If $i \in V$, this becomes a special case of case a.ii discussed above, and the perturbation has a positive probability of being realized. If $i \in \Gamma - V$, it is a special case of case a.i, and the perturbation has a positive probability of being lost.

Next, we assume that every server in the network has a buffer with a finite size.

b.i) Suppose that all customers in the network belong to a blocked tree with server i being the root of the tree, and $i \in V$. Denote the set of servers in the tree as U. In this case, we can prove that the perturbation in server i will be propagated to all customers in the tree, resulting in a case similar to a.ii). At beginning, two events may happen: (1) The customer being served in server i may enter a server, say server $i_0 \in \Gamma - V$, which is idle; or (2) The customer in server i may attempt to go to a server in the blocked tree. In both cases the perturbation will be propagated to the servers in a branch of the blocked tree. In the first case the perturbation will also be propagated to server i_0. Let U_1 be the subset of Γ such that the servers in U_1 obtain the perturbation through this propagation. Note that now the servers in U_1 are not blocked, but their buffers are full and each of them may become the root of a subtree. The situation for a subtree is different from that for the original tree, since there are customers in servers which are not in the subtree. Note that every server in $U_1 \cap U$ can reach server i without going through servers outside U_1. Because of the indecomposablity of the routing matrix, server i has a positive probability of reaching a server $j \in U - U_1 \cap U$ through a path denoted as $i \to i_1 \to i_2 \to \cdots \to i_l \to j$. Since $j \in U - U_1 \cap U$, server j belongs to a subtree blocked by one server, denoted as k, in $U_1 \cap U$. Denote this subtree as $j \to j_1 \to \cdots \to j_h \to k$, $k \in U_1 \cap U$. As mentioned above, there is a path inside U_1, denoted as $k \to k_1 \to k_2 \to \cdots \to k_g \to i$, which reaches server i. Thus, we have a loop of servers $i \to i_1 \to \cdots j \to j_1 \to \cdots k \to k_1 \to \cdots \to i$. Each server in the loop has a positive transition probability of reaching the server following it. Note that some servers may be on both path $i \to i_1 \to \cdots \to i_l$ and path $k_1 \to k_2 \to \cdots \to k_g$. Thus this loop may be divided into some subloops. (A pair of servers i and j such that $q_{i,j}q_{j,i} > 0$ is considered as a small loop.) Consider the subloop consisting of servers j to j_h and server k. If in the subloop there is a server which is not full, (e.g., if server i_0 is in this loop and $B_{i_0} > 1$, or $i_1 \notin U$) then the blocked chain may be released when a customer goes into this server. If all servers in this loop are full, then the servers in this loop may form a blocked loop, but this blocked loop may be released as soon as the deadlock loop appears. In

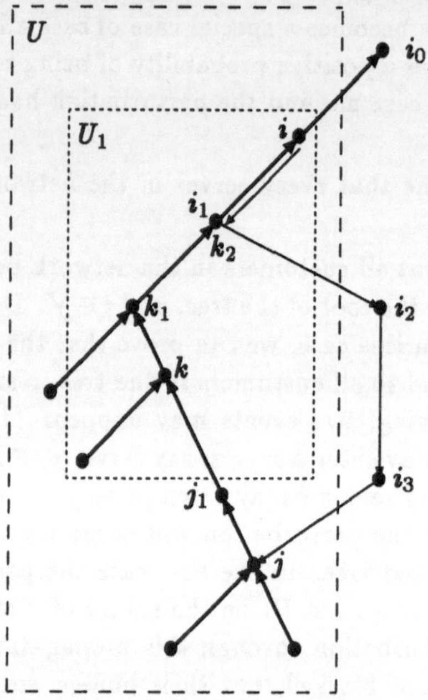

Figure 6.5: The Loop in the Proof of Lemma 6.7

both cases, the perturbation will be propagated to servers j and j_1 to j_h, and the probability that servers in U_1 and servers j and j_1 to j_h obtain the perturbation is positive. Next, we let $U_2 = U_1 + \{j, j_1, \cdots, j_h\}$. Repeating the above procedure for U_2 and all the remained sub-trees we conclude that the probability of the perturbation being realized is positive. In fact, this probability is one. Because all servers without the perturbation are blocked, after a server obtains the perturbation, it can never lose the perturbation.

Figure 6.5 shows a loop of servers consisting of two subloops, $i \to i_1 \to i$ and $i_1 \to i_2 \to \cdots \to j \to k \to k_2$.

b.ii) Suppose that every customer in the network belongs to a blocked tree with server i being the root of the tree, and $i \in \Gamma - V$. The same argument as b.i) shows that in this case the perturbation in any server in this tree will be lost. Thus, the probability of the perturbation being lost is one.

b.iii) Suppose that $B_i < N$ for all $i \in \Gamma$. We can find a blocked tree in the network and construct a feasible sequence which moves all customers into this tree. We start with any server, say server i_1. Since the routing matrix is indecomposable, there must be at least one server in $\Gamma - \{i_1\}$, denoted as i_2, such that $q_{i_2,i_1} > 0$. (Otherwise, by putting q_{i_1,i_1} at the lower-right corner of the matrix Q, Q will have a decomposable form.) If $b_{i_1} + b_{i_2} \geq N$, we have found the desirable tree. If not, we continue as follows. Let $V_1 = \{i_1, i_2\}$. Because of the indecomposability, there must exists at least one server in $\Gamma - V_1$, denoted as server i_3, such that either $q_{i_3,i_1} > 0$ or $q_{i_3,i_2} > 0$. (Otherwise, by reordering servers i_1 and i_2 as servers M and $M-1$, Q will have a decomposable form.) Now we have a tree consisting of servers i_1, i_2, and i_3. We continue the procedure until we find a tree consisting of g servers, i_1, i_2, $\cdots$, i_g, such that $\sum_{j=1}^{g} b_{i_j} > N$. Next, by Lemma 6.6 every state has a positive probability of reaching the state in which all customers in the network are in the blocked tree. The network then becomes in either case b.i) or case b.ii), depending on whether server i_1 has the perturbation or not. Therefore, the perturbation has a positive probability of being realized or lost.

Since the above discussion includes all possible cases of indecomposable networks, the proof of Lemma 6.7 is completed. □

Using Lemma 6.7 we can prove:

Theorem 6.4 *In a closed queuing network with finite buffer capacities, an irreducible routing matrix, and exponential service time distributions, a perturbation* $(\mathbf{x}, V)$ *will be realized or lost with probability one.*

The proof is similar to that of Theorem 6.1 and is omitted. Obviously, the theorem holds for networks in which the hazard rates of the service time distributions satisfy (6.8). The theorem says that such networks are irreducible in the sense defined in Definition 6.1.

6.2.3 The Sensitivity Analysis

Theorems

The realization facts satisfy Equations (6.10)-(6.12); Equation (6.9) has to be replaced by

$$\textit{If } n_i = 0 \textit{ or server } i \textit{ is blocked, then } c^{(f)}(\mathbf{x}, i) = 0. \tag{6.20}$$

Let $\mathbf{x}, \mathbf{y}$ be two states, and $q(\mathbf{x}, \mathbf{y})$ be the probability of the next state being $\mathbf{y}$ given that the current state is $\mathbf{x}$. If the transition from state $\mathbf{x}$ to $\mathbf{y}$ causes idle or blocked periods of servers in a set $U \subset \Gamma$ to be terminated by a customer in server k, we define $I_k(\mathbf{x}, \mathbf{y}) = U$ and $\chi_k(\mathbf{x}, \mathbf{y}) = 1$. Otherwise, $\chi_k(\mathbf{x}, \mathbf{y}) = 0$, and $I_k(\mathbf{x}, \mathbf{y}) = \emptyset$. Furthermore, we define $\mathbf{x}_{i,j}$ as the neighboring state that can be reached from $\mathbf{x}$ when a customer at server i completes its service and attemps to go to server j. The customer may or may not enter server j, depending on whether the buffer of server j is full or not. When there is a blocked chain in $\mathbf{x}$, $\mathbf{x}_{i,j}$ may reflect the changes of the status of all the servers in the chain. With these notations, we have the following theorem.

Theorem 6.5 *For networks with blocking, the realization factors satisfy*

$$\begin{aligned} c^{(f)}(\mathbf{x}, k) &= \sum_{i=1}^{M}\sum_{j=1}^{M} q(\mathbf{x}, \mathbf{x}_{i,j}) c^{(f)}(\mathbf{x}_{i,j}, k) \\ &+ \sum_{j=1}^{M} \{\chi_k(\mathbf{x}, \mathbf{x}_{k,j}) q(\mathbf{x}, \mathbf{x}_{k,j}) c^{(f)}(\mathbf{x}_{k,j}, I_k(\mathbf{x}, \mathbf{x}_{k,j})) \\ &\qquad + f(\mathbf{x}) - f(\mathbf{x}_{k,j})\}. \end{aligned} \tag{6.21}$$

Theorem 6.6 *For networks with blocking, the sample normalized derivative of $\eta_L^{(f)}$ converges to the following value:*

$$\lim_{L\to\infty} E\{\frac{\mu_v}{\eta_L^{(f)}}\frac{\partial \eta_L^{(f)}}{\partial \mu_v}\} = \sum_{all\ \mathbf{x}} p(\mathbf{x}) c^{(f)}(\mathbf{x}, v). \tag{6.22}$$

Like the multiclass network case, the sample normalized derivative may or may not converge to the normalized derivative of the steady-state performance. We shall discuss this in the next section.

Examples

Example 6.2 We study the throughput sensitivity in a cyclic network of three servers (i.e., $f \equiv 1$). The number of customers in the system is $N = 5$. The buffer sizes are $B_i = 3$, $i = 1, 2, 3$. The service time distributions are exponential with means $\bar{s}_1 = 5$, $\bar{s}_2 = 10$, and $\bar{s}_3 = 15$, respectively. Any server in the network can be blocked only by its preceding server. The system state can be represented by the numbers of the customers in each server and a symbol that indicates the blocked status of the servers. We use "+" to denote that a server is blocked. For instance, (3, 1, 1+) shows that the number of the customers in servers 1,2, and 3 are 3,1, and 1 respectively, and server 3 is blocked by server 1. All states and the transition diagram are shown in Figure 6.6. To save space, we only list below some representative

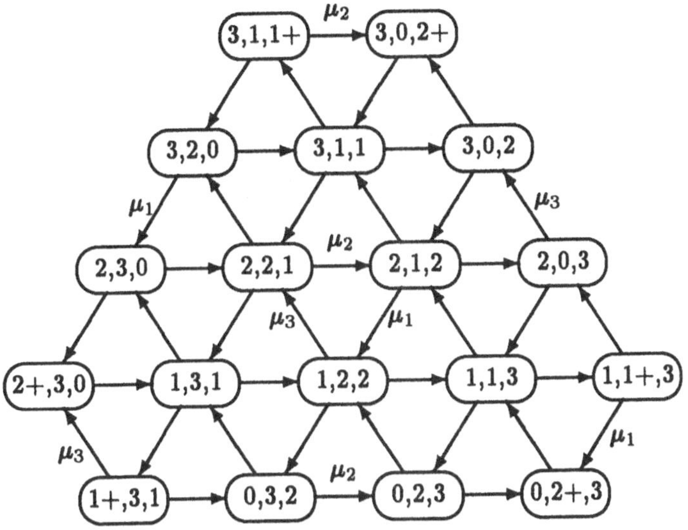

Figure 6.6: The State-Transition Diagram

equations in (6.21) for this system. It is not difficult to write all other equations by following the same model.

$$c[(0,3,2),\ 1] = 0; \tag{6.23}$$

$$(\mu_1 + \mu_2)c[(3,2,0),1] = \mu_1 c[(2,3,0),1] + \mu_2 c[(3,1,1),1]; \tag{6.24}$$

$$\begin{aligned} &(\mu_1 + \mu_2 + \mu_3)c[(3,1,1),1] \\ = \;& \mu_1 c[(2,2,1)1] + \mu_2 c[(3,0,2),1] + \mu_3 c[(3,1,1+),1]; \end{aligned} \tag{6.25}$$

$$\begin{aligned} &(\mu_1 + \mu_3)c[(3,0,2),1] \\ = \;& \mu_1 c[(2,1,2),1] + \mu_1 c[(2,1,2),2] + \mu_3 c[(3,0,2+),1]; \end{aligned} \tag{6.26}$$

$$\begin{aligned} &\mu_1 c[(3,0,2+),1] \\ = \;& \mu_1 c[(3,1,1),1] + \mu_1 c[(3,1,1),2] + \mu_1 c[(3,1,1),3] = \mu_1. \end{aligned} \tag{6.27}$$

No perturbation propagation is involved in (6.24) and (6.25). The second term on the right-hand side of (6.26) represents the perturbation propagation from server 1 to server 2 due to a termination of an idle period

$\mathbf{x}$	$p(\mathbf{x})$	$c(\mathbf{x},1)$	$c(\mathbf{x},2)$	$c(\mathbf{x},3)$
(3,1,1+)	0.0016	0.3950	0.6050	0.0000
(3,0,2+)	0.0031	1.0000	0.0000	0.0000
(3,2,0)	0.0049	0.0925	0.9075	0.0000
(3,1,1)	0.0074	0.2414	0.4053	0.3534
(3,0,2)	0.0069	0.5135	0.0000	0.4865
(2,3,0)	0.0128	0.0180	0.9820	0.0000
(2,2,1)	0.0172	0.0541	0.5413	0.4046
(2,1,2)	0.0237	0.1192	0.2321	0.6487
(2,0,3)	0.0167	0.2359	0.0000	0.7641
(2+,3,0)	0.0601	0.0000	1.0000	0.0000
(1,3,1)	0.0430	0.0088	0.5739	0.4174
(1,2,2)	0.0533	0.0201	0.2904	0.6894
(1,1,3)	0.0837	0.0378	0.1056	0.8567
(1,1+3)	0.0314	0.0590	0.0000	0.9410
(1+,3,1)	0.0516	0.0000	0.5796	0.4204
(0,3,2)	0.0949	0.0000	0.2993	0.7007
(0,2,3)	0.1547	0.0000	0.1162	0.8838
(0,2+,3)	0.3302	0.0000	0.0000	1.0000

Table 6.1: The Realization Probability in Example 6.2

of server 2. In (6.27), the customer in server 1 terminates both an idle period of server 2 and a blocked period of server 3. Thus the perturbation of server 1 is propagated to both servers 2 and 3. By property (6.12), $c[(3,0,2+),\ 1] = 1$.

There are a total of 72 equations for 54 variables $c(\mathbf{x},i)$; 54 of them are in the form of (6.23)-(6.27). 18 of them are in the form of (6.12). The values of these $c(\mathbf{x},i)$ are solved and listed in Table 6.1.

The steady-state distribution $p(\mathbf{x})$ can be obtained by solving the following equations:

$$-\tilde{q}(\mathbf{x})p(\mathbf{x}) + \sum_{\mathbf{y}\in\Phi-\mathbf{x}} \tilde{q}(\mathbf{x},\mathbf{y})p(\mathbf{y}) = 0, \qquad (6.28)$$

with

$$\tilde{q}(\mathbf{x}) = \sum_{all\ \mathbf{y}} \tilde{q}(\mathbf{x},\mathbf{y}),$$

where $\tilde{q}(\mathbf{x},\mathbf{y})$ is the transition rate of the Markov process $\mathcal{X}(t)$ and and

$$\sum_{\mathbf{x}\in\Phi} p(\mathbf{x}) = 1. \tag{6.29}$$

Note that

$$q(\mathbf{x},\mathbf{y}) = \frac{\tilde{q}(\mathbf{x},\mathbf{y})}{\tilde{q}(\mathbf{x})}.$$

The values of $p(\mathbf{x})$ are also listed in Table 6.1.

By (6.22), we can obtain the elasticities of η with respect to μ_i, $ELS(\mu_i)$, $i = 1, 2, 3$. They are

i	1	2	3
$ELS(\mu_i)$	0.0240	0.2215	0.7545

(6.30)

The steady-state throughput η can also be obtained by the classical method. In fact,

$$\eta = \sum_{all\ \mathbf{x}} \mu(\mathbf{x})p(\mathbf{x}),$$

where

$$\mu(\mathbf{x}) = \sum_{k=1}^{M} \epsilon(x_k)\mu_k,$$

with $\epsilon(x_k) = 0$ if $n_k = 0$ or server k is blocked, and $\epsilon(x_k) = 1$ otherwise (cf. (3.1) and (3.4)).

Thus,

$$\frac{\partial\eta}{\partial\mu_i} = \sum_{all\ \mathbf{x}} \epsilon(x_i)p(\mathbf{x}) + \sum_{all\ \mathbf{x}} \mu(\mathbf{x})\frac{\partial}{\partial\mu_i}p(\mathbf{x}). \tag{6.31}$$

From (6.28), $\frac{\partial}{\partial\mu_i}p(\mathbf{x})$ satisfies:

$$\begin{aligned} -\tilde{q}(\mathbf{x})\ \frac{\partial p(\mathbf{x})}{\partial\mu_i} &+ \sum_{\mathbf{y}\in\Phi-\mathbf{x}} \tilde{q}(\mathbf{x},\mathbf{y})\frac{\partial p(\mathbf{y})}{\partial\mu_i} \\ = \ & \frac{\partial\tilde{q}(\mathbf{x})}{\partial\mu_i}p(\mathbf{x}) - \sum_{\mathbf{y}\in\Phi-\mathbf{x}} \{\frac{\partial}{\partial\mu_i}\tilde{q}(\mathbf{x},\mathbf{y})\}p(\mathbf{y}). \end{aligned} \tag{6.32}$$

From (6.29), we have

$$\sum_{\mathbf{x}\in\Phi} \frac{\partial p(\mathbf{x})}{\partial\mu_i} = 0. \tag{6.33}$$

Equations (6.32) and (6.33) are solved for $\frac{\partial p(\mathbf{x})}{\partial\mu_i}$ for the system in this example. The results obtained by (6.31) are exactly the same as (6.30).

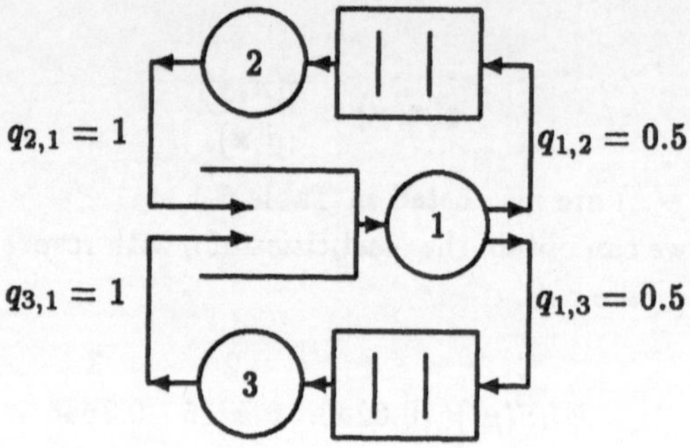

Figure 6.7: The Network in Example 6.3

Example 6.3 The network is shown in Figure 6.7. The buffer size of server 1 is infinite, and the buffer sizes of servers 2 and 3 are 3. The service time distributions are exponential with means $\bar{s}_1 = 5$, $\bar{s}_2 = 10$, and $\bar{s}_3 = 12$. There are 5 customers in the system. The system has altogether 19 states. We study the sensitivity of the throughput. Some equations of realization probabilities are:

$$(\mu_2 + \mu_3)c[(1_3, 1, 3), 1] = \mu_2 c[(2_3, 0, 3), 1] + \mu_3 c[(1, 1, 3), \emptyset],$$

$$\begin{aligned}(\mu_1 + \mu_3)c[(3, 0, 2), 1] &= 0.5\mu_1 c[(2, 1, 2), 1] + 0.5\mu_1 c[(2, 1, 2), 2] \\ &+ 0.5\mu_1 c[(2, 0, 3), 1] + c\mu_3 c[(4, 0, 1), 1].\end{aligned}$$

In the equations, 1_3 denotes that there are one customer in server 1 which is blocked by server 3. Solving the equations, we get the realization probabilities. From these values and the steady-state distribution we obtain the elasticities:

i	1	2	3
$ELS(i, \mu_i)$	0.2626	0.2770	0.4604

The classical Equations (6.31)-(6.33) yield the same results.

Other Sensitivities

As discussed in Section 3.5, the realization probability and the realization factor can be used to calculate the sensitivities due to any other changes in the service times. For example, they can be applied to calculate the sensitivity when each customer's service time increases by an equal amount. For networks with blocking, corresponding to Equations (3.84), we have

$$\lim_{L\to\infty} E\{\frac{\bar{s}_v}{\eta_L^{(f)}}\frac{\partial \eta_L^{(f)}}{\partial \bar{s}_v}\} = -\sum_{all\ \mathbf{x}} p(\mathbf{x})c_t^{(f)}(\mathbf{x}, v), \tag{6.34}$$

where the deivative should be understood in the sense that the service times are increased by equal amounts, and $c_t^{(f)}(\mathbf{x}, v)$ is the realization factor of a perturbation at server v's service completion time. $c_t^{(f)}(\mathbf{x}, v)$ can be determined in the same way as (3.85); i.e.,

$$\begin{aligned} c_t^{(f)}(\mathbf{x}, i) &= \sum_{j=1}^{M} q_{i,j}c_t^{(f)}(\mathbf{x}_{i,j}, i) \\ &+ \sum_{j=1}^{M} \chi_i(\mathbf{x}, \mathbf{x}_{i,j})q_{i,j}c_t^{(f)}(\mathbf{x}_{i,j}, \{I_i(\mathbf{x}, \mathbf{x}_{i,j})\}), \end{aligned} \tag{6.35}$$

where $c^{(f)}$ is the realization factor determined by Equations (6.12), (6.20), and (6.21). In the equation, $\mathbf{x}_{i,j}$ is a neighboring state of $\mathbf{x}$. For instance, in Example 6.3, if $\mathbf{x} = (2, 0, 3)$ then $\mathbf{x}_{1,3} = (2_3, 0, 3)$.

We make a convention that if there can not be any service completion of server i at state $\mathbf{x}$, then $c_t^{(f)}(\mathbf{x}, i) = 0$.

Example 6.4 The system is the same as in Example 6.2. The question is: what are the elasticities of the throughputs if the service time of each customer increases by the same amount $\Delta\bar{s}_i$? The realization probabilities $c_t(\mathbf{x}, i)$ for this system are calculated and listed in Table 6.2.

The limiting value of the sample elasticities obtained by (6.34) are:

i	1	2	3
ELS w.r.t. $\Delta\bar{s}_i$	0.0092	0.1245	0.5987

Note that these elasticities with respect to a fixed change in service time is smaller than that with respect to a change in mean service time obtained in Example 6.2. This has the same intuitive explanation as that in Example 3.8.

6.3 The Interchangeability

In the previous two sections, we have derived formulas for the limiting values of the sample derivatives of the performance for multiclass networks

x	$p(\mathbf{x})$	$c_t(\mathbf{x},1)$	$c_t(\mathbf{x},2)$	$c_t(\mathbf{x},3)$
(3,1,1+)	0.0016	0.0925	0.0000	0.0000
(3,0,2+)	0.0031	1.0000	0.0000	0.0000
(3,2,0)	0.0049	0.0180	0.7587	0.0000
(3,1,1)	0.0074	0.0541	0.0000	0.0000
(3,0,2)	0.0069	0.3513	0.0000	0.0000
(2,3,0)	0.0128	0.0000	0.9459	0.0000
(2,2,1)	0.0172	0.0088	0.2321	0.0000
(2,1,2)	0.0237	0.0201	0.0000	0.3534
(2,0,3)	0.0167	0.1434	0.0000	0.4865
(2+,3,0)	0.0601	0.0000	1.0000	0.0000
(1,3,1)	0.0430	0.0000	0.2904	0.0000
(1,2,2)	0.0533	0.0201	0.1056	0.4046
(1,1,3)	0.0837	0.0000	0.0000	0.6487
(1,1+,3)	0.0314	0.0000	0.0000	0.7641
(1+,3,1)	0.0516	0.0000	0.2993	0.0000
(0,3,2)	0.0949	0.0000	0.1162	0.4262
(0,2,3)	0.1547	0.0000	0.0000	0.7095
(0,2+,3)	0.3302	0.0000	0.0000	1.0000

Table 6.2: The Realization Probability in Example 6.4

and networks with blocking. In studying the sensitivity of steady-state performance, we first have to address the unbiasedness issue of the sample derivatives (cf. Lemmas 3.8 and 4.3). That is, we have to verify whether the following equation holds:

$$E\{\frac{\partial}{\partial\mu}[\eta_L^{(f)}(\mu,\xi)]|\mathbf{x}_0\} \stackrel{?}{=} \frac{\partial}{\partial\mu}\{E[\eta_L^{(f)}(\mu,\xi)|\mathbf{x}_0]\}, \tag{6.36}$$

where μ can be any service rate of interest (e.g., $\mu_{v;k}$ in multiclass networks), ξ is a random vector which determines all the random variables in the network, and $\mathbf{x}_0$ is the initial state of the network. The derivatives in the equation are understood as the one-sided derivative with respect to μ. In fact, as shown in Chapter 3, in many cases, the two one-sided derivatives may not equal.

We have proved that (6.36) holds for single-class networks with infinite buffer capacities; but we shall see that the equation does not hold for all the multiclass networks and networks with blocking. We first discuss some general concepts.

6.3.1 The Continuity of the Sample Functions

The continuity of the sample performance function $\eta_L^{(f)}$ is neither a sufficient nor a necessary condition for (6.36). However, as we see in Chapters 3 and 4, the continuity plays a crucial role in proving (6.36). After the continuity is established, the unbiasedness Equation (6.36) can usually be easily proved by checking some bounded conditions for the derivatives to apply the Lebesgue theorem; under many circumstances, these bounded conditions are satisfied. Thus, in most cases, the reason for (6.36) not to hold is the discontinuity of the sample function. The reference for this subsection is Cao [31].

To illustrate the idea, we give a simple example.

Example 6.5 In a simple lottery, a player draws a number from a random number generator creating uniformly distributed random variables over $[0, 1)$. If the number drawn is less than or equal to a parameter θ, the player receives one dollar, otherwise he receives nothing. In the example, we have $\xi \in [0, 1)$ and $0 < \theta < 1$. The sample performance function is

$$D(\theta,\xi) = \begin{cases} 1 & if\ \xi \le \theta, \\ 0 & if\ \xi > \theta. \end{cases}$$

The sample function (for a fixed ξ) has a jump at $\theta = \xi$. For this function, we have

$$E[D(\theta,\xi)] = \theta,$$

and

$$\frac{\partial}{\partial\theta}E[D(\theta,\xi)] = 1.$$

However, for any ξ,

$$\frac{\partial}{\partial\theta}D(\theta,\xi) = 0.$$

Therefore,

$$E\{\frac{\partial}{\partial\theta}[D(\theta,\xi)]\} \neq \frac{\partial}{\partial\theta}\{E[D(\theta,\xi)]\}.$$

The example clearly illustrate the reason why (6.36) does not hold when the sample function is discontinuous: the sample derivative $\frac{\partial}{\partial\theta}[D(\theta,\xi)]$ does not contain any information about the jumps of the sample function $D(\theta,\xi)$ in a neighborhood of θ. But the derivative of the mean performance function, $\frac{\partial}{\partial\theta}\{E[D(\theta,\xi)]\}$, certainly reflects the effect of the average jump heights in $[\theta,\theta+\Delta\theta]$. Thus, unless the average jump heights in $[\theta,\theta+\Delta\theta]$ is zero, the two sides of (6.36) will not equal.

Example 6.6 In this example, we change the reward function in Example 6.5 to the following

$$D(\theta,\xi) = \begin{cases} 1 & \textit{if } \xi \le \frac{1}{2} \textit{ and } \theta > 2\xi, \\ & \textit{or } \xi > \frac{1}{2} \textit{ and } \theta < 2\xi - 1, \\ 0 & \textit{otherwise.} \end{cases}$$

The values of the the function are illustrated in Figure 6.8.

For $\xi \le \frac{1}{2}$, the sample function jumps from 0 to 1 at 2ξ; for $\xi > \frac{1}{2}$, it jumps from 1 to 0 at $2\xi - 1$. We have

$$E[D(\theta,\xi)] = \frac{1}{2}.$$

Therefore,

$$E\{\frac{\partial}{\partial\theta}[D(\theta,\xi)]\} = \frac{\partial}{\partial\theta}\{E[D(\theta,\xi)]\} = 0.$$

The example shows that continuity is not necessary for (6.36). In the example, the average jump heights in $[\theta,\theta+\Delta\theta]$ is zero. In fact, the probability of a sample function jumping from 1 to 0 in $[\theta,\theta+\Delta\theta]$ is $\frac{1}{2}\Delta\theta$, and the probability of jumping from 0 to 1 in $[\theta,\theta+\Delta\theta]$ is also $\frac{1}{2}\Delta\theta$. In practical systems, this "average out" situtation seldon happens. Thus, the task

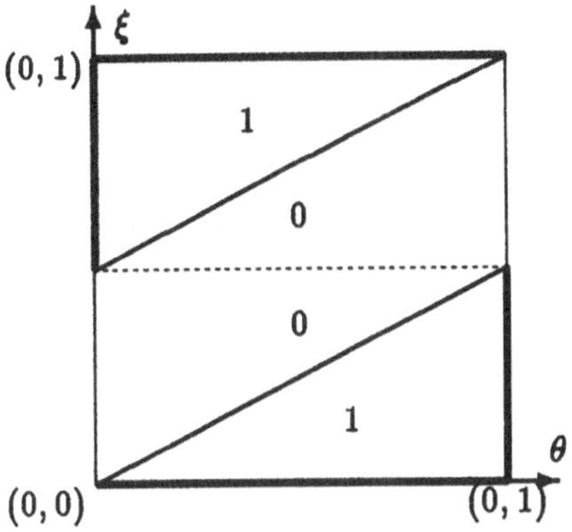

Figure 6.8: The Values of $D(\theta, \xi)$ in Example 6.6

of establishing (6.36) is to check the continuity of the sample performance function.

Nor is the continuity a sufficient condition for (6.36). This is illustrated in the next example.

Example 6.7 Let

$$D(\theta, \xi) = \begin{cases} 0 & if\ \xi \geq \theta, \\ \frac{1}{\sqrt{\xi}}(\sqrt{\theta} - \sqrt{\xi}) & if\ \xi < \theta. \end{cases}$$

The sample function (for a fixed ξ) are continuous at any θ. we have

$$E[D(\theta, \xi)] = \int_0^\theta \frac{1}{\sqrt{\xi}}(\sqrt{\theta} - \sqrt{\xi})d\xi = \theta$$

and

$$\frac{\partial}{\partial \theta} E[D(\theta, \xi)] = 1.$$

However, for any ξ, at $\theta = 0$ we have

$$\frac{\partial}{\partial \theta} D(\theta, \xi) = 0.$$

Therefore, at $\theta = 0$,

$$E\{\frac{\partial}{\partial \theta}[D(\theta, \xi)]\} \neq \frac{\partial}{\partial \theta}\{E[D(\theta, \xi)]\}.$$

In the example, the sample derivatives of $D(\theta, \xi)$ in $[0, \Delta\theta]$ are unbounded; this makes (6.36) not hold.

A sufficient condition for (6.36) is the uniform differentiability of the sample functions, which is defined below (see Cao [31]).

Assume that the one-side derivative $\frac{\partial}{\partial\theta}D(\theta, \xi)$ exists with probability one. Then

$$D(\theta + \Delta\theta, \xi) - D(\theta + \Delta\theta, \xi) = \frac{\partial D(\theta, \xi)}{\partial\theta}\Delta\theta + e(\Delta\theta, \xi) \qquad w.p.1,$$

with

$$\lim_{\Delta\theta\to 0} \frac{e(\Delta\theta, \xi)}{\Delta\theta} = 0 \qquad w.p.1.$$

Definition 6.2 *A set of functions $D(\theta, \xi)$, $\xi \in \Omega$, is said to be uniformly differentiable with probability one, if and only if for any $\epsilon > 0$, there exists a $\delta > 0$ such that if $|\Delta\theta| < \delta$, then $|\frac{e(\theta,\xi)}{\Delta\theta}| < \epsilon$ with probability one.*

For queuing networks, once the continuity of the sample path is proved, it is relatively easy to verify that the uniform bound exists. Then one can apply the Lebesgue theorem to prove the unbiasedness.

In the next subsections, we shall discuss the sample performance functions for multiclass networks and networks with blocking.

6.3.2 A Necessary Condition for Multiclass Networks

The Discontinuity

We first study a simple multiclass network (shown in Figure 6.9) for which (6.36) does not hold. There are two servers and two classes of customers in the network. Each class consists of only one customer. The class 1 customer goes back to server 1 immediately after it leaves the server; the class 2 customer goes from server 1 to server 2 and back.

Let the service time distributions be exponential, and let μ be the mean service rate of server 2 and λ be that of server 1 for both classes 1 and 2 customers. The system state can be denoted as (k_1, k_2), where k_i, $i = 1, 2$, is the class of the customer at position i in queue 1. $k_2 = 0$ indicates that there is only one customer in server 1. There are only three possible states: $(1, 2)$, $(2, 1)$, and $(1, 0)$.

Figure 6.10 illustrates a sample path of the network. Part a) of the figure shows the nominal path of the network, with T_0, T_1, and T_2 indicating the service completion times of the class 2 customer at both servers. The

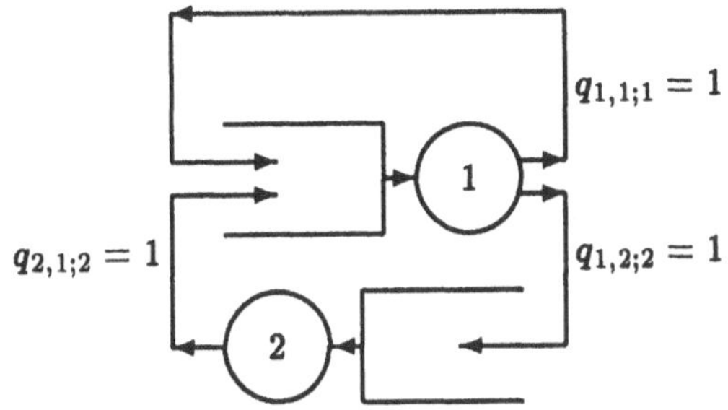

Figure 6.9: A Multiclass Network with Discontinuous Sample Functions

instants t_i, $i = 1, 2, 3, 4$, are the service completion times of the class 1 customer. At T_1, the class 2 customer moves from server 2 to server 1 and has to wait for the server to complete its service to the class 1 customer. $s_{1;2}$ is the service time of the class 2 customer at server 1. During the service period of the class 2 customer, the class 1 customer has to wait. $s_{1;1}$ is the service time of the class 1 customer after the class 2 customer leaves server 1. The figure shows that during $T_1 - T_0$, the class 1 customer feeds back at server 1 twice.

Part b) of Figure 6.10 shows the perturbed path of the network, in which the service time of the class 2 customer at server 2 increases by Δ, i.e., $T_1' = T_1 + \Delta$. In the nominal path, the class 2 customer arrives at server 1 before the class 1 customer feeds back at t_3 ($T_1 < t_3$); thus, the class 2 customer starts to receive service at t_3, and the class 1 customer has to wait until the service completion time T_2. In the perturbed path, because of the small delay of Δ, the class 2 customer arrives at server 1 after the class 1 customer feeds back at t_3 ($T_1' > t_3$); thus, the class 1 customer starts to receive another service at t_3, and the class 2 customer has to wait until t_4', when server 1 finishes its service to the class 1 customer. Therefore,

$$T_2' = t_3 + s_{1,1} + s_{1,2} = T_2 + s_{1;1}. \tag{6.37}$$

This clearly shows that T_2 is discontinuous with respect to the service rate of the class 2 customer.

The discontinuity of T_2 comes from the fact that the sample path is discontinuous in the following sense: a small change in the service time of the class 2 customer may cause a discontinuous change in the state

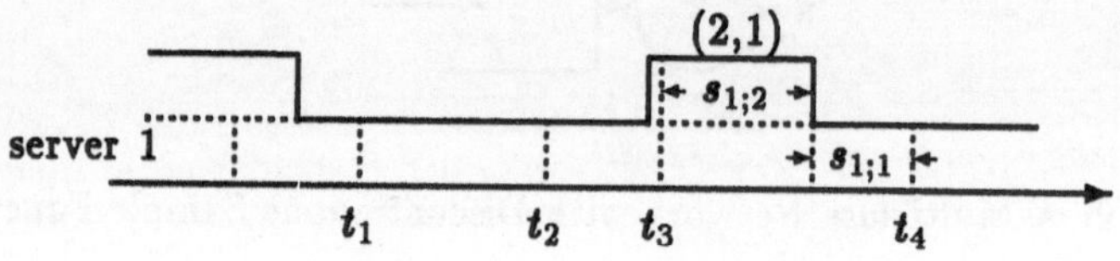

a) The Nominal Path

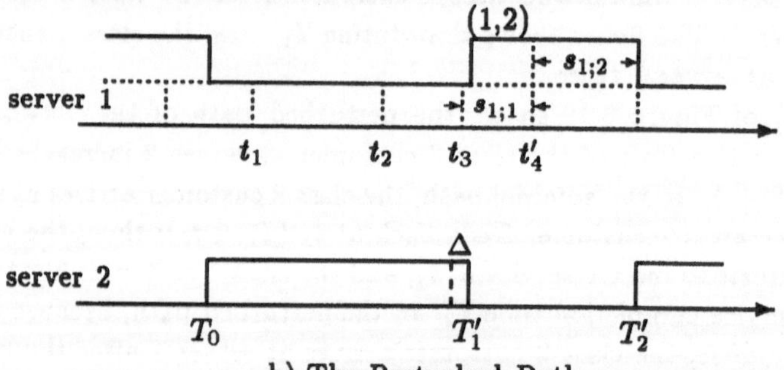

b) The Perturbed Path

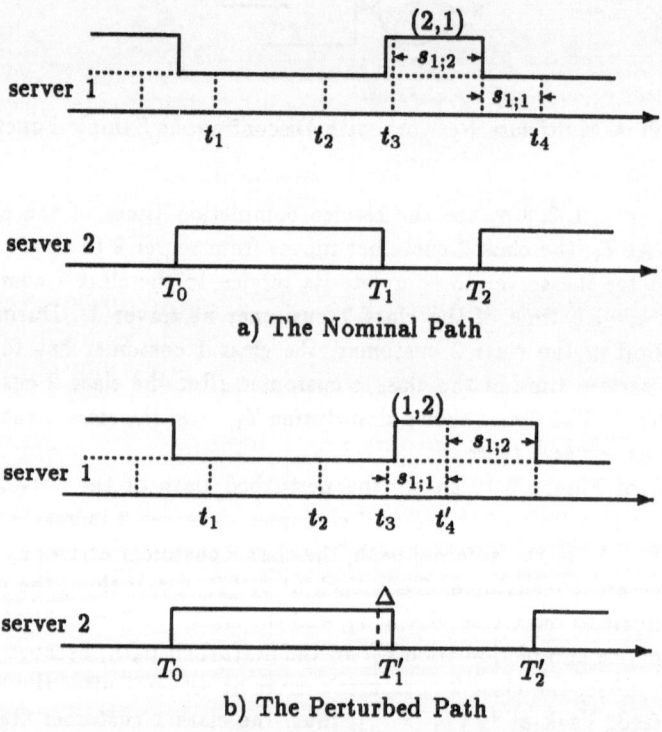

Figure 6.10: The Discontinuity of the Sample Path

sequence of the sample path. In fact, because of a small change in the service time, two transitions (T_1 and t_3 in Figure 6.10) may change their order of occurrence, which may result in the perturbed path reaching a different state from the nominal path after the two transitions. For example, after the service completion T_1 followed by service completion t_3, the system goes to state $(2,1)$,; while after t_3 followed by T_1', the system goes to state $(1,2)$.

More precisely, the sequence of states starting from T_0 for $T_1 < t_3$ is $(1,0)$, $(1,0)$, $(1,0)$, $(1,2)$, $(2,1)$, $(1,0)$, $\cdots$. As T_1 approaches t_3, the length of the state $(1,2)$, $t_3 - T_1$, decreases to zero. When $T_1 \uparrow t_3$, the sequence converges to $(1,0)$, $(1,0)$, $(1,0)$, $(2,1)$, $(1,0)$, $\cdots$. On the other hand, the state sequence for $T_1' > t_3$ is $(1,0)$, $(1,0)$, $(1,0)$, $(1,0)$, $(1,2)$, $(2,1)$, $(1,0)$, $\cdots$. As T_1' approaches t_3, the length of the last $(1,0)$ before the state $(1,2)$, $T_1' - t_3$, decreases to zero. When $T_1' \downarrow t_3$, the sequence converges to $(1,0)$, $(1,0)$, $(1,0)$, $(1,2)$, $(2,1)$, $(1,0)$, $\cdots$, which is different from the sequence obtained by $T_1 \uparrow t_3$.

In summary, the discontinuity is due to the fact that the change of the order of the occurrences of the two transitions may result in the system being in two different states. In this case, the sample path is discontinuous; hence the sample performance functions defined on the sample path are most likely discontinuous. This rule is general, and we shall apply this rule to discuss networks with blocking and generalized semi-Markov processes in the remainder of this chapter and the next chapter.

Example 6.8 The steady-state probabilities of the system in Figure 6.9 can be easily obtained by the Markov model.

$$p(1,0) = \frac{1}{1+2\frac{\mu}{\lambda}}$$

and

$$p(1,2) = p(2,1) = \frac{\frac{\mu}{\lambda}}{1+2\frac{\mu}{\lambda}}.$$

Thus,

$$\eta_{1;1} = [p(1,2) + p(1,0)]\lambda = \frac{\mu+\lambda}{1+2\frac{\mu}{\lambda}}$$

and

$$\eta_{1;2} = p(2,1)\lambda = \frac{\mu}{1+2\frac{\mu}{\lambda}}.$$

Therefore,

$$\frac{\mu}{\eta_{1;1}}\frac{\partial \eta_{1;1}}{\partial \mu} = -\frac{1}{(1+\frac{\lambda}{\mu})(1+2\frac{\mu}{\lambda})}$$

and

$$\frac{\mu}{\eta_{1;2}}\frac{\partial\eta_{1;2}}{\partial\mu} = -\frac{1}{1+2\frac{\mu}{\lambda}}.$$

On the orther hand, the sample derivatives are zero because any perturbation generated at server 2 can never be propagated to server 1, since there is no idle period at server 1. Thus, (6.36) does not hold for the network. This simple example was given in Cao [30]; some extensions were later studied in Heidelberger et al. [66].

A Necessary Condition

Observe that the right-hand side of (6.17) does not depend on j and h. That is, any fixed v and k, the limiting values

$$\lim_{L\to\infty} E\{\frac{\mu_{v;k}}{\eta_{j,h;L}}\frac{\partial\eta_{j,h;L}}{\partial\mu_{v;k}}\}$$

are all equal for all j and h. Thus, one necessary condition for the convergence of the sample elasticity of the throughput to the elasticity of the steady-state throughput is

$$\frac{\mu_{v;k}}{\eta_{j;h}}\frac{\partial\eta_{j;h}}{\partial\mu_{v;k}} \textit{ do not depend on } j \textit{ and } h.$$

This implies that

$$\frac{\eta_{j;h}(\mu_{v;k})}{\eta_{j';h'}(\mu_{v;k})} \textit{ is independent of } \mu_{v;k}. \tag{6.38}$$

This condition hold for single-class networks, where $\eta_{j,h} := \eta_j$, since $\frac{\eta_j}{y_j}$ does not depend on j, with y_j being the visit ratio to server j (see Section 2.3).

Condition (6.38) is very strong; but all is not lost. It is a necessary condition for two sample elasticities of $\eta_{j,h;L}$ and $\eta_{j',h';L}$, obtained on the same sample path, to converge simultanuously to the elasticities of their corresponding steady-state values. It is not necessary for a single sample elasticity of $\eta_{j,h;L}$ or $\eta_{j',h';L}$ to converge to the elasticity of its corresponding steady-state value.

Recall that a sample derivative is the derivative of a sample performance function $\eta_L^{(f)}(\theta,\xi)$ with ξ being fixed. The sample function $\eta_L^{(f)}(\theta,\xi)$ depends on how the representative random vector ξ is chosen. For different ξ, the form of the function is different. These different sample performance function gives the same value for the performance on the same sample path

(θ, ξ), but they give different values for the sample path with $(\theta + \Delta\theta, \xi)$. Thus, the sample derivatives based on different ξ are different. We call a ξ, together with the way to specify the sample path with ξ, a *representation* of the system (Glasserman [58]) and Cao [25]).

When (6.38) does not hold, it is possible that there are different representations, with each of them the sample elasticity of one $\eta_{j,h;L}$ converges to the elasticity of the steady-state throughput; but no representation allows such convergence for two different sample throughputs. Readers can find such an example in Cao [25].

6.3.3 A Sufficient Condition for Networks With Blocking

The Discontinuity

As explained in the last subsection, to check the discontinuity of sample functions, we need to examine whether a change in the occurrence of two transitions may lead to two completely different sequences of states on the sample paths.

Figure 6.11 illustrates a sample path of three servers in a network, where servers 2 and 3 are blocked *simultaneously* by server 1. Server 1 has a buffer capacity of 2. The figure shows that in the nominal path, server 2 is blocked at its service completion time t_2, and server 3, at a later time t_3. At $t_{1,1}$, server 1 completes the service to a customer. According to the first-blocked first-served principle, the customer in server 2 enters server 1 at $t_{1,1}$. Server 3 has to wait until the next service completion time $t_{1,2}$.

Now, suppose that server 2 gets a perturbation Δ, which may be due to the decreasing of the service rate of server 2, and that $t_2' = t_2 + \Delta > t_3$. In the perturbed path, server 3 is blocked by server 1 earlier than server 2. As a result, the blocked customer in server 3 enters server 1 at $t_{1,1}$, instead of $t_{1,2}$; and the blocked customer in server 2 has to wait until $t_{1,2}$. Thus, an infinitesimal perturbation may leads the system to a completely different sample path.

Figure 6.11 shows that if in a network there is a server that can simultaneously block two other servers, the sample functions are discontinuous. However, if in a network any server can block only one server, the order change shown in Figure 6.11 can never happen. It is easy to check that in such networks the sample functions are continuous. The cyclic networks with finite buffer capacities and the network shown in Figure 6.7 are examples of such networks.

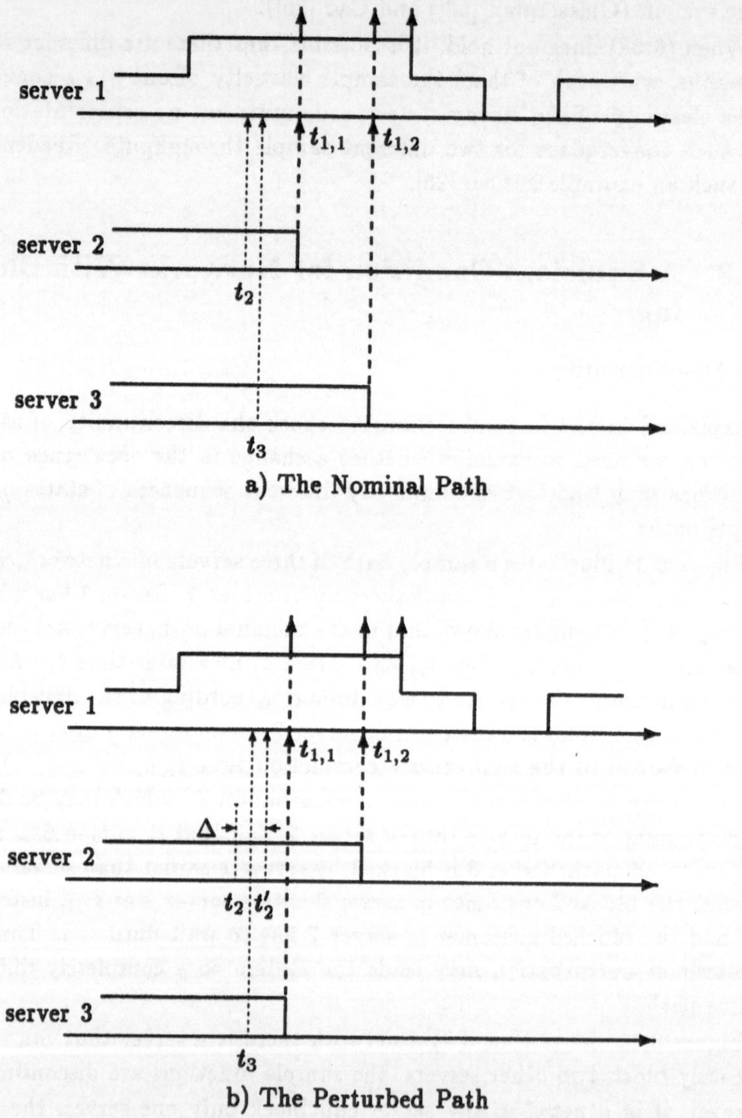

Figure 6.11: The Discontinuity Caused by Simultaneous blocking

The Sensitivity Equations

Furthermore, we can apply the same argument as that in Lemma 3.8 to prove that (6.36) holds for these networks. This is summarized in the following theorem.

Theorem 6.7 *In a network where no server can block more than one server simultaneously, the unbiasedness equation (6.36) holds.*

From this and Theorem 6.6, we can prove

Theorem 6.8 *In a network where no server can block more than one server simultaneously, we have*

$$\{\frac{\mu_v}{\eta^{(f)}}\frac{\partial \eta^{(f)}}{\partial \mu_v}\} = \sum_{all\ \mathbf{x}} p(\mathbf{x})c^{(f)}(\mathbf{x}, v). \tag{6.39}$$

To get some idea about the error of applying (6.39) to networks where simultaneous blocking occurs, we study the following example.

Example 6.9 The system contains 3 servers and 5 customers. The buffer sizes of servers 2 and 3 are 3, of server 1 is infinity. We study the network with four different routing matrices, shown in Table 6.3. The limiting values of the sample elasticities of the throughputs are calculated by Theorem 6.6, and the elasticities of the steady-state throughput are calculated by the Markov model (Equations (6.31)-(6.33)). These values are listed in Table 6.3 for comparison. Most errors are only at the level of 0.5%. Thus, (6.39) can be used to accurately estimate the elasticities of the steady-state throughputs in this network.

6.4 A Summary

We have studied two types of networks: multiclass networks and networks with blocking. The dynamics of these two types of networks are different from that of single-class infinite-buffer networks. The differences can be summarized as follows.

1. An infinitesimal perturbation may casue the perturbed sample path being significantly different from the nominal one. This is due to the fact that a small perturbation may lead to the change of the order of two transitions. In single-class infinite-buffer networks, the system reaches the same state after the order changes. However, in

No.		Routing Prob. $q_{i,j}$			$\bar{s}_i$	$\sum_{all\ \mathbf{x}} p(\mathbf{x})c(\mathbf{x},i)$	ELS_i
		j=1	2	3			
1	i=1	0.0	0.0	1.0	2	0.0002	0.0002
	2	0.0	0.0	1.0	2	0.0006	0.0006
	3	0.5	0.5	0.0	1	0.9993	0.9992
2	i=1	0.0	0.5	1.5	10	0.9653	0.9659
	2	0.9	0.0	0.1	5	0.0173	0.0170
	3	0.9	0.1	0.0	5	0.0173	0.0170
3	i=1	0.0	0.5	0.5	10	0.3105	0.3139
	2	0.5	0.0	0.5	10	0.3447	0.3430
	3	0.5	0.5	0.0	10	0.3447	0.3430
4	i=1	0.0	0.5	0.5	5	0.1276	0.1276
	2	0.7	0.0	0.3	10	0.2827	0.2823
	3	0.8	0.2	0.0	12	0.5898	0.5910

Table 6.3: The Error Due to Discontinuity

multiclass networks or networks with finite buffer capacities, the system may reach a different state after the order of the two transitions changes. Therefore, in these networks the sample performance function may be discontinuous, and the unbiasedness (6.36) may not hold. The formulas based on realization factors give the limiting values of the sample elastivcity; but they may not equal the elasticities of the steady-state performance.

We have shown that if in a network where no server can block more than one server simultaneously, the sample functions are continuous and the elasticities of the steady-state performance can be calculated by the realization factors. One example shows that the error is very small even if we apply the realization-factor formulas to networks with simultaneous blocking.

A necessary condition is given for the multiclass case. It is also pointed out that the system representation plays an important role. It is possible that the sample derivative is unbiased for one representation, but is not for another.

2. In multiclass networks, interaction between two servers may exist even the customers in each of these two servers can not reach the other. This may happen when the customers in these two servers visit a common third server. Thus, a perturbation may be realized

by the network even the routing matrix is decomposable. We define a network in which any perturbation will be realized or lost with probability one as an irreducible network. A necessary and sufficient condition is found for a multiclass network to be irreducible.

3. In networks with finite buffer capacities, perturbations may be propagated through blocking period. In some cases, there may be no idle periods. We proved that a network with an indecomposable routing matrix is irreducible; the proof is based on propagation through blocking.

Chapter 7

Sensitivity Analysis of Generalized Semi-Markov Processes

In the previous chapters, we studied various queuing networks. The state processes of these networks fit in a general model of stochastic processes, i.e., the generalized semi-Markov process (GSMP). In this chapter, we shall extend the concepts and results developed so far to GSMP. Since GSMP can be used to model many stochastic systems other than queuing systems, the results in this chapter extend the applicability of the theory.

The major differences between the sensitivity study of a queuing network and that of a GSMP are twofold. The first one is of conceptual nature. Perturbations in queuing networks are propagated among servers through idle or blocking periods. In GSMP, terminologies such as servers, customers, and idle or blocking periods do not exist; the perturbation propagation is based on an intrinsic property called *event coupling*. We shall characterize this property and study its relationship with perturbation realization. The second difference is technical in nature; unlike in the single-class queuing network case, GSMP may not be regenerative. Therefore, the proofs for GSMP, which are based on the event-coupling tree, are more general. Most of the material presented in this chapter is from Cao [18].

7.1 The Generalized Semi-Markov Processes

Generalized Semi-Markov Processes were introduced by Matthes [85] to study the insensitivity of steady-state distributions. In this chapter, we basically follow the standard description of a GSMP (e.g., see in Whitt [106]), except when there are two events whose lifetimes terminate at the same instant; in this case, our state transition rule is different from those in the previous papers (see this section and Section 7.3 for details).

Let Φ and $\mathcal{E}$ be subsets of positive integers. Φ represents the (phycical) state space of the GSMP. (The word "physical" is used to distinguish with the "mathematical" state defined later in Section 7.3; we shall omit the modifiers when there is no confusion.) At each state, some events may occur. $\mathcal{E}$ denotes the indices of all possible events that may occur during the evolution of a GSMP. All events that can occur in state $\mathbf{x} \in \Phi$ are denoted as set $B(\mathbf{x}) \subseteq \mathcal{E}$.

A GSMP stays in a state $\mathbf{x}$ until an event $\alpha \in B(\mathbf{x})$ triggers a transition. Let $q(\mathbf{x}'; \mathbf{x}, \alpha)$ be the probability of jumping to state $\mathbf{x}'$ when event α triggers a transition from state $\mathbf{x}$. Each event in $B(\mathbf{x})$ has a lifetime, and an event can trigger a transition only at the end of its lifetime. Associated with each event α, there is a clock whose reading is denoted as d_α. The clock runs at a speed $\sigma(\mathbf{x}, \alpha)$. If at time 0 the clock is set to d_α then at time τ the reading of the clock will be $d'_\alpha = d_\alpha - \sigma(\mathbf{x}, \alpha) \times \tau$. The lifetime of an event ends when the associated clock reading reaches zero. d_α is also called the *residual lifetime* of event α. Let $\mathbf{d}$ be a vector whose αth component is d_α. Also, let $R_+ = [0, \infty)$, $R_+^\infty = R_+ \times R_+ \times \cdots$, and

$$R_{\mathbf{x}} := \{\mathbf{d} \in R_+^\infty : d_\alpha > 0 \; if \; and \; only \; if \; \alpha \in B(\mathbf{x})\}.$$

$R_{\mathbf{x}}$ is the set of all possible readings of the clocks associated with the events that may occur in state $\mathbf{x}$. We assume $\sigma(\mathbf{x}, \alpha) > 0$ for all $\alpha \in B(\mathbf{x})$.

The event triggering the next transition is determined by the following rule. Suppose that at present the state is $\mathbf{x} \in \Phi$ and the clock reading vector is $\mathbf{d} \in R_{\mathbf{x}}$. Let

$$\tau^* = \tau^*(\mathbf{x}, \mathbf{d}) := min\{\tau_\alpha, : \; \alpha \in B(\mathbf{x}), \; d_\alpha - \tau_\alpha \times \sigma(\mathbf{x}, \alpha) = 0\},$$

$$d^*_\alpha = d^*_\alpha(\mathbf{x}, \mathbf{d}) := d_\alpha - \tau^* \times \sigma(\mathbf{x}, \alpha), \quad \alpha \in B(\mathbf{x}),$$

$$\alpha^* = \alpha^*(\mathbf{x}, \mathbf{d}) := \{\alpha \in B(\mathbf{x}) : d^*_\alpha(\mathbf{x}, \mathbf{d}) = 0\}.$$

The event $\alpha^*(\mathbf{x}, \mathbf{d})$ is the triggering event, and $\tau^*(\mathbf{x}, \mathbf{d})$ is the interval between the present time and the next transition. At the transition the system moves to state $\mathbf{x}'$ with probability $q(\mathbf{x}';\mathbf{x},\alpha)$. The above formula for α^* assumes that there is only one event α such that τ_α is the minimum. If there are two events, α^* and β^*, whose clock readings, $d^*_{\alpha^*}$ and $d^*_{\beta^*}$, both reach zero at τ^*, some additional rules should be applied to determine the state transition. Whitt [106] uses the event with the smallest index as the triggering event. Other rules such as the random selection can also be applied. However, in parametric study of a GSMP, the state transition based on a single triggering event usually results in discontinuity. To maintain the sample path continuity, it is convenient to define the state transition as the result of two consecutive transitions triggered by both events α^* and β^*. This point will be further discussed in Section 7.3.

The events associated with state $\mathbf{x}'$ are in set $B(\mathbf{x}')$. The clock readings after the transition are determined as follows. New clock readings are independently generated for each $\beta \in \mathcal{N}(\mathbf{x}',\mathbf{x},\alpha) := B(\mathbf{x}') - (B(\mathbf{x}) - \{\alpha\})$. The new clock reading for event $\beta \in \mathcal{N}(\mathbf{x}',\mathbf{x},\alpha)$ has a cumulative probability distribution (c.p.d.) $F(s;\mathbf{x}',\beta,\mathbf{x},\alpha)$. For events in both $B(\mathbf{x})$ and $B(\mathbf{x}')$ except for event α, the old clock readings are kept after the transition; i.e., for $\beta \in \mathcal{O}(\mathbf{x}',\mathbf{x},\alpha) := B(\mathbf{x}') \cap (B(\mathbf{x}) - \{\alpha\})$, set $d_\beta = d^*_\beta(\mathbf{x},\mathrm{d})$. For events in $B(\mathbf{x})$, but not in $B(\mathbf{x}')$, the clocks are set to quit; i.e., if $\beta \in (B(\mathbf{x}) - \{\alpha\}) - B(\mathbf{x}')$, then $d_\beta = \infty$ after the transition.

Let the GSMP be right-continuous, and let T_i, $\mathbf{x}_i$, β_i be the ith transition time, the ith state and the ith triggering event, respectively. Let $s_{\alpha,j}$ be the jth lifetime of event α. At T_i (after the transition), the service completion time for event $\alpha \in B(\mathbf{x}_i)$ is $\tau_{\alpha,i} = T_i + d_{\alpha,i}$, where $d_{\alpha,i}$ is the clocking reading (residual lifetime) of event α at T_i. We have $T_{i+1} = \tau_{\beta_{i+1},i}$. According to the clock setting rule, for $\alpha \in B(\mathbf{x}_{i+1})$ we have

$$\tau_{\alpha,i+1} = \begin{cases} \tau_{\alpha,i} & \textit{if } \alpha \in \mathcal{O}(\mathbf{x}_{i+1};\mathbf{x}_i,\beta_i) \\ \tau_{\beta_{i+1},i} + s_{\alpha,j} & \textit{if } \alpha \in \mathcal{N}(\mathbf{x}_{i+1};\mathbf{x}_i,\beta_i) \end{cases} \tag{7.1}$$

where $\tau_{\beta_{i+1},i} = T_{i+1}$ and $s_{\alpha,j}$ is the new lifetime of event α generated at T_{i+1}. The subscript j indicates that there are $j-1$ completions of event α before T_{i+1}. Equation (7.1) is similar to (3.9).

Let ξ represent the randomness of a sample path of the GSMP. The ith transition time, T_i, the ith state, $\mathbf{x}_i$, the ith triggering event, β_i, and the ith lifetime of α, $s_{\alpha,i}$, etc, are random variables depending on ξ and can be written as $T_i(\xi)$, $\mathbf{x}_i(\xi)$, $\beta_i(\xi)$, $s_{\alpha,i}(\xi)$, etc.

The process with $\sigma(\mathbf{x},\alpha) = 1$ for all $\mathbf{x}$ and α is called a *GSMP*, and the process with $\sigma(\mathbf{x},\alpha) \neq 1$ for some $\mathbf{x}$ and α is called a *GSMP with speeds*. To simplify the discussion, in this book we shall study only GSMP with $\sigma(\mathbf{x},\alpha) = 1$. The basic principles, however, apply to GSMP with speeds as well.

7.2 Event Coupling in a GSMP

As explained at the beginning of this chapter, in a GSMP, the event coupling is the concept that describes the interaction between events. In this section, we shall study this intrinstic property. The section consists of two main parts. In the first part, we classify, according to the event coupling properties, the GSMP into three classes, i.e., the strongly coupled, the hierachically coupled, and the decomposable GSMP. In a strongly coupled GSMP, every event at any state may affect every other event at any other state, either directly or through some other events. In a hierachically coupled GSMP, some events may affect others, but not vise versa. In a decomposable GSMP, the event-state set may be decomposed into subsets; the event-state in a subset may affect each other, but they cannot affect those in other subsets. In the second part of this section, we study the event-coupling on a sample path of a GSMP. The concept of event-coupling tree is introduced; based on it, the stably coupled GSMP are defined. The superposition and the exclusion of an event-coupling tree are investigated. These properties are the basis for the perturbation realization theory studied in the next section.

7.2.1 The Classification of Event Coupling

Thoughout this chapter, we make the following two assumptions:

Condition 7.1 *Every event in $B(\mathbf{x})$ may be a triggering event with a positive probability.*

Condition 7.2 *(Noninterruptive) For all $\mathbf{x} \in \Phi$, if $\alpha, \beta \in B(\mathbf{x})$, $\alpha \neq \beta$, and $q(\mathbf{x}'; \mathbf{x}, \alpha) > 0$, then $\beta \in B(\mathbf{x}')$.*

It is worthy mentioning that Condition 7.1 is nontrival. If the event lifetime distributions are finitely supported, the GSMP may reach some states at which the remaining lifetime of one event is definitely longer than the lifetimes of all the other events, even if that event is in $B(\mathbf{x})$ at the state. Condition 7.2 says that all events will die natually; i.e., the termination of any event will not kill other events. This Condition simplifies the discussion and is not crucial to the results of the chapter.

Define $\Sigma = \{(\mathbf{x}, \alpha) : \mathbf{x} \in \Phi,\ \alpha \in B(\mathbf{x}) \subseteq \mathcal{E}\}$ be the event-state space. For Jackson networks, $(\mathbf{x}, \alpha)$ takes the form $(\mathbf{n}, i)$, where i indicates that the event is the service completion of server i. We shall use $(\mathbf{x}, \alpha)$ to indicate that the event α is associated with state $\mathbf{x}$. For any two events $(\mathbf{x}, \alpha)$ and $(\mathbf{x}', \alpha')$, we define the *coupling coefficient*

$$\iota[(\mathbf{x}, \alpha); (\mathbf{x}', \alpha')] = \begin{cases} 1 & \textit{if } q(\mathbf{x}'; \mathbf{x}, \alpha) > 0 \textit{ and } \alpha' \in \mathcal{N}(\mathbf{x}'; \mathbf{x}, \alpha), \textit{ or} \\ & \alpha' = \alpha \textit{ and there is a } \beta \in B(\mathbf{x}) - \alpha \textit{ such} \\ & \textit{that } q(\mathbf{x}'; \mathbf{x}, \beta) > 0 \\ 0 & \textit{otherwise.} \end{cases}$$

If $\iota[(\mathbf{x}, \alpha); (\mathbf{x}', \alpha')] = 1$, we say that event $(\mathbf{x}', \alpha')$ is coupled to event $(\mathbf{x}, \alpha)$. The meaning of the coupling coefficient is quite simple. If $q(\mathbf{x}'; \mathbf{x}, \alpha) > 0$ and $\alpha' \in \mathcal{N}(\mathbf{x}'; \mathbf{x}, \alpha)$, then from Condition 7.1, α may trigger a transition from $\mathbf{x}$ to $\mathbf{x}'$. Thus, from the second part of (7.1), with α being the triggering event, the lifetime of $(\mathbf{x}', \alpha')$ is scheduled at the termination time of $(\mathbf{x}, \alpha)$. This means that if the termination time of $(\mathbf{x}, \alpha)$ is somehow postponed (or advanced) by a small amount, then the starting time and hence, the termination time of $(\mathbf{x}', \alpha')$, will also be postponed (or advanced) by the same amount. That is, a perturbation at $(\mathbf{x}, \alpha)$ may be *propagated* to $(\mathbf{x}', \alpha')$. We say that event $(\mathbf{x}, \alpha)$ *affects* $(\mathbf{x}', \alpha')$. Next, if $\alpha' = \alpha$, $\beta \in B(\mathbf{x}) - \alpha$, and $q(\mathbf{x}'; \mathbf{x}, \beta) > 0$, then by the first part of (7.1) and Condition 7.2, the event lifetime of α continues after the state transition triggered by β, and therefore, the changes in the termination time of $(\mathbf{x}, \alpha)$ will be passed to that of $(\alpha, \mathbf{x}')$. The above explanation shows that the coupling coefficient is simply a qualitative description of (7.1).

The matrix $\mathcal{J} = [\iota\{(\mathbf{x}, \alpha); (\mathbf{x}', \alpha')\}]_{(\mathbf{x}, \alpha), (\mathbf{x}', \alpha') \in \Sigma}$ is called the *(event-state) coupling matrix* of the GSMP. The coupling matrix contains all the information about which event can affect which other events.

Definition 7.1 *A GSMP is said to be strongly coupled, if its coupling matrix $\mathcal{J}$ is indecomposable.*

In a strongly coupled GSMP, all events in all states can affect each other, either directly or through some other events. Suppose a GSMP is not strongly coupled, then, by reordering the event-state pairs, $\mathcal{J}$ can be decomposed into the following form.

$$\mathcal{J} = \begin{bmatrix} \mathcal{J}_1 & O \\ \mathcal{J}_{2,1} & \mathcal{J}_2 \end{bmatrix},$$

where O is a matrix whose components are all zeros. Of course, $\mathcal{J}_1$ and $\mathcal{J}_2$ may be also decomposable.

Definition 7.2 *A GSMP is said to be hierachically coupled, if its coupling matrix $\mathcal{J}$ is decomposable and $\mathcal{J}_{2,1} \neq O$.*

In a hierachically coupled GSMP, events correponding to $\mathcal{J}_2$ can affect those corresponding to $\mathcal{J}_1$, but not vice versa. If $\mathcal{J}_{1,2} = O$, then the GSMP consists of two separate sets of event-states, the events in each set cannot affect those in the other set.

Example 7.1 We consider multiclass queuing networks with infinite buffer capacities and exponentially distributed service times. Let $\alpha_{(i,k)}$ be the event denoting the service completion of the customer (i, k). We assume that customers do not change their classes when traveling in the network and denote by $q_{i,j;k}$ the class k customers' routing probability from server i to j.

1. Consider the three-server two-class queuing network shown in Figure 6.1. The state process of this network is a GSMP which is strongly coupled because a change in the service completion time of any event can affect that of any other event through an idle period of a server.

2. Consider the two-server two-class queuing network shown in Figure 6.9. In the network class 1 customers never leave server 1 and hence it is never idle. Thus, a change in server 1's completion time may affect that of server 2 through its idle periods, but a change in server 2's completion time can never affect that of server 1. This is an example of a hierachically coupled GSMP.

3. Consider a two-server three-class network with one customer in each class and $q_{1,1;1} = 1$, $q_{2,2;2} = 1$, and $q_{1,2;3} = q_{2,1;3} = 1$. There are 4 states and 4 events for this system and 8 event-states. We shall not write out the 8×8 coupling matrix. It is, however, clear that an infinitesimal change in the service completion times of server 1, $\alpha_{(1,1)}$ and $\alpha_{(1,3)}$, cannot affect those of server 2, $\alpha_{(2,1)}$ and $\alpha_{(2,3)}$; and vice versa. Thus, $\mathcal{J}$ can be decomposed into two indecomposable matrices and $\mathcal{J}_{2,1} = O$.

It is worthwhile to note that the example shows that even when the network is physically connected the events can still be separated into two parts and the events in each of them do not affect those in the other.

In a hierachically coupled GSMP, the two matrices $\mathcal{J}_1$ and $\mathcal{J}_2$ may be further decomposable. Thus, the coupling matrix may have the following final form:

$$\mathcal{J} = \begin{bmatrix} \mathcal{J}_1 & O & O & \cdots & O \\ \mathcal{J}_{2,1} & \mathcal{J}_2 & O & \cdots & O \\ \cdot & \cdot & \cdot & \cdot & \cdot \\ \mathcal{J}_{J,1} & \mathcal{J}_{J,2} & \mathcal{J}_{J,3} & \cdots & \mathcal{J}_J \end{bmatrix},$$

where $\mathcal{J}_j$, $j = 1, \cdots, J$, are indecomposable and $\mathcal{J}_{j,i}$, $j > i$, may or may not be zero. Let Σ_j be the subset of Σ corresponding to $\mathcal{J}_j$, $j = 1, \cdots, J$. Since $\mathcal{J}_j$ is indecomposable, all the event-states $(\mathbf{x}, \alpha) \in \Sigma_j$ can be aggregated into one element, denoted as γ_j. The diagonal submatrix $\mathcal{J}_j$ can be replaced by one entry $\iota[\gamma_j; \gamma_j] = 1$, reflecting the fact that the events in Σ_j, can affect each other. For a pair of γ_i and γ_j, define $\iota[\gamma_j; \gamma_i] = 1$ if $\mathcal{J}_{j,i} \neq O$ and $\iota[\gamma_j; \gamma_i] = 0$ if $\mathcal{J}_{j,i} = O$. $\iota[\gamma_j; \gamma_i] = 1$ implies that there is at least one event $(\alpha_1, \mathbf{x}_1)$ in Σ_j and one event $(\alpha_2, \mathbf{x}_2)$ in Σ_i such that $\iota[(\alpha_1, \mathbf{x}_1); (\alpha_2, \mathbf{x}_2)] = 1$. Therefore, any changes in the completion times of any event $(\mathbf{x}, \alpha)$ in Σ_j can affect any event $(\mathbf{x}', \alpha')$ in Σ_i via a path $(\mathbf{x}, \alpha) \rightarrow (\mathbf{x}_1, \alpha_1) \rightarrow (\mathbf{x}_2, \alpha_2) \rightarrow (\mathbf{x}', \alpha')$. If $\iota[\gamma_j; \gamma_i] = 0$, then no event in Σ_j can affect directly the events in Σ_i. Of course, if $\iota[\gamma_j; \gamma_k] = 1$ and $\iota[\gamma_k; \gamma_i] = 1$, then an event in Σ_j can affect events in Σ_i via some events in Σ_k. In summary, the coupling matrix of a hierachical coupled GSMP can be reduced to a $J \times J$ matrix.

Finally, a strongly coupled GSMP should be distinguished from an *irreducible* GSMP, which means that every state $\mathbf{x} \in \Phi$ has a positive probability to reach any other state $\mathbf{x}' \in \Phi$. It is clear that a strongly coupled GSMP must be irreducible.

Lemma 7.1 *A strongly coupled GSMP is irreducible.*

The following example shows that, however, an irreducible GSMP may not be strongly coupled.

Example 7.2 Consider a GSMP with three states $\mathbf{x}$, $\mathbf{y}$ and $\mathbf{z}$, and three events α, β, and γ. The feasible event sets are $B(\mathbf{x}) = \{\alpha, \beta\}$, $B(\mathbf{y}) = \{\beta\}$, and $B(\mathbf{z}) = \{\alpha, \gamma\}$. The transition probabilities are $q(\mathbf{y}; \mathbf{x}, \alpha) = q(\mathbf{z}; \mathbf{x}, \beta) = q(\mathbf{z}; \mathbf{y}, \beta) = q(\mathbf{x}; \mathbf{z}, \gamma) = q(\mathbf{z}; \mathbf{z}, \alpha) = 1$. The GSMP is irreducible. We number the event-states as follows: $1 : (\mathbf{x}, \alpha)$, $2 : (\alpha, \mathbf{z})$, $3 : (\beta, \mathbf{x})$, $4 : (\beta, \mathbf{y})$, and $5 : (\gamma, \mathbf{z})$. Then the event coupling matrix of the GSMP is

$$\mathcal{J} = \left[\begin{array}{cc|ccc} 0 & 1 & 0 & 0 & 0 \\ 1 & 1 & 0 & 0 & 0 \\ \hline 0 & 0 & 0 & 1 & 0 \\ 0 & 1 & 0 & 0 & 1 \\ 0 & 0 & 1 & 0 & 1 \end{array}\right],$$

which is hierachically coupled.

7.2.2 The Event Coupling Tree

Now, we study the effect of event coupling on a sample path. Let

$$\begin{pmatrix} \mathbf{x}_0 \\ \beta_0 \end{pmatrix}, \begin{pmatrix} \mathbf{x}_1 \\ \beta_1 \end{pmatrix}, \cdots, \begin{pmatrix} \mathbf{x}_n \\ \beta_n \end{pmatrix} \tag{7.2}$$

represent a sequence of the state and triggering event pairs of a GSMP, with β_i, $i = 0, 1, \cdots, n$, denoting the triggering event from state $\mathbf{x}_i$ to $\mathbf{x}_{i+1}$. Consider an event $\alpha_0 \in B(\mathbf{x}_0)$ and let $\mathcal{E}_0 = \alpha_0$. We define for $i = 0, 1, \cdots, n-1$

$$\mathcal{E}_{i+1} = \begin{cases} \mathcal{E}_i & if\ \beta_i \notin \mathcal{E}_i \\ \{\mathcal{E}_i - \beta_i\} \cup \mathcal{N}(\mathbf{x}_{i+1}; \mathbf{x}_i, \beta_i) & if\ \beta_i \in \mathcal{E}_i \end{cases} \tag{7.3}$$

It is desirable to attach some physical meaning to help visualize the concept. Suppose that the initial event completion time of α_0 at state $\mathbf{x}_0$ is somehow delayed by a small amount Δ (i.e., $(\alpha_0, \mathbf{x}_0)$ obtains a *perturbation* Δ). $(\mathcal{E}_0, \mathbf{x}_0)$ denotes the set of events that have the perturbation. Because of event coupling described in (7.1), after a transition triggered by event β_0, the changes in the completion time of α_0 will affect those of the events in

$$\mathcal{E}_1 = \begin{cases} \alpha_0 & if\ \beta_0 \neq \alpha_0 \\ \mathcal{N}(\mathbf{x}_1; \mathbf{x}_0, \alpha_0) & if\ \beta_0 = \alpha_0. \end{cases}$$

More precisely, the completion times of the events in $\mathcal{E}_1$ will be delayed by the same amount Δ; i.e., the perturbation of $(\mathcal{E}_0, \mathbf{x}_0)$ will be *propagated* to $(\mathcal{E}_1, \mathbf{x}_1)$. The same explanation applies to $\mathcal{E}_i$, $i = 1, 2, \cdots, n$. $\mathcal{E}_i$ are called *perturbation sets,* and (7.3) is called the *perturbation propogation rule.* (7.3) can be viewed as a qualitative abstraction of (7.1). In (7.3), the actual values of the event lifetimes are ignored, and only the logical relation between event termination times are maintained.

The sequence $(\mathbf{x}_0, \mathcal{E}_0)$, $(\mathbf{x}_1, \mathcal{E}_1)$, $\cdots$, $(\mathbf{x}_n, \mathcal{E}_n)$, $\mathcal{E}_0 = \alpha_0$, is called an *event-coupling tree* with a root $(\mathbf{x}_0, \alpha_0)$. We can also define in a similar way an event-coupling tree with a root $(\mathbf{x}_0, \mathcal{E}_0)$ for any subset $\mathcal{E}_0 \subseteq B(\mathbf{x}_0)$. Based on any sample path of the GSMP and the definition in (7.3), a stochastic process $(\mathcal{E}_i, \mathbf{x}_i)$, called the *perturbation-state process,* can be constructed. The state space of this process is $2^{\mathcal{E}} \times \Phi$. The event-coupling tree is similar to the "tree representation" introduced in Suri [101].

Lemma 7.2 $\{(\mathbf{x}, B(\mathbf{x}))$, *all* $\mathbf{x}\}$ *and* $\{(\mathbf{x}, \emptyset)$, *all* $\mathbf{x}\}$ *are two absorbing sets of the perturbation-state process.*

Proof: If $\mathcal{E}_i = B(\mathbf{x}_i)$ on a sample path, then $\beta_i \in \mathcal{E}_i$ and the second part of (7.3) holds. Thus,

$$\begin{aligned}\mathcal{E}_{i+1} &= \{\mathcal{E}_i - \beta_i\} \cup \mathcal{N}(\mathbf{x}_{i+1};\mathbf{x}_i,\beta_i)\\ &= [B(\mathbf{x}_i) - \beta_i] \cup \{B(\mathbf{x}_{i+1}) - [B(\mathbf{x}_i) - \beta_i]\} = B(\mathbf{x}_{i+1}).\end{aligned}$$

From this, once the perturbation-state process reaches the set $\{(\mathbf{x}, B(\mathbf{x})),$ *all* $\mathbf{x}\}$, it stays in the set forever. Similarly, if $\mathcal{E}_i = \emptyset$, then $\mathcal{E}_{i+1} = \emptyset$. □

Theorem 7.1 *Superposition and exclusion properties hold for event-coupling trees. More precisely, let* $(\mathbf{x}_0, \mathcal{E}_0), (\mathbf{x}_1, \mathcal{E}_1), \cdots, (\mathbf{x}_n, \mathcal{E}_n)$; $(\mathbf{x}_0, \mathcal{E}_{1,0}), (\mathbf{x}_1, \mathcal{E}_{1,1}), \cdots, (\mathbf{x}_n, \mathcal{E}_{1,n})$; *and* $(\mathbf{x}_0, \mathcal{E}_{2,0}), (\mathbf{x}_1, \mathcal{E}_{2,1}), \cdots, (\mathbf{x}_n, \mathcal{E}_{2,n})$ *be the event-coupling trees on the same sample path* $\mathbf{x}_0, \mathbf{x}_1, \cdots, \mathbf{x}_n$ *with roots* $(\mathbf{x}_0, \mathcal{E}_0)$, $(\mathbf{x}_0, \mathcal{E}_{1,0})$, *and* $(\mathbf{x}_0, \mathcal{E}_{2,0})$, *respectively. We have*

1. *Superposition: If* $\mathcal{E}_0 = \mathcal{E}_{1,0} \cup \mathcal{E}_{2,0}$, *then* $\mathcal{E}_i = \mathcal{E}_{1,i} \cup \mathcal{E}_{2,i}$ *for all* $i \geq 0$; *and*

2. *Exclusion: If* $\mathcal{E}_{1,0} \cap \mathcal{E}_{2,0} = \emptyset$, *then* $\mathcal{E}_{1,i} \cap \mathcal{E}_{2,i} = \emptyset$ *for all* $i \geq 0$.

Proof: We prove the theorem by recursion.

1. Superposition: Suppose $\mathcal{E}_i = \mathcal{E}_{1,i} \cup \mathcal{E}_{2,i}$, which holds for $i = 0$. We discuss four cases. (i) $\beta_i \notin \mathcal{E}_{1,i} \cup \mathcal{E}_{2,i} = \mathcal{E}_i$. By the first part of (7.3), we have $\mathcal{E}_{i+1} = \mathcal{E}_i$, $\mathcal{E}_{1,i+1} = \mathcal{E}_{1,i}$, and $\mathcal{E}_{2,i+1} = \mathcal{E}_{2,i}$. Thus, the superposition holds for $i+1$. (ii) $\beta_i \in \mathcal{E}_{1,i} \cap \mathcal{E}_{2,i}$. Thus, $\beta_i \in \mathcal{E}_i$. By the second part of (7.3), we have

$$\mathcal{E}_{1,i+1} = (\mathcal{E}_{1,i} - \beta_i) \cup \mathcal{N}(\mathbf{x}_{i+1};\mathbf{x}_i,\beta_i),$$

and

$$\mathcal{E}_{2,i+1} = (\mathcal{E}_{2,i} - \beta_i) \cup \mathcal{N}(\mathbf{x}_{i+1};\mathbf{x}_i,\beta_i).$$

Thus,

$$\begin{aligned}\mathcal{E}_{1,i+1} \cup \mathcal{E}_{2,i+1} &= \{(\mathcal{E}_{1,i} - \beta_i) \cup (\mathcal{E}_{2,i} - \beta_i)\} \cup \mathcal{N}(\mathbf{x}_{i+1};\mathbf{x}_i,\beta_i)\\ &= \{(\mathcal{E}_{1,i} \cup \mathcal{E}_{2,i} - \beta_i)\} \cup \mathcal{N}(\mathbf{x}_{i+1};\mathbf{x}_i,\beta_i)\\ &= (\mathcal{E}_i - \beta_i) \cup \mathcal{N}(\mathbf{x}_{i+1};\mathbf{x}_i,\beta_i) = \mathcal{E}_{i+1}.\end{aligned}$$

(iii) $\beta_i \in \mathcal{E}_{1,i}$, $\beta_i \notin \mathcal{E}_{2,i}$. In this case, we have $\beta_i \in \mathcal{E}_i$, and

$$\mathcal{E}_{1,i+1} = (\mathcal{E}_{1,i} - \beta_i) \cup \mathcal{N}(\mathbf{x}_{i+1};\mathbf{x}_i,\beta_i), \quad \mathcal{E}_{2,i+1} = \mathcal{E}_{2,i}.$$

Again, from (7.3), we have

$$\begin{aligned}\mathcal{E}_{1,i+1} \cup \mathcal{E}_{2,i+1} &= \{\mathcal{E}_{1,i} - \beta_i\} \cup \mathcal{N}(\mathbf{x}_{i+1}; \mathbf{x}_i, \beta_i) \cup \mathcal{E}_{2,i} \\ &= \{\mathcal{E}_{1,i} \cup \mathcal{E}_{2,i} - \beta_i\} \cup \mathcal{N}(\mathbf{x}_{i+1}; \mathbf{x}_i, \beta_i) \\ &= \{\mathcal{E}_i - \beta_i\} \cup \mathcal{N}(\mathbf{x}_{i+1}; \mathbf{x}_i, \beta_i) = \mathcal{E}_{i+1}.\end{aligned}$$

(iv) $\beta_i \notin \mathcal{E}_{1,i}$, $\beta_i \in \mathcal{E}_{2,i}$. The proof is the same as (iii).

2. Exclusion: Suppose $\mathcal{E}_{1,i} \cap \mathcal{E}_{2,i} = \emptyset$, which holds for $i = 0$. Consider three cases. (i) $\beta_i \notin \mathcal{E}_{1,i} \cup \mathcal{E}_{2,i}$. We have $\mathcal{E}_{i+1} = \mathcal{E}_i$, $\mathcal{E}_{1,i+1} = \mathcal{E}_{1,i}$, and $\mathcal{E}_{2,i+1} = \mathcal{E}_{2,i}$. Thus, the exclusion property also holds for $i+1$. (ii) $\beta_i \in \mathcal{E}_{1,i}$. This implies $\beta_i \notin \mathcal{E}_{2,i}$. Thus, $\mathcal{E}_{2,i+1} = \mathcal{E}_{2,i}$ and $\mathcal{E}_{1,i+1} = \{\mathcal{E}_{1,i} - \beta_i\} \cup \mathcal{N}(\mathbf{x}_{i+1}; \mathbf{x}_i, \beta_i)$. Since $\mathcal{E}_{2,i} \in B(\mathbf{x}_i)$ and $\beta_i \notin \mathcal{E}_{2,i}$, we have $\mathcal{E}_{2,i} \in B(\mathbf{x}_i) - \beta_i$ and $\mathcal{E}_{2,i} \cap \mathcal{N}(\mathbf{x}_{i+1}; \mathbf{x}_i, \beta_i) = \emptyset$. Therefore,

$$\mathcal{E}_{1,i+1} \cap \mathcal{E}_{2,i+1} = \{(\mathcal{E}_{1,i} - \beta_i) \cup \mathcal{N}(\mathbf{x}_{i+1}; \mathbf{x}_i, \beta_i)\} \cap \mathcal{E}_{2,i} = \emptyset.$$

(iii) $\beta_i \in \mathcal{E}_{2,i}$. The proof is similar as that for (ii). □

From Theorem 7.1, trees with different roots do not overlap. Also, a tree that grows up on a set of different events is the union of the trees that grow up on each individual event in the set.

Corollary 7.1 *If $\mathcal{E}_{1,0} \subseteq \mathcal{E}_{2,0}$, then $\mathcal{E}_{1,i} \subseteq \mathcal{E}_{2,i}$ for all $i \geq 0$.*

Definition 7.3 *If on a sample path of a GSMP, the perturbation state eventually reaches the absorbing set $\{(\mathbf{x}, B(\mathbf{x})), \text{ all } \mathbf{x}\}$ (or $\{(\mathbf{x}, \emptyset), \text{ all } \mathbf{x}\}$); i.e.,*

$$\lim_{i\to\infty} \{\mathcal{E}_i - B(\mathbf{x}_i)\} = \emptyset \ (or \ \lim_{i\to\infty} \mathcal{E}_i = \emptyset),$$

then we say that the event-coupling tree prevails (or perishes), or the perturbation in $(\mathcal{E}_0, \mathbf{x}_0)$ is realized (or lost), on the sample path.

It is worthwhile to note that since $\mathcal{E}$ is a discrete space, the definition implies that $\mathcal{E}_i$ will reach its limit, either $B(\mathbf{x}_i)$ or $\emptyset$, in a finite number of steps.

Corollary 7.2 *If $\mathcal{E}_{0,1} \subseteq \mathcal{E}_{0,2}$ and the event-coupling tree rooted on $(\mathbf{x}_0, \mathcal{E}_{0,1})$ prevails on a sample path, then the event-coupling tree rooted on $(\mathbf{x}_0, \mathcal{E}_{0,2})$ prevails on the same sample path.*

Corollary 7.3 *If on a sample path a tree with a root $(\mathbf{x}_0, \alpha)$, $\alpha \in B(\mathbf{x}_0)$, prevails, then all the other trees with $(\mathbf{x}_0, \alpha')$, $\alpha' \in B(\mathbf{x}_0)$, $\alpha' \neq \alpha$ perish.*

The event-coupling tree provides a vital picture of the evolution of a perturbation. The perturbation realization, on the other hand, provides physical meanings to the concept. We shall use both terminologies in the remainder of this chapter.

Definition 7.4 *A GSMP is said to be stably coupled, if on a sample path, any perturbation* $(\mathcal{E}_0, \mathbf{x}_0)$ *will be either realized or lost with probability one. That is, if* $(\mathcal{E}_0, \mathbf{x}_0)$, $(\mathcal{E}_1, \mathbf{x}_1)$, $\cdots$, *is a tree rooted on* $(\mathcal{E}_0, \mathbf{x}_0)$, *then*

$$P\{\lim_{i\to\infty}[\mathcal{E}_i - B(\mathbf{x}_i)] = \emptyset\} + P[\lim_{i\to\infty}\mathcal{E}_i = \emptyset] = 1.$$

The state process of an irreducible multiclass queuing network, for example, is a stably coupled GSMP. In a stably coupled GSMP, if any set of events obtain a small perturbation, the GSMP will stabilize itself in the sense that eventually every event will obtain the same perturbation, or every event will lost the perturbation. In the former case, all the event completion times are perturbed by the same amount of time. In the latter, all the event completion times are not changed at all. In both cases, the relative positions of the event completion times are not changed. This shows that there is a strong connection in the system which synchronizes the event completion times.

Let us develop some sufficient conditions for the stably coupled GSMP. If $B(\mathbf{x})$ contains only one event, then $\mathbf{x}$ is called a *one-event state* (It is called a one-clock state and is used to prove the ergodicity of GSMP in Glasserman [57]). We have the following result.

Lemma 7.3 *If an irreducible GSMP has a one-event state* $\mathbf{x}$, *then the GSMP is stably coupled.*

Proof: Since the GSMP is irreducible, it will reach $\mathbf{x}$ with probability one. Let $\alpha = B(\mathbf{x})$ be the only event, and suppose that at the ith step the GMSP reaches $\mathbf{x}$. Then $\mathcal{E}_i \subseteq B(\mathbf{x}) = \alpha$. Thus, either $\mathcal{E}_i = \alpha = B(\mathbf{x})$ or $\mathcal{E}_i = \emptyset$. This proves the lemma. □

Although the lemma looks quite simple, it covers a considerably wide range of systems, including queuing networks in which every customer has a positive probability of visiting a common server Cao [16]. It is interesting to note that if an irreducible GSMP has a one-event state, the GSMP possesses a regenerative structure and is hence ergodic under some mild conditions (Glasserman [57]). For a GSMP which does not have a one-event state, the following Theorem helps to determine whether the GSMP is stably coupled.

Theorem 7.2 *An irreducible GSMP is stably coupled, if and only if there exists a sequence, denoted as (7.2), and an event* $\alpha_0 \in B(\mathbf{x}_0)$ *such that the corresponding event-coupling tree* $(\mathbf{x}_0, \alpha_0)$, $(\mathbf{x}_1, \mathcal{E}_1)$, $\cdots$, $(\mathbf{x}_n, \mathcal{E}_n)$ *prevails.*

Proof: First we prove the "if" part. Since the GSMP is irreducible, any sample path will, with probability one, visit the state $\mathbf{x}_0$ infinitely many times. Let p be the probability that the sequence (7.2) follows $\mathbf{x}_0$. By the assumption of the theorem, $p > 0$. This means that once the GSMP reaches $\mathbf{x}_0$, it will pass through the sequence in (7.2) with a positive probability and generate the event-coupling tree $(\mathbf{x}_0, \alpha_0)$, $(\mathbf{x}_1, \mathcal{E}_1)$, $\cdots$,$(\mathbf{x}_n, \mathcal{E}_n)$, which prevails. Thus, every sample path of the GSMP will, with probability one, pass through the event-coupling tree. Now consider any perturbation set $(\mathbf{x}'_0, \mathcal{E}'_0)$. Following (7.3), a tree $(\mathbf{x}'_0, \mathcal{E}'_0)$, $(\mathbf{x}'_1, \mathcal{E}'_1)$, $\cdots$, $(\mathbf{x}'_m, \mathcal{E}'_m)$, $\cdots$ will grow up on any sample path. With probability one, the tree will pass through $(\mathbf{x}_0, \alpha_0)$, $(\mathbf{x}_1, \mathcal{E}_1)$, $\cdots$, $(\mathbf{x}_n, \mathcal{E}_n)$. Suppose that at the jth transition, the tree reaches $\mathbf{x}_0$, i.e., $\mathbf{x}'_j = \mathbf{x}_0$, $\mathbf{x}'_{j+1} = \mathbf{x}_1$, $\cdots$, $\mathbf{x}'_{j+n} = \mathbf{x}_n$. If $\alpha_0 \in \mathcal{E}'_j$, then from Corollary 7.2, the event-coupling tree $(\mathbf{x}'_0, \mathcal{E}'_0)$, $\cdots$ prevails. If $\alpha_0 \notin \mathcal{E}_j$, then by Corollary 7.3, the tree perishes. In conclusion, any tree $(\mathbf{x}'_0, \mathcal{E}'_0)$, $(\mathbf{x}'_1, \mathcal{E}'_1)$, $\cdots$ will either prevail or perish with probability one.

The "only if" part is obvious for a GSMP in which there is at least one perturbation that is realized on a sample path. From Theorem 7.3 proved in the next section, this is true for all stably coupled GSMP. □

From Theorem 7.2, to determine that a GSMP is stably coupled, it is only necessary to find one sequence corresponding to an event-coupling tree that prevails. The event-coupling matrix $\mathcal{J}$ can be used to help in finding such a sequence. Lemma 7.3 can be viewed as a special case of Theorem 7.2. Theorem 6.1 for multiclass queuing networks and Theorem 6.4 are examples of the applications of Theorem 7.2.

The relationship between the stably coupled and the strongly couled GSMP will be discussed in the next section.

7.3 Sensitivity Analysis Based on Realization

Having introduced the concepts of event coupling and event-coupling trees, we are at a position to develop the sensitivity formulas for GSMP. The basic approach is the same as that for queuing network; but the asumptions for the analysis and the technicality in proof may be different. We shall only sketch these different parts.

7.3.1 Realization Factors

We use the elapsed time r_α, rather than the residual time d_α (or the clock reading), of an event α as a part of the mathematical state. If s_α is the event lifetime, then $r_\alpha = s_\alpha - d_\alpha$. Let $N = |\mathcal{E}|$ be finite, and $\mathbf{r}$ be an N-dimensional vector whose αth component is r_α. The system state can be denoted as $\mathcal{W}(t) = (\mathcal{X}(t), \mathcal{R}(t))$, where $\mathcal{X}(t)$ is the physical state of the GSMP, and $\mathcal{R}(t)$ is the vector of the elapsed lifetimes at time t. The state space of this stochastic process is $\Phi \times R_+^N$.

A close examination of the definition of a GSMP reveals that in order to use $\mathbf{w} = (\mathbf{x}, \mathbf{r})$ as a mathematical state, $F(s; \mathbf{x}', \beta, \mathbf{x}, \alpha)$ must not depend on α and $\mathbf{x}$. If the probability distribution depends on the previous event α and the previous state $\mathbf{x}$, then the distribution of the residual lifetime of an event will depend on the previous state. In such cases, the mathematical state should include the previous state and the previous event. This will make the results look unnecessarily complicated and we shall prefer not to do so.

Furthermore, we shall assume that $F(s; \mathbf{x}', \beta, \mathbf{x}, \alpha)$ does not depend on $\mathbf{x}'$ and denote it as $F(s; \mathbf{x}', \beta, \mathbf{x}, \alpha) = F_\beta(s)$. This assumption is not restrictive at all because one can always redefine β in a different $\mathbf{x}'$ as a new event so that each new event has its own probability distribution.

By Definition 7.4, in a stably coupled GSMP, a perturbation will be either realized or lost with probability one. Thus, for a stably coupled GSMP, we can define the realization probability $c(\mathbf{w}_0, \mathcal{E}_0)$ and the realization factor $c^{(f)}(\mathbf{w}_0, \mathcal{E}_0)$. Since the realization probability is defined only for stably coupled GSMP, all the results related to realization probabilities hold only for stably coupled GSMP. We shall not repeat this point in all our theorems.

Using the realization probability, we can easily establish the following lemma for the relationship between the strongly coupled and the stably coupled GSMP.

Lemma 7.4 *If* $c(\mathbf{x}_0, \mathbf{r}_0, \alpha_0) > 0$ *for all* $\mathbf{x}_0 \in \Phi$, *all* $\mathbf{r}_0$, *and all* $\alpha_0 \in B(\mathbf{x}_0)$, *then a stably coupled GSMP is strongly coupled.*

The lemma is self-explanatory. It is, however, worthwhile to note that if $c(\mathbf{x}_0, \mathbf{r}_0, \alpha_0) = 0$, then the event $(\mathbf{x}_0, \alpha_0)$ may not affect some events before it gets lost. The following example shows that the converse of Lemma 7.4 does not hold.

Example 7.3 Consider a GSMP with three states $\mathbf{x}$, $\mathbf{y}$ and $\mathbf{z}$, and three events α, β, and γ. The feasible event sets are $B(\mathbf{x}) = \{\alpha, \beta\}$, $B(\mathbf{y}) =$

$\{\beta, \gamma\}$, and $B(\mathbf{z}) = \{\gamma, \alpha\}$. The transition probabilities are $q(\mathbf{x}; \mathbf{x}, \beta) = q(\mathbf{y}; \mathbf{y}, \gamma) = q(\mathbf{z}; \mathbf{z}, \alpha) = 1$, and $q(\mathbf{y}; \mathbf{x}, \alpha) = q(\mathbf{z}; \mathbf{y}, \beta) = q(\mathbf{x}; \mathbf{z}, \gamma) = 1$. For this GSMP, all possible transitions and event-couplings can be illustrated by the following chart (the event above an arrow indicates the corresponding triggering event):

$$(\mathbf{x}, \alpha) \xrightarrow{\beta} (\mathbf{x}, \alpha) \xrightarrow{\alpha} (\gamma, \mathbf{y}) \xrightarrow{\gamma} (\gamma, \mathbf{y}) \xrightarrow{\beta} (\gamma, \mathbf{z}) \xrightarrow{\alpha} (\gamma, \mathbf{z}) \xrightarrow{\gamma}$$

$$(\beta, \mathbf{x}) \xrightarrow{\beta} (\beta, \mathbf{x}) \xrightarrow{\alpha} (\beta, \mathbf{y}) \xrightarrow{\gamma} (\beta, \mathbf{y}) \xrightarrow{\beta} (\alpha, \mathbf{z}) \xrightarrow{\alpha} (\alpha, \mathbf{z}) \xrightarrow{\alpha} (\mathbf{x}, \alpha).$$

The chart clearly indicates that the GSMP is strongly coupled but not stably coupled.

For further study, we require that all $F_\alpha(s)$ are absolutely continuous. Thus, the probability density functions $f_\alpha(s) = F'_\alpha(s)$ exist and $|f_\alpha(s)| < \infty$, $s < \infty$. The hazard rate of the lifetime of event α is

$$g_\alpha(r_\alpha) = \frac{f_\alpha(r_\alpha)}{1 - F_\alpha(r_\alpha)}.$$

We state our requirements for the distribution functions of the event lifetimes in the following condition.

Condition 7.3 *$F(s; \mathbf{x}', \beta, \mathbf{x}, \alpha) = F_\beta(s)$ does not depend on α, $\mathbf{x}$, and $\mathbf{x}'$; all $F_\beta(s) < \infty$ for $s < \infty$ and are absolutely continuous.*

For any N-dimensional vector $\mathbf{r}$ and any set $\mathcal{E}_0 \subseteq \mathcal{E}$, we define a vector $\mathbf{r}_{-\mathcal{E}_0}$ with components

$$\{\mathbf{r}_{-\mathcal{E}_0}\}_\alpha = \begin{cases} \{\mathbf{r}\}_\alpha & \text{if } \alpha \notin \mathcal{E}_0, \\ 0 & \text{if } \alpha \in \mathcal{E}_0. \end{cases}$$

Let $\Delta\mathbf{t}$ be a vector whose components are all Δt's. The sample performance function is defined in the same way as that for queuing networks, with f being a performance function. We assume $|f| < H < \infty$ holds.

With these notations and using the superposition and exclusion properties of the event-coupling tree, we can prove the following lemma and theorem.

Lemma 7.5

$$c^{(f)}(\mathcal{E}_0, \mathbf{x}, \mathbf{r}) = 0, \text{ if } \mathcal{E}_0 \notin B(\mathbf{x}). \tag{7.4}$$

$$c^{(f)}(B(\mathbf{x}), \mathbf{x}, \mathbf{r}) = f(\mathbf{x}). \tag{7.5}$$

If $\mathcal{E}_1 \cap \mathcal{E}_2 = \emptyset$ and $\mathcal{E}_1 \cup \mathcal{E}_2 = \mathcal{E}_3$, $\mathcal{E}_i \subseteq \mathcal{E}$, $i = 1, 2, 3$, then

$$c^{(f)}(\mathcal{E}_1, \mathbf{x}, \mathbf{r}) + c^{(f)}(\mathcal{E}_2, \mathbf{x}, \mathbf{r}) = c^{(f)}(\mathcal{E}_3, \mathbf{x}, \mathbf{r}). \tag{7.6}$$

and

$$\sum_{\beta \in B(\mathbf{x})} c^{(f)}(\beta, \mathbf{x}, \mathbf{r}) = f(\mathbf{x}). \tag{7.7}$$

Theorem 7.3 *Under Conditions 7.2 and 7.3, we have*

$$\begin{aligned} \sum_{\beta \in B(\mathbf{x})} \frac{\partial c^{(f)}(\alpha, \mathbf{x}, \mathbf{r})}{\partial r_\beta} &= c^{(f)}(\alpha, \mathbf{x}, \mathbf{r}) \sum_{\beta \in B(\mathbf{x})} g_\beta(r_\beta) \\ &- \sum_{\mathbf{x}' \in \Phi} \sum_{\beta \in B(\mathbf{x}) - \alpha} g_\beta(r_\beta) q(\mathbf{x}'; \mathbf{x}, \beta) c^{(f)}(\alpha, \mathbf{x}', \mathbf{r}_{-\mathcal{N}(\mathbf{x}';\mathbf{x},\beta)}) \\ &- \sum_{\mathbf{x}' \in \Phi} g_\alpha(r_\alpha) q(\mathbf{x}'; \mathbf{x}, \alpha) \{ \sum_{\gamma \in \mathcal{N}(\mathbf{x}';\mathbf{x},\alpha)} c^{(f)}(\gamma, \mathbf{x}', \mathbf{r}_{-\mathcal{N}(\mathbf{x}';\mathbf{x},\alpha)}) \\ &+ f(\mathbf{x}) - f(\mathbf{x}') \} \end{aligned} \tag{7.8}$$

for any $\alpha \in \mathcal{E}$, $\mathbf{x} \in \Phi$, and $\mathbf{r}$.

Applying (7.4)-(7.8) to closed queuing networks with a single class of customers, we can obtain the results in Chapter 5 as a special case.

7.3.2 Sample Sensitivities

The perturbation generation function is the same as that for queuing networks. Thus, the generation function for event α is

$$G_\alpha(r, \theta) = \frac{\partial}{\partial r} \left\{ \frac{\partial F_\alpha^{-1}(\rho, \theta)}{\partial \theta} \Big|_{\rho = F_\alpha(r, \theta)} \right\}.$$

Just like the network case, we need the following bounded condition.

Condition 7.4 *There is a positive number $K > \infty$ such that in a neighborhood of θ, $\theta \in [\theta_a, \theta_b]$, $|G_\alpha(r, \theta)| < K$ holds for all $r \geq 0$.*

To study the convergence of the sample derivative, we need the ergodicity of the GSMP.

Condition 7.5 *The GSMP is ergodic with a unique steady-state probability density function $p(\mathbf{w}) = p(\mathbf{x}, \mathbf{r})$.*

The ergodicity of a GSMP per se is a complex topic and has been discussed in many papers (e.g, Borovkov [6], [7] and [8], Whitt [106], Glynn [60], and Sigman [97] and [98]).

Theorem 7.4 *Under Conditions 7.3-7.5, the normalized sample derivative of the system performance,* $\frac{1}{\eta_L^{(I)}}\frac{\partial \eta_L^{(f)}(\xi,\theta)}{\partial \theta}$, *converges with probability one to the expected value of the product of the perturbation generation function and the realization factor. That is,*

$$\lim_{L\to\infty} \frac{1}{\eta_L^{(I)}} \frac{\partial \eta_L^{(f)}(\xi,\theta)}{\partial \theta} = E\{G_\alpha[r_\alpha(t),\theta]c^{(f)}(\alpha,\mathbf{x},\mathbf{r})\}$$

$$= \sum_{all\ \mathbf{x}} \int_{R_+^N} G_\alpha[r_\alpha(t),\theta]c^{(f)}(\alpha,\mathbf{x},\mathbf{r})p(\mathbf{x},\mathbf{r})d\mathbf{r}, \qquad w.p.1. \quad (7.9)$$

Proof: The approach of the proof is similar to that of Theorem 5.2 in Chapter 5. However, in a GSMP, there may not exist a one-event state such as $(N, 0, \cdots, 0)$ for closed networks discussed in Chapter 5. The first step in the proof is to establish

$$\lim_{L\to\infty} \frac{1}{T_L} \int_0^{T_L} G_\alpha[r_\alpha(t),\theta]\{c_L^{(f)}[\alpha, X(t), R(t), \xi, t]$$
$$-c^{(f)}[\alpha, X(t), R(t), \xi, t]\}dt = 0 \qquad w.p.1.$$

By Theorem 7.2, for a stably coupled GSMP, there exists a sequence $\mathcal{S}$, denoted in (7.2), such that the perturbation in $(\alpha_0, \mathbf{x}_0)$ is realized on the sequence. Let $L_0 = 0$ and $L_d = min\{l > L_{d-1},\ X_l = \mathbf{x}_0, X_{l+1} = \mathbf{x}_1, \cdots, X_{l+n} = \mathbf{x}_n\}$, with $\{\mathbf{x}_0, \mathbf{x}_1, \cdots, \mathbf{x}_n\}$ forming the sequence $\mathcal{S}$. Then $E(L_{d+1} - L_d) = mp < \infty$, $d = 1, 2, \cdots$, where m is the expected value of the number of steps between two consecutive visits to $\mathbf{x}_0$ and $p > 0$ is the probability that the sequence $\mathcal{S}$ follows $\mathbf{x}_0$. Since the GSMP is irreducible, $m < \infty$. Note that any perturbation generated before T_{L_d} will be either realized or lost when the GSMP reaches X_{L_d+n} at T_{L_d+n}. (If α_0 belongs to the perturbation set at T_{L_d}, then the perturbation will be realized; otherwise, it will be lost.) The remainder of the proof is the same as that for Theorem 5.2 and is omitted here. □

7.3.3 Sensitivity of Steady-State Performance

As explained in Chapter 5, the essential fact for the steady-state performance sensitivity is the unbiasedness of the sample derivatives, which largely depends on the continuity of the sample function.

We first give a condition under which the sample paths, and hence the sample functions, are continuous with respect to a parameter of a GSMP.

First, under Condition 7.3, the sample function $\eta_L^{(f)}$ is piecewise differentiable. The discontinuity occurs only when on (ξ,θ) there are two events, α and β, whose lifetimes terminate at the same time. Suppose that on $(\xi,\theta+\Delta\theta)$, $\Delta\theta > 0$, α terminates before β and α triggers a state transition from $\mathbf{x}_1$ to $\mathbf{x}_2$ and β, from $\mathbf{x}_2$ to $\mathbf{x}_3$. The GSMP will remain in the sequence $\mathbf{x}_1$, $\mathbf{x}_2$, $\mathbf{x}_3$ as $\Delta\theta \downarrow 0$. Thus, if we define the state transition at (ξ,θ), triggered by two events α and β simultaneously, as from $\mathbf{x}_1$ to $\mathbf{x}_3$, then the sample function $\eta_L^{(f)}$ is right-continuous at θ. Now, consider the sample path $(\xi,\theta+\Delta\theta)$ with $\Delta\theta < 0$, on which β terminates before α. Let $\mathbf{x}_1$, $\mathbf{x}_4$, $\mathbf{x}_5$ be the state sequence. As $\Delta\theta \uparrow 0$, the GSMP remains in the same sequence. Thus, $\eta_L^{(f)}$ is left-continuous only if we define the state transtion on (ξ,θ), triggered by both α and β simultaneously, to be from $\mathbf{x}_1$ to $\mathbf{x}_5$. Therefore, it is clear that the continuity of $\eta_L^{(f)}$ at θ holds if and only if $\mathbf{x}_5 = \mathbf{x}_3$.

The above discussion shows that for the purpose of sensitivity study, it is necessary to define the state transition at the time when two event lifetimes terminate simulaneously as the result of two consecutive transitions triggered by each event separately.

Condition 7.6 *On any sample path (ξ,θ) of a GSMP, the state of the GSMP after two transitions that occur simultaneously does not depend on the order of these two transitions.*

This Condition is equivalent to the *commuting condition* in Glasserman [57].

Lemma 7.6 *Under Conditions 7.3, 7.4, and 7.6, we have*

$$E\{\frac{\partial}{\partial\theta}\eta_L^{(f)}(\xi,\theta)|\mathbf{x}_0,\mathbf{r}_0\} = \frac{\partial}{\partial\theta}E\{\eta_L^{(f)}(\xi,\theta)|\mathbf{x}_0,\mathbf{r}_0\}.$$

Proof: From

$$\eta_L^{(f)} = \frac{1}{L}\sum_{l=0}^{L-1} f(X_l)(T_{l+1}-T_l)$$

and (5.20), we have

$$\begin{aligned} |\frac{\partial\eta_L^{(f)}}{\partial\theta}| &\le \frac{1}{L}\sum_{l=0}^{L-1}|f(X_l)|\{|\frac{\partial T_{l+1}}{\partial\theta}|+|\frac{\partial T_l}{\partial\theta}|\} \\ &< HK\frac{1}{L}\sum_{l=0}^{L-1}(T_{l+1}+T_l) < 2HKT_L. \end{aligned}$$

Under Condition 7.6, $\eta_L^{(f)}$ is a continuous function of θ. The above equation holds for both the right- and left-sided derivatives. The lemma then follows from the monotonicity of $T_L(\theta, \xi)$ and the same argument as for Lemma 5.4. (The monotonicity of the sample function is obvious if the perturbation generation function has the same sign for all r. For a discussion of the monotonicity of the mean, see Glasserman and Yao [59]). □

The following Condition is the same as Condition 5.4 in Chapter 5. In the theorem, $\pi(\mathbf{x}, \mathbf{r})$ is the steady-state distribution seen at the transition instants, which satisfies

$$\sum_{all\ \mathbf{x}} \int_{R_{\mathbf{x}}} \pi(\mathbf{x}, \mathbf{r}) d\mathbf{r} = 1. \tag{7.10}$$

Condition 7.7 *There exist two real numbers $r^* > 0$ and $K^* > 0$ such that $|\frac{\partial \pi(\mathbf{x},\mathbf{r})}{\partial \theta}| < K^* \pi(\mathbf{x}, \mathbf{r})$ for all $\mathbf{r} \in R^M_{r^*-}$, and the integration in (7.10) uniformly converges in a neighborhood of θ.*

Finall, we have

Theorem 7.5 *Under Conditions 7.3-7.7, the sample derivative of the system performance with respect to a service distribution parameter, based on a single sample path in $[0, T_L]$, converges to the derivative of the steady-state performance with probability one as the length of the sample path goes to infinity; moreover, the normalized derivative of the steady-state performance, $\frac{1}{\eta^{(I)}} \frac{\partial \eta^{(f)}}{\partial \theta}$, equals the expected value of the product of the perturbation generation function and the realization factor. That is,*

$$\begin{aligned}
\frac{1}{\eta^{(I)}} \frac{\partial \eta^{(f)}}{\partial \theta} &= \lim_{L \to \infty} \frac{1}{\eta_L^{(I)}} \frac{\partial \eta_L^{(f)}(\xi, \theta)}{\partial \theta} \\
&= E[G_\alpha(r_\alpha, \theta) c^{(f)}(\alpha, \mathbf{x}, \mathbf{r})] \qquad w.p.1 \\
&= \sum_{all\ \mathbf{x}} \int_{R_+^N} G_\alpha(r_\alpha, \theta) c^{(f)}(\mathbf{x}, \mathbf{r}, \alpha) p(\mathbf{x}, \mathbf{r}) d\mathbf{r}.
\end{aligned}$$

Chapter 8

Operational Sensitivity Analysis

In literature, there is an alternative approach to the study of queuing nwteorks, i.e., the operational analysis (Buzen and Denning [13], Denning and Buzen [53], and Dallery and David [51] and [52]). The basic principle of operational analysis is to analyze the behaviour of a queuing network during a finite period of time. The behaviour is characterized by means of operational variables, which are defined as the average of certain quantities over a finite period of time. Under some assumptions pertaining to the operational variables (i.e., the average behaviour of a system), relationships among performance parameters and the basic operational variables can be established. These assumptions are called *operational assumptions*. Especially, under some assumptions the proportion of the time that a system stays in a given state can be expressed as a product of the parameters relative to each station. This is a counterpart of the product-form solution in the stochastic approach.

In Dallery and Cao [50], the operational analysis is extended to study the asymptotic behaviour of a system, and the operational variables are defined as the average over an infinite period of time. In the paper, the product-form solution (Baskett et al, [4]), the arrival theorem (Lavenberg and Reiser [79] and Sevcik and Mitrani [95]), the Norton's aggregation theorem (Balsamo and Iazeolla [3] and Chandy et al. [37]), and Marie's approximate method (Marie [84]) are established using the operational setting.

The system dynamics reflects the system's behaviour on a sample path.

In particular, the realization probability measures the average behaviour of a perturbation on a sample path. Thus, it is natural that the sensitivity analysis based on the realization probability has an operational version. The objective of this chapter is to develop an operational approach to sensitivity analysis (shortly named as "operational sensitivity analysis") of closed queuing networks (Cao and Dallery [32]).

In Section 8.1, we give a brief review of the operational analysis. A diagram is given to depict the relation among the existing methods in studying queuing networks. In Section 8.2, we develop the basic theory for the operational sensitivity analysis. We shall study only the fundamental case, i.e., the sensitivity of throughput in a closed Jackson network. Operational formulas are derived for realization ratios and throughput derivatives using operational assumptions. The formulas have the same form as their stochastic counterparts.

8.1 Operational Variables and Operational Assumptions

In operational analysis, no probability and stochastic concepts are involved; we analyze the relationship among operational variables under some operational assumptions . Operational variables are the averages of certain quantities on a sample path; operational assumptions specify certain system behaviour through the operational variables. All operational variables can be directly measured by observing a sample path, and the operational assumptions can be directly verified for each sample path.

The following are some notation used by Denning and Buzen [53]. They will be also used in the next section to derive the results of operational sensitivity analysis of queuing networks.

$\mathbf{n}, \mathbf{m}$:	states of the system;
T:	total observation time;
$T(\mathbf{n})$:	total time during which the state of the system is $\mathbf{n}$;
$C(\mathbf{n}, \mathbf{m})$:	number of state transitions from $\mathbf{n}$ to $\mathbf{m}$;
$C(\mathbf{n})$:	number of times the system leaves state $\mathbf{n}$;

The first assumption used in operational analysis is (to be consistent with the terminologies used in the literature, in this chapter we shall use the word "assumption" instead of "condition.")

Assumption 8.1 *(Flow Balance): The system state at the initial time is* *the same as the state at the final time; i.e.,* $\mathcal{N}(0) = \mathcal{N}(T)$.

This assumption simply means that the number of entries to every state is the same as the number of exits from that state during the observation period. From this assumption, we can derive the following state balance equations.

$$\sum_{\mathbf{m}} \{p(\mathbf{m}) \frac{C(\mathbf{m}, \mathbf{n})}{T(\mathbf{m})}\} = p(\mathbf{n}) \sum_{\mathbf{m}} \frac{C(\mathbf{n}, \mathbf{m})}{T(\mathbf{n})}. \tag{8.1}$$

where $p(\mathbf{n}) = T(\mathbf{n})/T$ is the proportion of time the system is in state $\mathbf{n}$. It is obviously the counterpart of the state probability.

To simplify these equations, we need more notations and assumptions. Let

$\mathbf{n}_{i,j}$:	the state $(n_1, ..., n_i - 1, ..., n_j + 1, ..., n_M)$;
$C_{i,j}$:	number of customer transitions from server i to server j;
C_i:	number of customers served by server i;
$C_{i,j}(\mathbf{n})$:	number of customer transitions from server i to server j given that the system state was $\mathbf{n}$ before the transition;
$C_i(\mathbf{n})$:	number of customer completions at server i when the system state is $\mathbf{n}$;
$C_i(n_i)$:	number of customer completions at server i when n_i customers are present;
$T_i(n_i)$:	total time during which there are n_i customers in server i;

In the above definitions, we have

$$C_i = \sum_{j=1}^{M} C_{i,j},$$

and

$$C_i(\mathbf{n}) = \sum_{j=1}^{M} C_{i,j}(\mathbf{n}).$$

Now we define some other operational quantities:

$$q_{i,j}(\mathbf{n}) = \frac{C_{i,j}(\mathbf{n})}{C_i(\mathbf{n})}, \quad q_{i,j} = \frac{C_{i,j}}{C_i}; \tag{8.2}$$

$$s_i(\mathbf{n}) = \frac{T(\mathbf{n})}{C_i(\mathbf{n})}, \quad s_i(n_i) = \frac{T(n_i)}{C_i(n_i)}. \tag{8.3}$$

$q_{i,j}(\mathbf{n})$ is the routing frequency of a customer when the system state is $\mathbf{n}$. $q_{i,j}$ is the routing frequency. $s_i(\mathbf{n})$ and $s_i(n_i)$ are called *global pseudo-service time* and *local pseudo-service time* of server i, respectively. They are generally different from the real service time. For more details, see Dallery and David [51] and [52].

In order to get the product-form solution, we need three more assumptions.

Assumption 8.2 *(One-Step Behaviour): At any time instant only one customer transition may occur.*

Assumption 8.3 *(Routing Homogeneity): The routing frequencies $q_{i,j}(\mathbf{n})$ are independent of the system's state; i.e., $q_{i,j}(\mathbf{n}) = q_{i,j}$.*

Assumption 8.4 *(Global Pseudo-service Time Homogeneity): The global pseudo-service times at server i depend only on n_i; i.e., $s_i(\mathbf{n}) = s_i(n_i)$.*

Under Assumptions 8.1-8.4, for any strongly connected networks, the solution to the state balance equation (8.1) takes the following product form

$$p(\mathbf{n}) = \frac{1}{G_M(N)} \prod_{i=1}^{M} \prod_{n=0}^{n_i} y_i s_i(n), \tag{8.4}$$

where y_i is the visit ratio to server i; i.e., the solution to (2.43) with $q_{i,j}$ specified by (8.2), and $G_M(N)$ is a normalizing constant.

Note that in this form the pseudo-service time $s_i(n)$ does not necessary equal the mean service time of server i. Thus in order to use (8.4) to evaluate system performance without requiring an observation, another assumption is needed.

Assumption 8.5 *(Homogeneous Behaviour): The local pseudo-service time equals the mean service time; i.e., $s_i(n_i) = s_i$.*

Using this assumption we can replace $s_i(n)$ in (8.4) by the mean service time s_i. Then formula (8.4) looks exactly the same as the product-form solution for Jackson networks.

Discussions about Assumptions 8.1-8.5 have been appeared in several papers (Dallery and David [51] and [52], and Denning and Buzen [53]). Assumptions 8.1 and 8.2 are hardly ever restrictive. Assumptions 8.3 and 8.4 are often approximatively satisfied by queuing networks in which no blocking occurs between two stations. Assumption 8.5 is considered as the most restrictive assumption of operational analysis. However, from the results shown in Suri [100], the system performance is usually robust to the violation of this assumption.

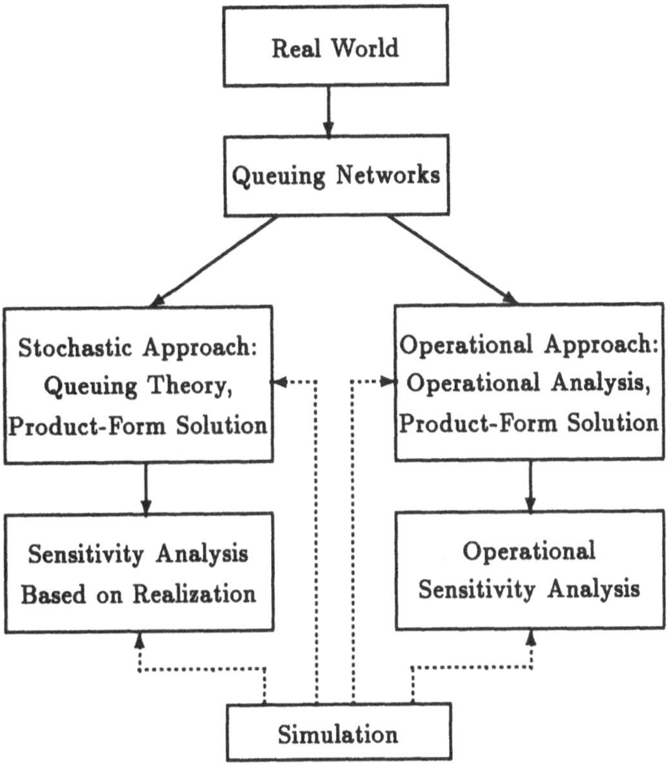

Figure 8.1: The Structure of Analysis

Structure of the Analysis

Figure 8.1 shows the basic structure of the analysis of queuing networks. A real world problem is first modelled mathematically as a queuing network. Then this queuing network can be analyzed using either the stochastic or the operational approach. To obtain the values of the derivatives of the system performance using the stochastic approach, there also exist two ways: by either directly taking derivatives from formulas if they are available, or by using sensitivity formulas based on realization probabilities. Counterparts of these two choices also exist for the operational approach. We shall develop operational sensitivity formulas in the next section.

For most systems the equations derived by either the stochastic or the operational approach are not easy to solve. Therefore, one has to run the system or a simulation in order to get a sample path of the system. The values measured on this sample path are operational variables such as the "proportion of time the system is in a given state" and the "operational realization ratio" (defined below). For a stochastic system, they are estimates of stochastic quantities such as the "state probability' and the "realization ratio." From these values we can calculate, for example, the sensitivities of system throughputs.

8.2 Operational Sensitivity Analysis

In this section, we apply the operational approach to the sensitivity of the throughput in single-class closed networks. We derive formulas using operational variables; the conditions for these formulas can be verified on a sample path. The approach helps us to understand why these conditions are needed.

8.2.1 Realization Ratios

First, we study the behaviour of a single perturbation on a sample path. Recall that in Chapter 3, we proved $S_{L,min} > 0$ with probability one. This corresponds to Assumption 8.2, the one-step behaviour assumption in operational approach. This assumption is needed to avoid possible confusion in perturbation propagation. For example, suppose that there is only one customer in server i at time instant t_- and the following two events happen simultaneously at time t: The customer in server i leaves this server and at the same time a customer arrives to server i from another server. In this case one can hardly determine whether propagation rules in Chapter 3 should be applied.

Let $X_1, X_2, \cdots, X_L$ be the sequence of states observed. X_L is the state before the last completion; i.e., the observation period is $(0, T_L]$ $(T = T_L)$. We also assume that the state preceding X_1 is known as $X_0 = X_l$ (flow balance). We define the following new notations:

X_l: the lth state of the path;
$L_{\mathbf{n}}$: the set of indices at which the system enters state $\mathbf{n}$;
$L_{\mathbf{m},\mathbf{n}}$: the set of indices at which the system enters state $\mathbf{n}$ and the previous state is $\mathbf{m}$;

The precise definitions are:

$$L_{\mathbf{n}} = \{\, l \in \{1, 2, ..., L\} : X_l = \mathbf{n}\}$$

and

$$\mathcal{L}_{\mathbf{m},\mathbf{n}} = \{\, l \in \{1, 2, ..., L\} : \; X_l = \mathbf{n}, X_{l-1} = \mathbf{m}\}.$$

$\mathcal{L}_{\mathbf{n}}$ and $\mathcal{L}_{\mathbf{m},\mathbf{n}}$ are subsets of the set of all positive integers. $\mathcal{L}_{\mathbf{m},\mathbf{n}}$ is a subset of $\mathcal{L}_{\mathbf{n}}$. The number of integers in set $\mathcal{L}_{\mathbf{n}}$ is $C(\mathbf{n})$, and the number of integers in set $\mathcal{L}_{\mathbf{m}.\mathbf{n}}$ is $C(\mathbf{m}, \mathbf{n})$.

We use $(\mathbf{n}, V, l)$ to denote that $X_l = \mathbf{n}$ and at the lth transition the servers in set $V \subseteq \Gamma$ have a perturbation, $l < L$. We are interested in the final effect of this perturbation on an "output" server, denoted as server M for convenience. That is, we study the sensitivity of the throughput of server M.

Definition 8.1 *The perturbation in state* $(\mathbf{n}, V, l)$ *is realized at server* M *at time* T_L*, if server* M *possesses the perturbation at time* T_L.

Definition 8.2 *(Realization Index) The realization index of the perturbation* $(\mathbf{n}, V, l)$ *at server* M *is:*

$$RI_{L,M}(\mathbf{n}, V, l) = \begin{cases} 1 & \text{if the perturbation is realized at server } M \text{ at } T_L; \\ 0 & \text{otherwise.} \end{cases} \tag{8.5}$$

Definition 8.3 *(Realization Ratio) The realization ratio of the perturbation* $(\mathbf{n},\ V)$ *in* $[0, T_L]$ *is*

$$R(\mathbf{n},\ V) = \frac{\sum_{l \in \mathcal{L}_{\mathbf{n}}} RI_{L,M}(\mathbf{n},\ V,\ l)}{C(\mathbf{n})}. \tag{8.6}$$

The operational variable "realization ratio" is the proportion of the perturbation $(\mathbf{n}, V)$ that are propagated to the output server at the end of the observation period. This operational concept is defined on a finite sample path of a system.

It is straigtforward that properties similar to (3.14)-(3.17) hold for realization ratios. In particular, we have

Lemma 8.1

$$\textit{If } n_i = 0, \textit{ then } R(\mathbf{n}, i) = 0; \tag{8.7}$$

$$R(\mathbf{n}, \Gamma) = 1; \tag{8.8}$$

If $V_1 \cap V_2 = \emptyset$ *and* $V_1 \cup V_2 = V_3$*, then*

$$R(\mathbf{n}, V_1) + R(\mathbf{n}, V_2) = R(\mathbf{n}, V_3); \tag{8.9}$$

and

$$\sum_{k=1}^{M} R(\mathbf{n}, k) = 1. \tag{8.10}$$

Proof: Equations (8.7) and (8.8) follows directly the definition. Equation (8.10) is a direct consequence of (8.9) and (8.8). The proof of (8.9) is similar to that of (3.16) and (3.24). Just note that (3.25) holds for $k = M$. Thus,

$$RI_{L,M}(\mathbf{n}, V_1, l) + RI_{L,M}(\mathbf{n}, V_2, l) = RI_{L,M}(\mathbf{n}, V_3, l), \qquad 0 > l < L.$$

Equation (8.9) follows from this and (8.6). □

Realization Ratio Equations

We are ready to derive the operational identity for operational realization ratios. First we define

Definition 8.4 *The realization ratio of the perturbation* $(\mathbf{n}, V)$ *given that the previous state is* $\mathbf{m}$ *is*

$$R_{\mathbf{m}}(\mathbf{n}, V) = \frac{\sum_{l \in L_{\mathbf{m},\mathbf{n}}} RI_{L,M}(\mathbf{n}, V, l)}{C(\mathbf{m}, \mathbf{n})}. \tag{8.11}$$

To establish the relation between realization ratios, we also need one more assumption.

Assumption 8.6 *(Realization Homogeneity to the Previous State) For any* $\mathbf{n}$ *and* V, $R_{\mathbf{m}}(\mathbf{n}, V)$ *is independent of* $\mathbf{m}$.

This assumption is intuitively reasonable. It says that, on the average, the realization of a perturbation at state $\mathbf{n}$ is independent of the preceding state. The stochastic counterpart of this assumption is the Markov property of the trajectory. Under this assumption we have

$$\begin{aligned} R_{\mathbf{m}}(\mathbf{n}, V) &= \frac{\sum_{l \in L_{\mathbf{m},\mathbf{n}}} RI_{L,M}(\mathbf{n}, V, l)}{C(\mathbf{m}, \mathbf{n})} \\ &= \frac{\sum_{\mathbf{m}} \sum_{l \in L_{\mathbf{m},\mathbf{n}}} RI_{L,M}(\mathbf{n}, V, l)}{\sum_{\mathbf{m}} C(\mathbf{m}, \mathbf{n})}. \end{aligned}$$

If the flow-balance Assumption 8.1 holds, then $C(\mathbf{n}) = \sum_{\mathbf{m}} C(\mathbf{m}, \mathbf{n})$. Thus,

$$R_{\mathbf{m}}(\mathbf{n}, V) = \frac{\sum_{l \in L_{\mathbf{n}}} RI_{L,M}(\mathbf{n}, V, l)}{C(\mathbf{n})} = R(\mathbf{n}, V), \qquad \textit{for all } \mathbf{n}. \tag{8.12}$$

Because $X_0 = X_L$, $C(\mathbf{m}, \mathbf{n})$ equals the number of integers in $L_{\mathbf{m},\mathbf{n}}$. If X_0 is not available, then the only error in (8.12) would be at most a $+1$ term missing in the numerator. This error is not significant if state $\mathbf{n} = X_1$ is visited frequently during the observation period. We shall see that the assumption $X_0 = X_L$ is also not essential. The corresponding situation in the stochastic approach is that a regenerative period is considered.

To prove the main theorem we need one more assumption.

Assumption 8.7 *(Realization Balance)*

$$RI_{L,M}(X_0, V, 0) = RI_{L,M}(X_L, V, L) \qquad \textit{for all } V.$$

We define

$$q(\mathbf{n}, \mathbf{m}) = \frac{C(\mathbf{n}, \mathbf{m})}{C(\mathbf{n})}.$$

This is the state transition frequency from state $\mathbf{n}$ to state $\mathbf{m}$. Now we can prove the following theorem.

Theorem 8.1 *Under the assumptions of realization homogeneity, flow balance, and realization balance, the realization ratios satisfy the following equations:*

$$R(\mathbf{n}, V) = \sum_{\mathbf{m}} R(\mathbf{m}, \ U)q(\mathbf{n}, \mathbf{m}) \quad \textit{for all } \mathbf{n}, \tag{8.13}$$

where $U = \{k :$ *server k has the perturbation after the transition* $(\mathbf{n}, V) \rightarrow \mathbf{m}\}$.

Proof: Let us first consider a state $\mathbf{n} \neq X_0$. We have

$$\begin{aligned} R(\mathbf{n}, V) &= \frac{\sum_{l \in L_{\mathbf{n}}} RI_{L,M}(\mathbf{n}, V, l)}{C(\mathbf{n})} \\ &= \frac{\sum_{\mathbf{m}} \sum_{l \in L_{\mathbf{n},\mathbf{m}}} RI_{L,M}(\mathbf{n}, V, l-1)}{C(\mathbf{n})} \end{aligned}$$

In the last equation $L_{\mathbf{n}}$ is decomposed into subsets $L_{\mathbf{n},\mathbf{m}}$. The perturbation at state $X_{l-1} = \mathbf{n}$ will be propagated to state $X_l = \mathbf{m}$. Now if U is the set of servers having the perturbation after a transition from state $\mathbf{n}$ with a perturbation set V to state $\mathbf{m}$, then the propagation rules imply $RI_{L,M}(\mathbf{n}, V, l-1) = RI_{L,M}(\mathbf{m}, U, l)$, $l \in L_{\mathbf{n},\mathbf{m}}$. Thus,

$$R(\mathbf{n}, V) = \frac{\sum_{\mathbf{m}} \sum_{l \in L_{\mathbf{n},\mathbf{m}}} RI_{L,M}(\mathbf{m}, \ U, \ l)}{C(\mathbf{n})}$$

$$= \sum_{\mathbf{m}} \left\{ \frac{\sum_{l \in L_{\mathbf{n},\mathbf{m}}} RI_{L,M}(\mathbf{m}, U, l)}{C(\mathbf{n}, \mathbf{m})} \right\} \frac{C(\mathbf{n}, \mathbf{m})}{C(\mathbf{n})}$$
$$= \sum_{\mathbf{m}} R_{\mathbf{n}}(\mathbf{m}, U) q(\mathbf{n}, \mathbf{m})$$
$$= \sum_{\mathbf{m}} R(\mathbf{m}, U) q(\mathbf{n}, \mathbf{m}). \tag{8.14}$$

The last equation is due to the realization homogeneity.

For $\mathbf{n} = X_0 = X_L$, the last state X_L is not included in the above term $\sum_{l \in L_{\mathbf{n},\mathbf{m}}} RI_{L,M}(\mathbf{m}, U, l)$, since T_{L+1} is not in the observation period. Also, the summation $\sum_{l \in L_{\mathbf{n}}}$ does not count T_0. Therefore,

$$R(X_0, V) = \frac{\sum_{\mathbf{m}} \sum_{l \in L_{\mathbf{n},\mathbf{m}} - \{L\}} RI_{L,M}(\mathbf{m}, U, l)}{C(\mathbf{n})} + \frac{RI_{L,M}(X_L, U, L)}{C(\mathbf{n})}$$
$$= \frac{\sum_{\mathbf{m}} \sum_{l \in L_{\mathbf{n},\mathbf{m}} - \{L\} + \{0\}} RI_{L,M}(\mathbf{m}, U, l)}{C(\mathbf{n})}$$

By the flow balance and realization balance assumptions, $X_0 = X_L$ and $RI_{L,M}(X_L, U, L) = RI_{L,M}(X_0, U, 0)$. Substituting this into the above equation, we have

$$R(X_0, V) = \frac{\sum_{\mathbf{m}} \sum_{l \in L_{\mathbf{n},\mathbf{m}}} RI_{L,M}(\mathbf{m}, U, l)}{C(\mathbf{n})}.$$

This is the same as the first equation in (8.14), which directly leads to (8.13). □

The realization balance assumption may appear strong. However, if this assumption and the assumption $X_0 = X_L$ are not satisfied, the error would be at most a +1 term missing on the numerator of $R(X_0, V)$. Again this is not significant if $C(\mathbf{n})$ is not too small. From a stochastic point of view, the realization balance assumption appears reasonable, since the first and the last states are the same. The restriction of this assumption in operational analysis comes from the fact that only a finite period of observation is considered. Another related issue is that some perturbation generated near the end of the observation which is not propagated to the output server at the end of this observation might be propagated to that server if we observed longer. Nevertheless, the operational analysis reflects the system behaviour in a finite observed period, and the boundary effect decreases as the length of the observation period increases.

Equation (8.13) is somewhat similar to the state space balance equations in (8.1). Theorem 8.1 and Equations (8.7)-(8.10) are sufficient to determine the realization ratio by the transition rate $q(\mathbf{n}, \mathbf{m})$. The only essential assumption needed is the assumption of realization homogeneity, which is in a sense weaker than the Markov assumption. Hence one can expect that the operational equations (8.13) apply to more general systems than their stochastic counterpart (3.19).

Realization Ratio Equations for Homogeneous Behaviour Systems

By the one-step behaviour assumption, state $\mathbf{n}$ can only transit to a neighboring state $\mathbf{n}_{i,j}$ in one step. Therefore, (8.13) is equivalent to

$$R(\mathbf{n}, V)C(\mathbf{n}) = \sum_{i,j} R(\mathbf{n}_{i,j}, U)C_{i,j}(\mathbf{n}),$$

where

$$C(\mathbf{n}) = \sum_{i:n_i \neq 0} C_i(\mathbf{n}) = \sum_i \epsilon(n_i)C_i(\mathbf{n}).$$

Dividing both sides of the above equation by $T(\mathbf{n})$, we obtain

$$R(\mathbf{n}, V)\sum_i \{\epsilon(n_i)\frac{C_i(\mathbf{n})}{T(\mathbf{n})}\} = \sum_{i,j}\{R(\mathbf{n}_{i,j}, U)\frac{C_{i,j}(\mathbf{n})}{C_i(\mathbf{n})}\frac{\epsilon(n_i)C_i(\mathbf{n})}{T(\mathbf{n})}\}.$$

By Assumptions 8.3 and 8.4, this is equivalent to

$$R(\mathbf{n}, V)\sum_i \frac{\epsilon(n_i)}{s_i(n_i)} = \sum_{i,j} R(\mathbf{n}_{i,j}, U)\frac{\epsilon(n_i)q_{i,j}}{s_i(n_i)}. \tag{8.15}$$

Now consider the case $V = k$ in Theorem 8.1; i.e., only the kth server is perturbed. The above equation becomes

$$R(\mathbf{n}, k)\sum_i \frac{\epsilon(n_i)}{s_i(n_i)} = \sum_{i,j} R(\mathbf{n}_{i,j}, U)\frac{\epsilon(n_i)q_{i,j}}{s_i(n_i)}, \tag{8.16}$$

where U is the set of servers that have the perturbation after a customer transition from server i to server j, given that before this transition server k has the perturbation. So if $i = k$ and $n_j = 0$ then $U = \{k, j\}$, otherwise $U = \{k\}$. Note that by convention if $n_i = 0$ then $R(\mathbf{n}, i) = 0$ for all i.

Thus if $i = k$, and $n_k = 1$ we can still consider that $k \in U$ since after the customer transition $n_k = 0$. Hence,

$$R(\mathbf{n}, k) \sum_{i=1}^{M} \frac{\epsilon(n_i)}{s_i(n_i)} = \sum_{i,j:i \neq k \ or \ n_j \neq 0} \{\epsilon(n_i) \frac{q_{i,j}}{s_i(n_i)} R(\mathbf{n}_{i,j}, k)\}$$
$$+ \sum_{j=1}^{M} [1 - \epsilon(n_j)] R(\mathbf{n}_{k,j}, \{j, k\}) \frac{q_{k,j}}{s_k(n_k)}.$$

By Lemma 8.1, we have

$$R(\mathbf{n}_{k,j}, \{j, k\}) = R(\mathbf{n}_{k,j}, j) + R(\mathbf{n}_{k,j}, k).$$

Substituting this into the second term in the above equation and combining with the first term, we obtained

$$R(\mathbf{n}, k) \sum_{i=1}^{M} \frac{\epsilon(n_i)}{s_i(n_i)} = \sum_{i,j} \epsilon(n_i) \frac{q_{i,j}}{s_i(n_i)} R(\mathbf{n}_{i,j}, k) + \sum_{j=1}^{M} [1 - \epsilon(n_j)] \, R(\mathbf{n}_{k,j}, j) \frac{q_{k,j}}{s_k(n_k)}. \tag{8.17}$$

If, furthermore, we take the assumption of homogeneous behaviour (Assumption 8.5), i.e., $s_i(n_i) = s_i$ for all n_i, then (8.17) appears exactly the same as (3.19).

8.2.2 Operational Sensitivity Formulas

The realization ratio describes the way in which a single perturbation at state $\mathbf{n}$ will affect the throughput of the output server. Using the realization ratio, we shall derive in this subsection the operational formulas for the sensitivity of the throughput with respect to the changes in service times. Since operational analysis does not use any stochastic concept, the "mean" service time does not make sense. Thus, we assume that the service rate of a server, say server v, changes. This is equivalent to that every service time at server v changes by an amount proportional to its service time. Let S be the generic variable for the service time. We assume $\Delta S = \gamma S$. We choose γ to be so small such that γT_L is less than $S_{L,min}$.

Let $S_l(\mathbf{n})$ denote the length of the lth state $X_l = \mathbf{n}$. We have the following equation which is similar to (3.38):

$$\Delta T_L = \sum_{all \ l} \gamma S_l(\mathbf{n}) RI_{L,M}(\mathbf{n}, v, l).$$

The throughput of server M in $[0, T_L]$ is defined as

$$\eta_{L,M} = \frac{C_M}{T},$$

with $T = T_L$. Thus,

$$\begin{aligned} \frac{S}{\eta_{L,M}} \frac{\partial \eta_{L,M}}{\partial S} &= \lim_{\gamma \to 0} \frac{1}{\gamma} \frac{\Delta \eta_{L,M}}{\eta_{L,M}} = - \lim_{\gamma \to 0} \frac{1}{\gamma} \frac{\Delta T_L}{T} \\ &= - \sum_{all\ l} \frac{S_l(\mathbf{n})}{T} RI_{L,M}(\mathbf{n}, v, l) \\ &= - \sum_{\mathbf{n}} \sum_{l \in L_{\mathbf{n}}} \{ \frac{S_l(\mathbf{n})}{T} RI_{L,M}(\mathbf{n}, i, l) \}. \end{aligned} \tag{8.18}$$

To continue the analysis, we need one more assumption.

Assumption 8.8 *(Realization Homogeneity to State Lengths) The average length of those states* $\mathbf{n}$ *whose realization indices are 1 equals the average length over all states* $\mathbf{n}$.

To express this in a formula, we have

$$\frac{\sum_{l \in L_{\mathbf{n}}} S_l(\mathbf{n}) RI_{L,M}(\mathbf{n}, v, l)}{\sum_{l \in L_{\mathbf{n}}} RI_{L,M}(\mathbf{n}, v, l)} = \frac{\sum_{l \in L_{\mathbf{n}}} S_l(\mathbf{n})}{C(\mathbf{n})}. \tag{8.19}$$

In the above formula, $\sum_{l \in L_{\mathbf{n}}} RI_{L,M}(\mathbf{n}, v, l)$ is the number of states $\mathbf{n}$ at which the realization indices are 1. By definition,

$$T(\mathbf{n}) = \sum_{l \in L_{\mathbf{n}}} S_l(\mathbf{n}). \tag{8.20}$$

Substituting (8.19), (8.20) and (8.6) into (8.18) yields the equation in the following theorem.

Theorem 8.2 *Under the assumption of the realization homogeneity to state lengths, it holds*

$$\begin{aligned} \frac{S}{\eta_{L,M}} \frac{\partial \eta_{L,M}}{\partial S} &= - \sum_{\mathbf{n}} \{ \frac{1}{T} \frac{T(\mathbf{n})}{C(\mathbf{n})} \sum_{l \in L_{\mathbf{n}}} RI_{L,M}(\mathbf{n}, v, l) \} \\ &= - \sum_{\mathbf{n}} p(\mathbf{n}) R(\mathbf{n}, v). \end{aligned} \tag{8.21}$$

Equation (8.21) is the operational counterpart of (3.67).

Assumption 8.8 is not very restrictive. If a system's state process satisfies the Markov property, then the realization ratio should depend only on the state of the system, and should be independent of the length of the state. Thus, Assumption 8.8 holds for these systems. Just like other operational assumptions, this assumption describes the average behaviour of a system. Therefore, it may apply to some systems for which the Markov property does not hold. Assumption 8.6 asserts that realization ratios do not depend on the previous state, while Assumption 8.8 requires that on the average they do not depend on the length of the current state.

Discussion

We have developed the operational sensitivity analysis for single-class closed queuing networks. The main results are the operational equations (8.13), (8.17) and (8.21). To develop (8.13), essentially we need only one assumption, i.e., the assumption of realization homogeneity to the previous state (Assumption 8.6). For (8.21), we need only the assumption of realization homogeneity to the state lengths (Assumption 8.8). Both assumptions correspond to the Markov property in the stochastic approach. These two assumptions describe the average behaviour of a system. Apparently, the Markov property of a system implies more than these two assumptions. In this sense, operational analysis tells us precisely what features of the Markov property are used to develop each equation. It shows that Equations (8.13) and (8.21) only require that on the average the realization ratio does not depend on the previous state or the length of the current state, respectively. Equation (8.17) requires three more assumptions which are the same as those leading to the product-form solution of the steady-state queue-length distribution. The stochastic counterpart of (8.17) is the realization probability equation for closed Jackson networks.

If a system satisfies the Markov property, then Assumptions 8.3, 8.4, and 8.5 hold, as the length of the observation period goes to infinity. Thus, for systems with the Markov property, the product-form solution, the realization equation (8.13), and the sensitivity equation (8.21) hold. Therefore, compared with stochastic approach, operational analysis describes the special features required by each formula, respectively, through different operational assumptions. In this sense, the Markov property is too strong for each formula. Figure 8.2 illustrates this idea. Of course, the above discussion is only a descriptive explanation which indicates some advantages of the operational approach.

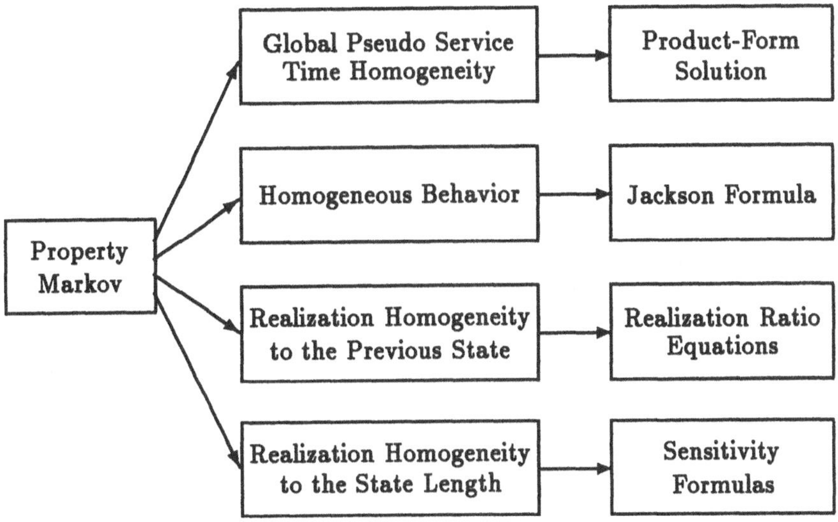

Figure 8.2: The Relation between the Markov Property and the Formulas

Compared with the stochastic approach, operational sensitivity analysis has all the advantages that operational analysis possesses. For example, no information about distributions is needed. However, neither (8.13) nor (8.17) is easy to solve. Therefore, simulation is necessary to estimate the sensitivity. It is worthwhile to note that (8.18) does not require any assumptions; thus it applies to any general systems. We shall not continue the discussion about the comparison between the stochastic and the operational approaches; interested readers are referred to Sevcik and Klawe [94] and the papers cited at the beginning of this chapter.

Chapter 9

Explicit Formulas, Algorithms, and Applications to Optimization

Using the dynamic point of view, we have developed a new approach to the sensitivity analysis of queuing networks. The main concept introduced in this book is the realization probability (or the realization factor), which measures the final effect of a perturbation on a performance measure and can be viewed as the fundamental "building block" of the sensitivity analysis. We have shown that the steady-state performance sensitivity can be expressed as the mean of the product of the perturbation generation function and the realization factor; the generation function is simply a constant if the service time distribution is exponential. A set of linear equations has been developed for the realization factors for different types of networks.

In Chapter 3, we solved the realization probability equations for the symmetric Jackson networks. In general, however, no closed-form solution has been found. Thus, numerical methods or simulation are needed to obtain the performance sensitivities. Many algorithms have been developed to estimate the performance sensitivity by using a single sample path (Glasserman [57] and Ho and Cao [68]); these algorithms can be viwed as efficient ways to estimate the mean of the product of the generation function and the realization factors.

In this chapter, we show that using the aggregation method, one can obtain explicit formulas for the performance sensitivities of load-dependent closed Jackson networks by solving the realization factor equations for a two-server network. The formulas are expressed in a form similar to Buzen's algorithm and hence are computationally tractable. The computational complexity of the algorithm is comparable to that of the formulas obtained by directly taking the derivative of the product-form formulas (see, e.g., Williams and Bhandiwad [105], Tay and Suri [103], Liu and Nain [80], and Cao and Ma [35]).

The explicity formulas obtained in this chapter have a very interesting form; i.e., the performance derivatives with respect to a service rate can be factored into a product of a positive quantity and a quantity whose sign depends only on other service rates of the networks. This result has an important application in optimization. By this result, some optimal control policies (such as the bang-bang control and the threshold policy), which are otherwise difficult to establish, become intuitively clear. We shall use this result to develop some new optimal control policies.

Another result derived in this chapter, which is of interest on its own, is the invariance property of the performance sensitivity for Norton's aggregation (see Theorem 9.2). It is well known that in a closed network consisting of M load-dependent servers, servers 1 to $M-1$ can be replaced by an equivalent server, resulting in an equivalent two-server network (the equivalent server and server M) in which server M has the same marginal steady-state distribution as in the original network. In this chapter, we prove that the sensitivities of the performance measures with respect to the mean service rate of server M in the equivalent network are the same as those in the original network, provided that the performance functions of the equivalent network are properly chosen.

In Section 9.1, we first solve analytically the realization factor equations for a two-server cyclic network and obtain the performance sensitivity formulas for the two-server network. In Section 9.2, we prove the sensitivity-invariance property for Norton's aggregation; we use the property to obtain the sensitivity formulas for M-server load-dependent networks. In Section 9.3, we develop convolution algorithms for the sensitivity formulas; these algorithms are in a similar form as Buzen's (Buzen [14]). The results in these three sections are based on Cao and Ma [33]. In Section 9.4, we discuss the application of these formulas to optimization problems (see Ma and Cao [83] and [82]).

9.1 A Two-Server Cyclic Network

The Realization Factor Equations

Consider a two-server cyclic network in which $M = 2$, $q_{1,1} = q_{2,2} = 0$, $q_{1,2} = q_{2,1} = 1$. The service requirement of each server is exponentially distributed with mean 1, and the service rates depend on the numbers of customers in each server. We denote the service rates of servers 1 and 2 as μ_{n_1} and λ_{n_2}, respectively. Since $n_1 + n_2 = N$, we can denote the system state as $\mathbf{n} = n_2$. The realization factor equations (4.14) and (4.17) become

$$c^{(f)}(N,1) = c^{(f)}(0,2) = 0, \tag{9.1}$$

and

$$c^{(f)}(k,1) + c^{(f)}(k,2) = f(k), \qquad k = 1,2,\cdots,N, \tag{9.2}$$

where $f(k) = f[(N-k,k)]$ is the performance function. Let $k = N$ in (9.2); we have

$$c^{(f)}(N,2) = f(N). \tag{9.3}$$

Specifying (4.19) to the two-server network, together with (9.1) and (9.3), gives the following matrix equation:

$$A\mathbf{x} = \mathbf{b}, \tag{9.4}$$

where

$$A = \begin{bmatrix} \lambda_1 + \mu_{N-1} & -\mu_{N-1}\frac{\lambda_1}{\lambda_2} & 0 & 0 & \cdots \\ -\lambda_2\frac{\mu_{N-2}}{\mu_{N-1}} & \lambda_2 + \mu_{N-2} & -\mu_{N-2}\frac{\lambda_2}{\lambda_3} & 0 & \cdots \\ \cdots & \cdots & \cdots & \cdots & \cdots \\ \cdots & -\lambda_k\frac{\mu_{N-k}}{\mu_{N-k+1}} & \lambda_k + \mu_{N-k} & -\mu_{N-k}\frac{\lambda_k}{\lambda_{k+1}} & \cdots \\ \cdots & \cdots & \cdots & \cdots & \cdots \\ \cdots & 0 & -\lambda_{N-2}\frac{\mu_2}{\mu_3} & \lambda_{N-2} + \mu_2 & -\mu_2\frac{\lambda_{N-2}}{\lambda_{N-1}} \\ 0 & \cdots & \cdots & -\lambda_{N-1}\frac{\mu_1}{\mu_2} & \lambda_{N-1} + \mu_1 \end{bmatrix}$$

in which the entries represented by dots are all zeros, and

$$\mathbf{x} = (x_1, x_2, \cdots, x_{N-1})^T,$$

with $x_k = c^{(f)}(k,2)$, $k = 1,2,\cdots,N-1$, and T denoting transpose, and

$$\mathbf{b} = (\lambda_1 a_1, \lambda_2 a_2, \cdots, \lambda_{N-1} a_{N-1})^T,$$

where

$$a_k = f(k) - \frac{\mu_{N-k}}{\mu_{N-k+1}} f(k-1), \qquad k = 1,2,\cdots,N-2, \tag{9.5}$$

and

$$a_{N-1} = \frac{\mu_1}{\lambda_N} f(N) + [f(N-1) - \frac{\mu_1}{\mu_2} f(N-2)]. \tag{9.6}$$

The solution

Let $\mathbf{b}_i = (0, \cdots, 0, \lambda_i a_i, 0, \cdots, 0)^T$, and denote the solution to the equation $A\mathbf{x}_i = \mathbf{b}_i$ as $\mathbf{x}_i = (x_{1,i}, x_{2,i}, \cdots, x_{N-1,i})^T$. Then the solution to (9.4) is

$$x_k = \sum_{i=1}^{N-1} x_{k,i}, \qquad k = 1, 2, \cdots, N-1. \tag{9.7}$$

We shall first solve the equation

$$A\mathbf{x}_{N-1} = \mathbf{b}_{N-1}. \tag{9.8}$$

Note that $(b_{N-1})_k = 0$ for $k = 1, 2, \cdots, N-2$. Working on the equations in (9.8) from the first one downward, we get

$$\frac{x_{k,N-1}}{x_{k+1,N-1}} = \frac{\lambda_k \alpha_k}{\lambda_{k+1} \alpha_{k+1}}, \qquad k = 1, 2, \cdots, N-2, \tag{9.9}$$

where $\alpha_1 = 1$ and

$$\alpha_{k+1} = 1 + \sum_{i=0}^{k-1} \prod_{j=0}^{i} \frac{\lambda_{k-j}}{\mu_{N-k+j}}, \qquad k = 1, 2, \cdots, N-1. \tag{9.10}$$

The last equation in (9.8) is

$$-\lambda_{N-1} \frac{\mu_1}{\mu_2} x_{N-2,N-1} + (\lambda_{N-1} + \mu_1) x_{N-1,N-1} = \lambda_{N-1} a_{N-1}. \tag{9.11}$$

From (9.9) and (9.11), we get

$$x_{k,N-1} = a_{N-1} \frac{\lambda_k}{\mu_1} \frac{\alpha_k}{\alpha_N}, \qquad k = 1, 2, \cdots, N-1.$$

Next, we solve for the equation

$$A\mathbf{x}_1 = \mathbf{b}_1. \tag{9.12}$$

Working on the equations in (9.12) from the last one upward, we get

$$\frac{x_{k,1}}{x_{k-1,1}} = \frac{\mu_{N-k} \beta_{N-k}}{\mu_{N-k+1} \beta_{N-k+1}}, \qquad k = 2, \cdots, N,$$

where $\beta_1 = 1$ and

$$\beta_{k+1} = 1 + \sum_{i=0}^{k-1} \prod_{j=0}^{i} \frac{\mu_{k-j}}{\lambda_{N-k+j}}. \tag{9.13}$$

From this equation and the last equation in (9.12), we obtain

$$x_{k,1} = a_1 \frac{\mu_{N-k}}{\mu_{N-1}} \frac{\beta_{N-k}}{\beta_N}, \qquad k = 1, 2, \cdots, N-1.$$

Now, consider the equation

$$A\mathbf{x}_i = \mathbf{b}_i. \tag{9.14}$$

Using the same method as for (9.8) and (9.12), we get

$$\frac{x_{j,i}}{x_{j-1,i}} = \frac{\lambda_j \alpha_j}{\lambda_{j-1}\alpha_{j-1}}, \qquad j = 2, 3, \cdots, i,$$

or

$$x_{1,i} : x_{2,i} : \cdots : x_{i-1,i} : x_{i,i} = \lambda_1\alpha_1 : \lambda_2\alpha_2 : \cdots : \lambda_{i-1}\alpha_{i-1} : \lambda_i\alpha_i. \tag{9.15}$$

Similarly, we have

$$\frac{x_{j,i}}{x_{j+1,i}} = \frac{\mu_{N-j}\beta_{N-j}}{\mu_{N-j-1}\beta_{N-j-1}}, \qquad j = i, i+1, \cdots, N-2,$$

or

$$\begin{aligned} & x_{i,i} : x_{i+1,i} : \cdots : x_{N-2,i} : x_{N-1,i} \\ = \; & \mu_{N-i}\beta_{N-i} : \mu_{N-i-1}\beta_{N-i-1} : \cdots : \mu_2\beta_2 : \mu_1\beta_1. \end{aligned} \tag{9.16}$$

The ith equation in (9.14) is

$$-\lambda_i \frac{\mu_{N-i}}{\mu_{N-i+1}} x_{i-1,i} + (\lambda_i + \mu_{N-i}) x_{i,i} - \mu_{N-i} \frac{\lambda_i}{\lambda_{i+1}} x_{i+1,i} = \lambda_i a_i. \tag{9.17}$$

Solving (9.15)-(9.17), we obtain

$$\left(-\frac{\mu_{N-i}}{\lambda_i} \frac{\lambda_{i-1}}{\mu_{N-i+1}} \frac{\alpha_{i-1}}{\alpha_i} + 1 + \frac{\mu_{N-i}}{\lambda_i} - \frac{\mu_{N-i-1}}{\lambda_{i+1}} \frac{\beta_{N-i-1}}{\beta_{N-i}}\right) x_{i,i} = a_i. \tag{9.18}$$

With (9.10) and (9.13), it is easy to check that the following two equations hold:

$$\frac{\mu_{N-i-1}}{\lambda_{i+1}} \beta_{N-i-1} = \beta_{N-i} - 1, \tag{9.19}$$

and

$$\frac{\lambda_i}{\mu_{N-i}}\alpha_i = \alpha_{i+1} - 1. \tag{9.20}$$

Using these two equations, we can reduce (9.18) to

$$\begin{aligned} x_{i,i} &= a_i \frac{\lambda_i}{\mu_{N-i}} \frac{\alpha_i \beta_{N-i}}{\alpha_{i+1} + \beta_{N-i} - 1} \\ &= a_i \frac{\alpha_i \beta_{N-i}}{\alpha_i + \beta_{N-i+1} - 1}. \end{aligned} \tag{9.21}$$

Therefore, based on (9.15)-(9.16), the solution to (9.14) is

$$x_{k,i} = \begin{cases} a_i \frac{\lambda_k}{\lambda_i} \frac{\alpha_k \beta_{N-i}}{\alpha_i + \beta_{N-i+1} - 1}, & k = 1, 2, \cdots, i; \\ a_i \frac{\mu_{N-k}}{\mu_{N-i}} \frac{\alpha_i \beta_{N-k}}{\alpha_i + \beta_{N-i+1} - 1}, & k = i+1, \cdots, N-1. \end{cases} \tag{9.22}$$

It is interesting to note that although (9.8) and (9.12) take forms different from (9.14), the solution to (9.14) reduces to those to (9.8) and (9.12) when $i = 1$ and $i = N - 1$, respectively. Finally, using (9.7), we get the solution to (9.4) as follows:

$$\begin{aligned} c^{(f)}(k, 2) &= \lambda_k \alpha_k \sum_{i=k}^{N-1} \frac{a_i}{\lambda_i} \frac{\beta_{N-i}}{\alpha_i + \beta_{N-i+1} - 1} \\ &+ \mu_{N-k}\beta_{N-k} \sum_{i=1}^{k-1} \frac{a_i}{\mu_{N-i}} \frac{\alpha_i}{\alpha_i + \beta_{N-i+1} - 1} \\ & \qquad k = 1, 2, \cdots, N-1. \end{aligned} \tag{9.23}$$

The Sensitivity Formula

The steady-state probabilities are (Kleinrock [76])

$$p(0) = \Big(1 + \sum_{j=1}^{N} \prod_{i=1}^{j} \frac{\mu_{N-i+1}}{\lambda_i}\Big)^{-1}, \tag{9.24}$$

and

$$p(k) = p(0) \prod_{i=1}^{k} \frac{\mu_{N-i+1}}{\lambda_i}, \qquad k = 1, 2, \cdots, N. \tag{9.25}$$

Specifying (4.46) to the two-server cyclic network, we obtain easily the normalized derivatives of the customer-average performance with respect to the service rates of server 2 for the two-server cyclic queuing network.

Theorem 9.1 *In a two-server cyclic network, the normalized derivatives of the customer-average performance measure with respect to the mean service rates of server 2 are*

$$\frac{\lambda_k}{\eta^{(f)}}\frac{\partial \eta^{(f)}}{\partial \lambda_k} = -p(k)c^{(f)}(k,2), \qquad k = 0, \cdots, N, \tag{9.26}$$

where $p(k)$, $k = 0, 1, 2, \cdots, N$, *are the steady-state probabilities shown in (9.24) and (9.25), and* $c^{(f)}(k,2)$, $k = 0, 1, 2, \cdots, N$, *are the realization factors shown in (9.1), (9.3) and (9.23). In the formulas,* a_i, α_i, *and* β_i, $i = 1, 2, \cdots, N$, *are defined in (9.5), (9.6), (9.10) and (9.13).*

The derivatives of the time-average performance measure with respect to the service rates of server 2 are

$$\lambda_k \frac{\partial \eta_T^{(f)}}{\partial \lambda_k} = p(k)\{\eta_T^{(f)} c(k,2) - c^{(f)}(k,2)\}, \qquad k = 0, \cdots, N. \tag{9.27}$$

Example 9.1 As a special case, we consider load-independent servers; i.e., $\mu_k = \mu$ and $\lambda_k = \lambda$, for all k. Let $\rho = \frac{\mu}{\lambda} = \frac{1}{\sigma}$. Suppose $\lambda \neq \mu$. Then

$$p(k) = \frac{\rho^k(1-\rho)}{1-\rho^{N+1}} \qquad k = 0, 1, \cdots, N.$$

For system throughput, $f = 1$ for all n. We have $a_k = f(k) - f(k-1) = 0$, $k = 1, 2, \cdots, N-2$, and $a_{N-1} = \frac{\mu}{\lambda} = \rho$. Therefore, from (9.23), we have

$$c(k,2) = \frac{a_{N-1}\alpha_k}{\alpha_{N-1} + \beta_2 - 1} = \frac{\alpha_k}{\alpha_N}, \qquad k = 1, 2, \cdots, N-1.$$

and

$$\begin{aligned} \alpha_k &= 1 + \sigma + \sigma^2 + \cdots + \sigma^{k-1} \\ &= \frac{1-\sigma^k}{1-\sigma} = \frac{1}{\rho^{k-1}}\frac{1-\rho^k}{1-\rho}, \qquad k = 1, 2, \cdots, N. \end{aligned}$$

Thus the realization factors for the throughput of the two-server network are

$$c(k,2) = \rho^{N-k}\frac{1-\rho^k}{1-\rho^N}, \qquad k = 1, 2, \cdots, N.$$

For such a network, the elasticity of the throughput with respect to λ is

$$\frac{\lambda}{\eta}\frac{\partial \eta}{\partial \lambda} = \sum_{k=1}^{N} p(k)c(k,2)$$

$$= \sum_{k=1}^{N} \frac{\rho^N(1-\rho)(1-\rho^k)}{(1-\rho^N)(1-\rho^{N+1})}$$

$$= \frac{N\rho^N - (N+1)\rho^{N+1} + \rho^{2N+1}}{(1-\rho^N)(1-\rho^{N+1})}.$$

This formula can be easily verified by directly taking the derivative of the throughput

$$\begin{aligned} \eta &= [1-p(0)]\mu \\ &= \frac{\rho - \rho^{N+1}}{1-\rho^{N+1}}\mu \\ &= \lambda \frac{1-\rho^N}{1-\rho^{N+1}}. \end{aligned}$$

Finally, if $\lambda = \mu$, then $p(k) = \frac{1}{N+1}$ and $c(k,2) = \frac{k}{N}$. The system reduces to the symmetric network discussed in Examples 3.1 and 3.3, and $\frac{\lambda}{\eta}\frac{\partial \eta}{\partial \lambda} = \frac{1}{2}$.

9.2 Explicit Formulas for Load-Dependent Closed Jackson Networks

9.2.1 Norton's Aggregation for Sensitivities

We have solved the performance sensitivity problem for a two-server network. To extend the results to M-server networks, we can use the aggregation technique. From Chandy et al. [37], the first $M-1$ servers in the network can be replaced by an equivalent, load-dependent server so that the resultant two-server network has the same marginal steady-state distribution as the original M-server network (see Section 2.3). In this section, we shall prove that the performance function of the two-server equivalent network can be so chosen that the performance sensitivities with respect to server M's mean service rates are the same for both the equivalent and the original networks.

By the aggregation results, the marginal probability of server M, $p(n_M) = \sum_{n_1,\cdots,n_{M-1}} p(n_1, n_2, \cdots, n_M)$ equals the marginal probability of the server with the same service rates as server M in a two-server cyclic queuing network in which the other server has the following service rates:

$$\mu_n = y_M \frac{G_{M-1}(n-1)}{G_{M-1}(n)}$$

$$= y_M \frac{\sum_{n_1+\cdots+n_{M-1}=n-1} \prod_{i=1}^{M-1} y_i^{n_i}/A_i(n_i)}{\sum_{n_1+\cdots+n_{M-1}=n} \prod_{i=1}^{M-1} y_i^{n_i}/A_i(n_i)},$$
$$n = 1, 2, \cdots, N. \qquad (9.28)$$

We shall call the two-server cyclic network, with one server having service rates $\mu_{M,n}$ and the other μ_n, $n = 1, 2, \cdots, N$, the equivalent network. Next, let us define the performance function of the equivalent network. For the original network, we have

$$\begin{aligned} \eta_T^{(f)} &= \sum_{all\ \mathbf{n}} f(\mathbf{n})p(\mathbf{n}) \\ &= \sum_{n_M} \{ \sum_{n_1,\cdots,n_{M-1}} f(n_1, n_2, \cdots, n_M)p(n_1, n_2, \cdots, n_M)\} \\ &= \sum_{n_M} f'(n_M)p(n_M), \end{aligned} \qquad (9.29)$$

where

$$\begin{aligned} p(n_M) &= \sum_{n_1,\cdots,n_{M-1}} p(n_1, n_2, \cdots, n_M) \\ &= \frac{y_M^{n_M}}{A_M(n_M)} \frac{G_{M-1}(N - n_M)}{G_M(N)} \end{aligned}$$

is the marginal probability of the number of customers in server M (Buzen [14]), and

$$\begin{aligned} f'(n_M) &= \sum_{n_1,\cdots,n_{M-1}} f(n_1, \cdots, n_M) \frac{p(n_1, \cdots, n_M)}{p(n_M)} \\ &= \frac{1}{G_{M-1}(N - n_M)} \sum_{n_1,\cdots,n_{M-1}} f(n_1, \cdots, n_M) \prod_{i=1}^{M-1} \frac{y_i^{n_i}}{A_i(n_i)} \\ &= E[f(n_1, n_2, \cdots, n_M)|n_M] \end{aligned} \qquad (9.30)$$

is the conditional mean of the performance measure, given that the number of customers in server M is n_M. We define $f'(n_M)$, $n_M = 0, 1, 2, \cdots, N$, as the performance function of the equivalent network. By (9.29), the time-average performance $\eta_T^{(f)}$ is the same for both the original and the equivalent networks. The customer-average performance thus defined for the equivalent network, $\eta_e^{(f)} = \eta_T^{(f)}/\eta_e$, is different from that of the original network because the throughput η_e is different from that of the original network, η. Despite this, we have the following theorem.

Theorem 9.2 *(Norton's Theorem for Sensitivity) The elasticities of the performance measures with respect to $\mu_{M,n}$ (while keeping μ_n fixed) for the equivalent network are equal to those (while keeping $\mu_{i,n}$, $i = 1, 2, \cdots, M-1$, fixed) for the original network; i.e.,*

$$\frac{\mu_{M,k}}{\eta_T^{(f)}} \left\{ \frac{\partial \eta_T^{(f)}}{\partial \mu_{M,k}} \right\}_{fix\ \mu_n} = \frac{\mu_{M,k}}{\eta_T^{(f)}} \left\{ \frac{\partial \eta_T^{(f)}}{\partial \mu_{M,k}} \right\}_{fix\ \mu_{i,n},\ i=1,..,M-1} \tag{9.31}$$

and

$$\frac{\mu_{M,k}}{\eta_e^{(f)}} \left\{ \frac{\partial \eta_e^{(f)}}{\partial \mu_{M,k}} \right\}_{fix\ \mu_n} = \frac{\mu_{M,k}}{\eta^{(f)}} \left\{ \frac{\partial \eta^{(f)}}{\partial \mu_{M,k}} \right\}_{fix\ \mu_{i,n},\ i=1,..,M-1} . \tag{9.32}$$

Proof: We first prove (9.31). By (2.44), the conditional probability of $(n_1, \cdots, n_{M-1})$ given n_M is

$$\begin{aligned} p(n_1, n_2, \cdots, n_M | n_M) &= \frac{p(n_1, \cdots, n_M)}{p(n_M)} \\ &= \frac{1}{G_{M-1}(N - n_M)} \prod_{i=1}^{M-1} \frac{y_i^{n_i}}{A_i(n_i)}. \end{aligned} \tag{9.33}$$

From (9.30) and (9.33), the conditional mean $f'(n_M)$ does not depend on $\mu_{M,k}$, $k = 0, 1, 2, \cdots, N$. Therefore, by (9.29), we have

$$\begin{aligned} \left\{ \frac{\partial \eta_T^{(f)}}{\partial \mu_{M,k}} \right\}_{fix\ \mu_{i,n}} &= \sum_{all\ \mathbf{n}} f(\mathbf{n}) \left\{ \frac{\partial p(\mathbf{n})}{\partial \mu_{M,n}} \right\}_{fix\ \mu_{i,n}} \\ &= \sum_{n_M} f'(n_M) \left\{ \frac{\partial p(n_M)}{\partial \mu_{M,n}} \right\}_{fix\ \mu_{i,n}} . \end{aligned}$$

By Norton's theorem (Chandy et al. [37]) and the fact that μ_n in the equivalent network does not depend on $\mu_{M,k}$, we have

$$\left\{ \frac{\partial p(n_M)}{\partial \mu_{M,n}} \right\}_{fix\ \mu_{i,n}} = \left\{ \frac{\partial p(n_M)}{\partial \mu_{M,n}} \right\}_{fix\ \mu_n} .$$

Thus,

$$\begin{aligned} \left\{ \frac{\partial \eta_T^{(f)}}{\partial \mu_{M,k}} \right\}_{fix\ \mu_{i,n}} &= \sum_{n_M} f'(n_M) \left\{ \frac{\partial p(n_M)}{\partial \mu_{M,n}} \right\}_{fix\ \mu_n} \\ &= \left\{ \frac{\partial \eta_T^{(f)}}{\partial \mu_{M,k}} \right\}_{fix\ \mu_n} . \end{aligned}$$

This leads to (9.31). For (9.32), we have $\eta^{(f)} = \frac{\eta_T^{(f)}}{\eta}$ and $\eta_e^{(f)} = \frac{\eta_T^{(f)}}{\eta_e}$. Thus,

$$\frac{\mu_{M,k}}{\eta^{(f)}} \frac{\partial \eta^{(f)}}{\partial \mu_{M,k}} = \frac{\mu_{M,k}}{\eta_T^{(f)}} \frac{\partial \eta_T^{(f)}}{\partial \mu_{M,k}} - \frac{\mu_{M,k}}{\eta} \frac{\partial \eta}{\partial \mu_{M,k}}. \tag{9.34}$$

A similar equation holds for the equivalent network. Therefore, (9.32) holds if

$$\frac{\mu_{M,k}}{\eta_e} \left\{ \frac{\partial \eta_e}{\partial \mu_{M,k}} \right\}_{fix\ \mu_n} = \frac{\mu_{M,k}}{\eta} \left\{ \frac{\partial \eta}{\partial \mu_{M,k}} \right\}_{fix\ \mu_{i,n}}. \tag{9.35}$$

Now, according to the definition,

$$\begin{aligned} \eta &= \sum_{\mathbf{n}} p(\mathbf{n})\mu(\mathbf{n}) = \sum_{\mathbf{n}} [\sum_{i=1}^{M} \epsilon(n_i)\mu_{i,n_i}] p(\mathbf{n}) \\ &= \sum_{i=1}^{M} \{\sum_{\mathbf{n}} [\epsilon(n_i)\mu_{i,n_i} p(\mathbf{n})]\} \\ &= \sum_{i=1}^{M} \sum_{k=1}^{N} \mu_{i,k} p(n_i = k) \\ &= \sum_{i=1}^{M} \eta_i, \end{aligned}$$

where $p(n_i = k) = \sum_{n_i=k} p(\mathbf{n})$ and $\eta_i = \sum_{k=1}^{N} \mu_{i,k} p(n_i = k)$ is the steady-state throughput of server i in the original network. We have

$$\begin{aligned} \eta_M &= \sum_{k=1}^{N} \mu_{M,k} p(n_M = k) \\ &= \sum_{k=1}^{N} y_M \frac{G_{M-1}(N-k)}{G_M(N)} \frac{y_M^{k-1}}{A_M(k-1)} \\ &= y_M \frac{G_M(N-1)}{G_M(N)}. \end{aligned}$$

Similarly,

$$\eta_i = y_i \frac{G_M(N-1)}{G_M(N)}.$$

Thus,

$$\eta = \frac{G_M(N-1)}{G_M(N)} (\sum_{i=1}^{M} y_i).$$

On the other hand, in the equivalent network,

$$\eta_e = \sum_{i=1}^{2} \eta_{e,i},$$

where $\eta_{e,i}$ is the throughput of server i, $i = 1, 2$, in the equivalent network. By the network structure, we have $\eta_{e,1} = \eta_{e,2}$. By Norton's theorem,

$$\eta_{e,2} = \eta_M = y_M \frac{G_M(N-1)}{G_M(N)}.$$

Therefore,

$$\frac{\eta}{\eta_e} = \frac{\sum_{i=1}^{M} y_i}{2y_M} = \alpha,$$

where α is a constant, independent of $\mu_{M,k}$. From this, it is easy to check that (9.35) holds; thus, (9.32) holds, and the proof is completed. □

It should be noted that the aggregation of the performance sensitivity does not yield any savings on computation. This is clear from (9.30), which shows that $G_{M-1}(N - n_M)$, etc, are needed for calculating $f'(n_M)$. In fact, the same statement is true for Norton's equivalent in the sense of marginal probability distributions.

9.2.2 The Explicit Sensitivity formulas

Theorem 9.2 shows that we can obtain the performance sensitivity of an M-server network via the equivalent two-server network. Therefore, the performance sensitivity formulas of an M-server network can be obtained by combining Theorems 9.1 and 9.2. In this section, we shall derive the performance sensitivity formulas by using these two theorems and Buzen's algorithm for steady-state probabilities.

Let $p_n(n_M = k)$, $k \le n$, be the probability of $n_M = k$ in a network with a population of n. We have (Buzen [14])

$$p_n(n_M = k) = \frac{y_M^k}{A_M(k)} \frac{G_{M-1}(n-k)}{G_M(n)}. \tag{9.36}$$

Then the equivalent service rates expressed in (9.28) can be written as

$$\mu_n = \mu_{M,N-n} \frac{p_{N-1}(n_M = N-n)}{p_{N-1}(n_M = N-n-1)}, \qquad n = 1, 2, \cdots, N-1. \tag{9.37}$$

In the equivalent two-server network, server M corresponds to server 2. Thus, we let $\lambda_i = \mu_{M,i}$ for $i = 0, 1, 2, \cdots, N$. By (9.37), we have

$$\begin{aligned}\prod_{j=0}^{i} \frac{\lambda_{k-j}}{\mu_{N-k+j}} &= \prod_{j=0}^{i} \frac{p_{N-1}(n_M = k-j-1)}{p_{N-1}(n_M = k-j)} \\ &= \frac{p_{N-1}(n_M = k-i-1)}{p_{N-1}(n_M = k)}.\end{aligned}$$

From this equation and (9.10), we get

$$\begin{aligned}\alpha_{k+1} &= 1 + \sum_{i=0}^{k-1} \frac{p_{N-1}(n_M = k-i-1)}{p_{N-1}(n_M = k)} \\ &= \frac{p_{N-1}(n_M \le k)}{p_{N-1}(n_M = k)}.\end{aligned} \tag{9.38}$$

Similarly,

$$\begin{aligned}\prod_{j=0}^{i} \frac{\mu_{k-j}}{\lambda_{N-k+j}} &= \prod_{j=0}^{i} \frac{p_{N-1}(n_M = N-k+j)}{p_{N-1}(n_M = N-k+j-1)} \\ &= \frac{p_{N-1}(n_M = N-k+i)}{p_{N-1}(n_M = N-k-1)}.\end{aligned}$$

From this equation and (9.13), we get

$$\begin{aligned}\beta_{k+1} &= 1 + \sum_{i=0}^{k-1} \frac{p_{N-1}(n_M = N-k+i)}{p_{N-1}(n_M = N-k-1)} \\ &= \frac{p_{N-1}(n_M \ge N-k-1)}{p_{N-1}(n_M = N-k-1)}.\end{aligned} \tag{9.39}$$

Thus,

$$\begin{aligned}\beta_{N-i+1} + \alpha_i &= \frac{p_{N-1}(n_M \ge i-1)}{p_{N-1}(n_M = i-1)} + \frac{p_{N-1}(n_M \le i-1)}{p_{N-1}(n_M = i-1)} \\ &= \frac{1 + p_{N-1}(n_M = i-1)}{p_{N-1}(n_M = i-1)}.\end{aligned}$$

Therefore,

$$\begin{aligned}\frac{\beta_{N-i}}{\beta_{N-i+1} + \alpha_i - 1} &= p_{N-1}(n_M = i-1)\beta_{N-i} \\ &= \frac{p_{N-1}(n_M = i-1)}{p_{N-1}(n_M = i)} p_{N-1}(n_M \ge i),\end{aligned}$$

and

$$\begin{aligned}\frac{\alpha_i}{\beta_{N-i+1}+\alpha_i-1} &= p_{N-1}(n_M=i-1)\alpha_i \\ &= p_{N-1}(n_M\le i-1).\end{aligned}$$

Using (9.23) and the above results, we obtain the realization factor for the equivalent network as

$$\begin{aligned}c^{(f)}(k,2) &= \frac{\mu_{M,k}}{p_{N-1}(n_M=k-1)} \\ &\times \left\{p_{N-1}(n_M\le k-1)\sum_{i=k}^{N-1}\frac{a_i}{\mu_{M,i}}\frac{p_{N-1}(n_M=i-1)}{p_{N-1}(n_M=i)}p_{N-1}(n_M\ge i)\right. \\ &+ \left. p_{N-1}(n_M\ge k)\sum_{i=1}^{k-1}\frac{a_i}{\mu_{M,i}}\frac{p_{N-1}(n_M=i-1)}{p_{N-1}(n_M=i)}p_{N-1}(n_M\le i-1)\right\}.\end{aligned} \tag{9.40}$$

It is interesting to note that $c^{(f)}(k,2)$ can be determined by the steady-state probabilities of the same system but with only $N-1$ customers. In (9.40), from (9.5) and (9.6),

$$\begin{aligned}a_i &= f'(i)-\frac{\mu_{N-i}}{\mu_{N-i+1}}f'(i-1) \\ &= f'(i)-\frac{G_{M-1}(N-i-1)G_{M-1}(N-i+1)}{[G_{M-1}(N-i)]^2}f'(i-1), \\ &\quad i=1,2,\cdots,N-2,\end{aligned} \tag{9.41}$$

$$\begin{aligned}a_{N-1} &= \frac{y_M}{\mu_{M,N}G_{M-1}(1)}f'(N) \\ &+ \{f'(N-1)-\frac{G_{M-1}(2)}{[G_{M-1}(1)]^2}f'(N-2)\},\end{aligned} \tag{9.42}$$

and

$$p_{N-1}(n_M=k)=\frac{y_M^k}{A_M(k)}\frac{G_{M-1}(N-k-1)}{G_M(N-1)}. \tag{9.43}$$

Now we can prove the following theorem.

Theorem 9.3 *In a load-dependent closed Jackson queuing network, the normalized derivatives of the customer-average performance measure with* *respect to the mean service rates of server M are*

$$\begin{aligned}
&\frac{\mu_{M,k}}{\eta^{(I)}}\frac{\partial \eta^{(f)}}{\partial \mu_{M,k}} \\
&= -y_M\frac{G_M(N-1)}{G_M(N)} \times \left\{ \left[\sum_{j=0}^{k-1}\frac{y_M^j}{A_M(j)}\frac{G_{M-1}(N-j-1)}{G_M(N-1)}\right]\right. \\
&\quad \times \sum_{i=k}^{N-1}\frac{a_i}{y_M}\frac{G_{M-1}(N-i)}{G_{M-1}(N-i-1)}\left[\sum_{j=i}^{N-1}\frac{y_M^j}{A_M(j)}\frac{G_{M-1}(N-j-1)}{G_M(N-1)}\right] \\
&\quad + \left[\sum_{j=k}^{N-1}\frac{y_M^j}{A_M(j)}\frac{G_{M-1}(N-j-1)}{G_M(N-1)}\right]\sum_{i=1}^{k-1}\frac{a_i}{y_M}\frac{G_{M-1}(N-i)}{G_{M-1}(N-i-1)} \\
&\quad \left.\times \left[\sum_{j=0}^{i-1}\frac{y_M^j}{A_M(j)}\frac{G_{M-1}(N-j-1)}{G_M(N-1)}\right]\right\}, \\
&\qquad\qquad k = 1,2,\cdots,N-1, \qquad (9.44)
\end{aligned}$$

and

$$\frac{\mu_{M,N}}{\eta^{(I)}}\frac{\partial \eta^{(f)}}{\partial \mu_{M,N}} = -\frac{y_M^N}{A_M(N)G_M(N)}f'(N),$$

where a_i, $i = 1,2,\cdots,N-1$, are defined by (9.41)-(9.42). The elasticities of the time-average performance measure with respect to the mean service rates of server M are

$$\frac{\mu_{M,k}}{\eta_T^{(f)}}\frac{\partial \eta_T^{(f)}}{\partial \mu_{M,k}} = \frac{\mu_{M,k}}{\eta^{(f)}}\frac{\partial \eta^{(f)}}{\partial \mu_{M,k}} + \frac{\mu_{M,k}}{\eta}\frac{\partial \eta}{\partial \mu_{M,k}}. \qquad (9.45)$$

The formulas for the elasticity of the time-average performance are the same as (9.44) except that in determining a_i, $i = 1,2,\cdots,N-1$, $f'(i)$ are replaced by $[f'(i)/\eta_T^{(f)}] - 1$, $i = 1,2,\cdots,N$.

Proof: The first part of the theorem is a direct consequence of Theorem 9.2 and Equations (9.36), (9.40) and (9.43). To prove the second part, note that (9.45) is the same as (9.34) and can be written as

$$\frac{\mu_{M,k}}{\eta_T^{(f)}}\frac{\partial \eta_T^{(f)}}{\partial \mu_{M,k}} = \frac{1}{\eta_T^{(f)}}\frac{\mu_{M,k}}{\eta^{(I)}}\frac{\partial \eta^{(f)}}{\partial \mu_{M,k}} - \frac{\mu_{M,k}}{\eta^{(I)}}\frac{\partial \eta^{(I)}}{\partial \mu_{M,k}}.$$

By the linearity of all the equations involved, we can see that a_i corresponding to $[f'(i)/\eta_T^{(f)}] - 1$, $i = 1,2,\cdots,N$, lead to the right-hand side of the above equation. □

(i,j)	1	2	3
1	0.2	0.35	0.45
2	0.1	0.2	0.7
3	0.4	0.3	0.3

Table 9.1: The Routing Matrix in Example 9.2

k	$\mu_{1,k}$	$\mu_{2,k}$	$\mu_{3,k}$
1	4.7	3.4	5.0
2	3.0	3.0	4.6
3	5.0	2.9	3.0
4	4.0	4.0	3.6
5	2.3	4.0	3.0
6	1.8	3.0	4.6
7	2.1	2.3	3.0
8	3.1	1.2	4.0
9	2.4	3.0	5.0
10	5.0	3.0	3.0

Table 9.2: The Service Rates in Example 9.2

To illustrate the correctness of our formulas, we now present a numerical example. The result of the example verifies the formulas.

Example 9.2 Consider a queuing network with 3 servers and 10 customers. The system has a total of 66 states. The routing probability matrix and the load-dependent service rates are arbitrarily chosen and are listed in Tables 9.2 and 9.1, respectively. The performance function of interest, f, is taken as $f(\mathbf{n}) = f(n_1, n_2, n_3) = n_1^2$.

We consider the customer-average performance measure associated with f. Applying the normalized derivatives (9.44) in Theorem 9.3 to each server node, we obtain in Table 9.3 the normalized derivatives of the customer-average performance measure with respect to $\mu_{i,k}$, $\frac{\mu_{i,k}}{\eta^{(I)}}\frac{\partial \eta^{(J)}}{\partial \mu_{i,k}}$, for $i = 1, 2, 3$ and $k = 1, 2, \cdots, 10$.

To verify the results, we also calculate these normalized derivatives by directly solving the flow balance equations of the steady-state probabilities. The values thus obtained are exactly the same as those listed in Table 9.3.

k	$\frac{\mu_{1,k}}{\eta^{(f)}}\frac{\partial\eta^{(f)}}{\partial\mu_{1,k}}$	$\frac{\mu_{2,k}}{\eta^{(f)}}\frac{\partial\eta^{(f)}}{\partial\mu_{2,k}}$	$\frac{\mu_{3,k}}{\eta^{(f)}}\frac{\partial\eta^{(f)}}{\partial\mu_{3,k}}$
1	-3.171112	0.818023	1.039880
2	-4.463585	1.103940	1.884079
3	-4.853974	1.063887	2.463057
4	-4.708748	0.965881	2.992016
5	-4.453705	0.771188	3.165722
6	-3.968324	0.641260	2.913746
7	-3.164469	0.525189	2.395751
8	-2.064210	0.410976	1.534830
9	-1.214777	0.176114	0.661650
10	-0.344556	0.000000	0.000000

Table 9.3: A Numerical Example for the Explicit Sensitivity Formula

9.3 The Convolution Algorithm for the Sensitivities

In this section, we discuss the computational complexity of the derived formulas. We identify a wide class of performance functions for which the computational complexity can be reduced to that of Buzen's algorithm. Then, we compare the formulas of this paper with the formulas that are obtained by directly taking the directives of the product-form solutions.

The Computation of a_i

Applying the formulas (9.44) requires computing a_i for all $i = 1, 2, \cdots, N-1$, which from (9.41)-(9.42) in turn requires calculating the conditional mean $f'(i)$ for all $i = 1, 2, \cdots, N$. For any arbitrary performance function f, this usually requires $K = \frac{(M+N-1)!}{(M-1)!N!}$ (i.e., the number of states) summations, as indicated in (9.30). This is approximately equivalent to the same computation as calculating $E[f] = \eta_T^{(f)}$. However, in many cases, the computational effort can be reduced. For example, to compute the throughput sensitivity, no conditional computation for the mean values is needed since $f'(n_M) \equiv 1$ for all n_M.

In what follows, we show that for a wide range of performance functions, the idea of Buzen's algorithm about computing the $G_m(n)$'s for the $p(\mathbf{n})$'s can be easily adapted to compute f', and the order of computational complexity in computing f' will thus reduce to that of Buzen's algorithm.

To that end, for some integer $W \geq 1$, let $g_{w,i}$ be a mapping from

$\{0, 1, \cdots, N\}$ to $R = (-\infty, \infty)$ for each $w = 1, 2, \cdots, W$ and $i = 1, 2, \cdots, M$. Consider the following class of performance functions f,

$$f(\mathbf{n}) = f(n_1, \cdots, n_M) = \sum_{w=1}^{W} f_w(n_1, \cdots, n_M), \tag{9.46}$$

where

$$f_w(n_1, \cdots, n_M) = \prod_{i=1}^{M} g_{w,i}(n_i) \qquad w = 1, 2, \cdots, W. \tag{9.47}$$

With f given by (9.46), $f'(n_M)$ in (9.30) becomes

$$\begin{aligned} f'(n_M) &= \frac{1}{G_{M-1}(N - n_M)} \sum_{w=1}^{W} \sum_{n_1, \cdots, n_{M-1}} f_w(n_1, \cdots, n_M) \prod_{i=1}^{M-1} \frac{y_i^{n_i}}{A_i(n_i)} \\ &= \frac{1}{G_{M-1}(N - n_M)} \sum_{w=1}^{W} g_{w,M}(n_M) \sum_{n_1, \cdots, n_{M-1}} \prod_{i=1}^{M-1} \frac{g_{w,i}(n_i) y_i^{n_i}}{A_i(n_i)}. \end{aligned} \tag{9.48}$$

To compute the right-hand side of (9.48), define, for each $w = 1, 2, \cdots, W$,

$$G_{w,m}(n) := \sum_{n_1 + \cdots + n_m = n} \prod_{i=1}^{m} \frac{g_{w,i}(n_i) y_i^{n_i}}{A_i(n_i)},$$

$$m = 1, 2, \cdots M - 1, \quad n = 1, 2, \cdots, N \tag{9.49}$$

with $G_{w,m}(0) = 1$. Observe that for each w, the $G_{w,m}(n)$'s can be recursively computed via the recursion

$$G_{w,m}(n) = \sum_{k=0}^{n} \frac{g_{w,m}(k) y_m^k}{A_m(k)} G_{w,m-1}(n - k), \tag{9.50}$$

with $G_{w,1}(n) = \frac{g_{w,1}(n) y_1^n}{A_1(n)}$ for all n and $G_{w,m}(0) = 1$ for all m. This is exactly the same as Buzen's algorithm for $G_m(n)$ (Buzen [14]) except for the additional weighting factor $g_{w,m}(k)$. Thus, with the outputs from this adaptation of Buzen's algorithm, the value $f'(n_M)$ of (9.48) is now easily determined to be

$$f'(n_M) = \frac{1}{G_{M-1}(N - n_M)} \sum_{w=1}^{W} g_{w,M}(n_M) G_{w,M-1}(N - n_M). \tag{9.51}$$

The computational complexity originally involved in the K summations in (9.30) now reduces to the complexity involved in applying the convolution algorithm W times.

Note that the class of performance functions discussed above covers a wide range of performance measures. For example, if we choose $f(\mathbf{n}) = n_1$ (this corresponds to $W = 1$, $g_{1,1}(n_1) = n_1$ and $g_{1,i}(n_i) = 1$ for $i = 2, \cdots, M$), we are interested in the response time for server 1. If we choose $f(\mathbf{n}) = \epsilon(n_1)$ (this corresponds to $W = 1$, $g_{1,1}(n_1) = \epsilon(n_1)$ and $g_{1,i}(n_i) = 1$ for $i = 2, \cdots, M$), we are interested in server 1's utilization. Similarly, if we choose $f(\mathbf{n}) = \sum_{i=1}^{M} h_i(n_i)$ (this corresponds to $W = M$, and for each $w = 1, \cdots, M$ and $i = 1, \cdots, M$, $g_{w,i}(n_i) = h_i(n_i)$ if $i = w$ and $g_{w,i}(n_i) = 1$ otherwise), then we are somewhat interested in the performance measure associated with additive holding costs in all the servers. The interested reader should also find many other types of performance measures that fit into the description in (9.46) and (9.47).

The Comparison with Direct Formulas

As shown in (2.51), the direct formulas (based on product-form solutions) for the elasticity of the customer-average performance measure with respect to a service rate of a server have the following form:

$$\frac{\mu_{i,k}}{\eta^{(f)}} \frac{\partial \eta^{(f)}}{\partial \mu_{i,k}} = p_{N-1}(n_i \geq k) - \frac{E[f|n_i \geq k] p_N(n_i \geq k)}{E[f]}, \tag{9.52}$$

for all $i = 1, 2, \cdots, M$ and $k = 1, 2, \cdots, N$. With a general f, (9.52) covers many special cases such as the derivatives of throughput, queue lengths, and utilization with repect to the mean service times.

The direct formulas have a completely different form from those derived here based on realization factors. It is simple to see that the computational effort required by the direct formulas is in the same order as that for $E[f]$ and $E[f|n_i \geq k]$, which requires $K = \frac{(M+N-1)!}{(M-1)!N!}$ (i.e., the number of states) summations. For the same class of performance functions (9.46) and (9.47), the computational effort can be reduced in the same way as (9.51) to that of Buzen's algorithm (Cao and Ma [35]). Thus, the computational complexities of both approaches are comparable to each other.

9.4 Decentralized Optimization of A Closed Jackson Network

In this section, we use the performance sensitivity formulas developed in the previous sections to solve the decentralized optimization problem for closed Jackson networks. We assume that the service rates at each server of a closed Jackson network are controllable. We seek an optimal control policy that minimizes certain performance measures. The proof is based on an intrinsic property of the performance sensitivity formulas: The derivative of a performance measure with respect to a service rate can be factored into a product of a positive term and a quantity that does not depend on the service rate.

Compared with the traditional method based on linear programming (Yao and Schechner [107]), the method based on sensitivity formulas is more direct and hence offers a more intuitive picture. Furthermore, the results obtained apply to more general cost functions. Especially, these results indicate that the bang-bang optimal policy may apply to some servers but may not apply to others. The work in this section is reported in Ma and Cao [82]. Motivated by the approach, we also developed a solution to the decentralized control problem by using directly the derivatives of the product-form solution (see Ma and Cao [81]).

9.4.1 The Problem and the Solution

The Problem

In the decentralized control problem, we assume that the service rates of each server in a closed Jackson network are controllable in a decentralized manner; i.e., the service rate of a server can be chosen according to some policy based on the number of customers in the server. Let $[u_i, U_i]$, $0 \leq u_i < U_i$, $i = 1, 2, \cdots, M$, be the interval from which the service rate of server i can be chosen. Let $\mu_{i,0} = 0$.

We consider stationary control policies. A stationary policy v is specified by a mapping $\Phi \rightarrow [u_1, U_1] \times \cdots \times [u_M, U_M]$, with $v(\mathbf{n}) = \{\mu_{1,n_1}, \mu_{2,n_2}, \cdots, \mu_{M,n_M}\}$. Under policy v, the service rate of server i is the ith components of $v(\mathbf{n})$. Denote the set of all stationary policies by Υ. Under each policy, the network is a load-dependent Jackson network.

The performance measure to be optimized is the average of some cost functions. Let $f(\mathbf{n}, v(\mathbf{n}))$ be the instantaneous cost incurred when the system state is $\mathbf{n}$ and the servers operate at their respective service rates specified by $v(\mathbf{n})$. Define the long-run customer-average expected cost under the policy $v \in \Upsilon$ as

$$J(v) = \lim_{L\to\infty} E\left\{\frac{1}{L}\int_0^{T_L} f(\mathbf{N}(t), v(\mathbf{N}(t)))dt\right\}. \tag{9.53}$$

We consider only the customer average; the problem for time average is similar. The state process $\mathbf{N}(t)$ depends on the policy v.

The decentralized optimization problem can now be stated as: to find an optimal policy $v \in \Upsilon$ that minimizes $J(v)$. The optimal policy may not be unique.

For some problems the instantaneous cost consists of two parts: the instantaneous holding cost and the instantaneous operating cost. The former is defined as a mapping $h : \Phi \to R$, which specifies the cost incurred for storing the customers in the system. The latter is defined as a mapping $g : [u_1, U_1] \times \cdots \times [u_M, U_M] \to R$, which represents the operating cost for servers to provide serices at their respective service rates. In this case,

$$f\{\mathbf{N}(t), v(\mathbf{N}(t))\} = h[\mathbf{N}(t)] + g[v(\mathbf{N}(t))].$$

The Solution

We assume that the routing probability matrix $[q_{i,j}]$ is irreducible. Thus, if $u_i > 0$ for all i, then the state process under any policy is irreducible. If $u_i = 0$ for some i, then the state process under the policies that choose $\mu_{i,n_i^*} = 0$ may contains several recurrent subsets. As discussed in Section 4.1, under such a policy the states with $n_i < n_i^*$ are transient. Thus, for the stationary performance, such a policy is equivalent to a policy of the form

$$\mu_{i,n_i}\begin{cases} = 0 & n_i \le n_i^*; \\ > 0 & n_i > n_i^*. \end{cases} \tag{9.54}$$

Any policy in Υ can be considered as having form (9.54) for some $n_i^* \ge 0$, $i = 1, 2, \cdots, M$.

The solution to the optimization problem is stated in Theorem 9.4; which is based on the following Lemma.

Lemma 9.1 *Under any policy $v \in \Upsilon$, the derivative of $J(v)$ with respect to a service rate $\mu_{i,k}$ is the product of a non-negative term and a quantity whose value is independent of $\mu_{i,k}$, if and only if the cost function $f(\mathbf{n}, v(\mathbf{n}))$ at $n_i = k$ is linear in the service rate $\mu_{i,k}$.*

Lemma 1 immediately implies that if the condition in the lemma is satisfied, then the sign (positive, negative, or zero) of the derivative of $J(v)$ with respect to $\mu_{i,k}$ is independent of $\mu_{i,k}$. The proof of Lemma 1 is in the next subsection.

Theorem 9.4 *Suppose that the cost function $f(\mathbf{n}, v(\mathbf{n}))$ at $n_i = k$ is linear in a service rate $\mu_{i,k}$, $i \in \Gamma$. Then the optimal policy for this service rate is the bang-bang policy; i.e., the optimal service rate of $\mu_{i,k}$ is either the maximum value U_i or the minimum value u_i. If $u_i = 0$ and $\mu_{i,k} = 0$ is the optimal value, then the bang-bang policy is equivalent to a threshold policy in which $\mu_{i,n_i} = 0$ for $n_i \leq k$.*

Proof. First, we observe from the product-form solution that the performance measure $J(v)$ is a continuous function of $\mu_{i,k}$, even at $\mu_{i,k} = 0$. Second, from Lemma 9.1, any policy with $\mu_{i,k} \in (u_i, U_i)$ cannot be better than the policy with $\mu_{i,k}$ being one of the boundary values. The theorem follows immediately. □

If the linearity condition holds for all i and k, then by applying Theorem 9.4 to each $\mu_{i,k}$, we have the following corollary.

Corollary 9.1 *If the cost function $f(\mathbf{n}, v(\mathbf{n}))$ is linear in all service rates μ_{i,n_i}, $i = 1, 2, ..., M$, $n_i = 1, \cdots, N$, then the optimal policy is a bang-bang policy for all servers. If $u_i = 0$ for all i, then the optimal policy is a threshold policy with thresholds n_i^*, $i = 1, 2, ..., M$, such that $\mu_{i,n_i} = 0$ for $n_i \leq n_i^*$ and $\mu_{i,n_i} = U_i$ for $n_i > n_i^*$.*

The Form of the Cost Function

From Theorem 9.4, the optimal policy for choosing a service rate $\mu_{i,k}$ is bang-bang, if the cost function takes the form:

$$f(\mathbf{n}, v(\mathbf{n}))|_{n_i=k} = \phi_1(\mathbf{n}, v_{-i}(\mathbf{n}))|_{n_i=k} + \mu_{i,k}\phi_2(\mathbf{n}, v_{-i}(\mathbf{n}))|_{n_i=k} \quad (9.55)$$

where the notation

$$v_{-i}(\mathbf{n}) := \{\mu_{1,n_1}, ..., \mu_{i-1,n_{i-1}}, \mu_{i+1,n_{i+1}}, ..., \mu_{M,n_M}\},$$

is used and ϕ_1 and ϕ_2 are any functions. Note that $f(\mathbf{n}, v(\mathbf{n}))$ at $n_i = k' \neq k$ may be of any form.

In the case of Corollary 9.1, where the linearity condition of the cost function holds for each service rate, it follows readily from (9.55) that the cost function may take the following general form:

$$f(\mathbf{n}, v(\mathbf{n})) = h(\mathbf{n}) + \sum_{k=1}^{K} \prod_{i \in \Gamma_k} g_{i,k}(\mathbf{n}) \mu_{i,n_i}, \tag{9.56}$$

where $\Gamma_k \subseteq \Gamma$ is any subset of Γ and h and $g_{i,k}$'s are any functions on Φ. The function h is the holding cost and the other term in (9.56) is the operating cost.

The optimality results for threshold policies were established in Yao and Schechner [107] by linear programming for the holding cost function

$$h(\mathbf{n}) = \sum_{i=1}^{M} h_i(n_i),$$

and the operating cost function

$$g(\mathbf{n}, v(\mathbf{n})) = \sum_{i=1}^{M} g_i(n_i) \mu_{i,n_i}.$$

9.4.2 The Proof of the Lemma

The Sensitivity Formulas

Denote f_v as the composite mapping of f and v; i.e., set $f_v(\mathbf{n}) = f(\mathbf{n}, v(\mathbf{n}))$. From (9.53),

$$J(v) = \sum_{\mathbf{n}} f(\mathbf{n}, v(\mathbf{n})) \frac{\pi(\mathbf{n})}{\mu(\mathbf{n})} = \sum_{\mathbf{n}} f_v(\mathbf{n}) \frac{\pi(\mathbf{n})}{\mu(\mathbf{n})} := \eta^{(f_v)}. \tag{9.57}$$

If $f(\mathbf{n}, v(\mathbf{n})) = I(\mathbf{n}) = 1$ for all $\mathbf{n}$ in Φ, then

$$\eta^{(I)} = \sum_{\mathbf{n}} \frac{\pi(\mathbf{n})}{\mu(\mathbf{n})} = \frac{1}{\eta}$$

where η is the steady-state system throughput under the policy v. In this case, the policy tries to maximize the throughput under the constraints of the operating costs.

From (4.47), we have, for any function h on Φ that does not depend on the service rates,

$$\frac{\mu_{i,k}}{\eta^{(I)}} \frac{\partial \eta^{(h)}}{\partial \mu_{i,k}} = - \sum_{\mathbf{n} \in \Phi_0, n_i = k} p(\mathbf{n}) c^{(h)}(\mathbf{n}, i), \tag{9.58}$$

where Φ_0 is the irreducible set corresponding to the set of service rates. Because $f_v(\mathbf{n}) = f(\mathbf{n}, v(\mathbf{n}))$ depends on $\mu_{i,k}$, Equation (9.58) is not directly applicable. However, from (9.57), we have

$$\frac{\partial \eta^{(f_v)}}{\partial \mu_{i,k}} = \sum_{\mathbf{n}} f(\mathbf{n}, v(\mathbf{n})) \frac{\partial \frac{\pi(\mathbf{n})}{\mu(\mathbf{n})}}{\partial \mu_{i,k}} + \sum_{\mathbf{n}} \frac{\pi(\mathbf{n})}{\mu(\mathbf{n})} \frac{\partial f(\mathbf{n}, v(\mathbf{n}))}{\partial \mu_{i,k}}.$$

The first term on the right-hand side of the equation can be viewed as the partial derivative of $\eta^{(f_v)}$ by treating the mapping f_v as being *independent* of $\mu_{i,k}$, and thus is given by (9.58). Therefore, we have

$$\begin{aligned} \frac{\mu_{i,k}}{\eta^{(I)}} \frac{\partial J(v)}{\partial \mu_{i,k}} &= \frac{\mu_{i,k}}{\eta^{(I)}} \frac{\partial \eta^{(f_v)}}{\partial \mu_{i,k}} \\ &= - \sum_{\mathbf{n} \in \Phi_0, n_i = k} p(\mathbf{n}) \{ c^{(f_v)}(\mathbf{n}, i) - \mu_{i,k} \frac{\partial f(\mathbf{n}, v(\mathbf{n}))}{\partial \mu_{i,k}} \}. \end{aligned} \tag{9.59}$$

In (9.59), the equality $p(\mathbf{n}) = \eta \frac{\pi(\mathbf{n})}{\mu(\mathbf{n})}$ is used.

To prove Lemma 9.1, we shall factorize the derivative in such a way that $\mu_{i,k}$ appears only in a term that is always nonnegative. In (9.59), $p(\mathbf{n}) \geq 0$; thus, we shall work on the term in the braces. We shall use the explicit formula for realization factors.

The Two-Server Cyclic Network

We first study the two-server cyclic network discussed in Section 9.1. From (9.59), we have

$$\begin{aligned} \frac{\lambda_k}{\eta^{(I)}} \frac{\partial J(v)}{\partial \lambda_k} &= \frac{\lambda_k}{\eta^{(I)}} \frac{\partial \eta^{(f_v)}}{\partial \lambda_k} \\ &= -p(k) \{ c^{(f_v)}(k, 2) - \lambda_k \frac{\partial f(k, \mu_{N-k}, \lambda_k)}{\partial \lambda_k} \} \end{aligned} \tag{9.60}$$

for $k = 1, 2, ..., N$.

The realization factors given in Section 9.1 correspond to the situation where $n_i^* = 0$ for $i = 1, 2$. Observe that even if n_1^* or $n_2^* > 0$ in v, the solution for realization factors will still have the same form as in (9.5), (9.6), (9.10), (9.13), and (9.23), with obvious modification in the indices from $\{0, 1, ..., N-1\}$ to $\{n_2^*, n_2^*+1, ..., N-1-n_1^*\}$. All the structural results obtained will be the same. Therefore, for simplicity, we shall proceed our derivations with the assumption that $n_i^* = 0$ for $i = 1, 2$.

The proof of Lemma 9.1 for $k = N$ is obvious: Note $c^{(f_v)}(N,2) = f_v(N) = f(N, \mu_0, \lambda_N)$, and thus from (9.60),

$$\frac{\partial J(v)}{\partial \lambda_N} = -\frac{p(N)}{\eta \lambda_N}\{f(N,\mu_0,\lambda_N) - \lambda_N \frac{\partial f(N,\mu_0,\lambda_N)}{\partial \lambda_N}\}.$$

The two terms in the braces on the right-hand side of the equation have their terms containing λ_N cancelled if and only if $f(N, \mu_0, \lambda_N)$ is linear in λ_N.

To prove the lemma for $k = 1, 2, \cdots, N-1$, we first observe that in (9.23),

$$\alpha_i + \beta_{N-i+1} - 1 = D_N\Big(\prod_{j=N-i+1}^{N-1} \mu_j\Big)^{-1}\Big(\prod_{j=i}^{N-1} \lambda_j\Big)^{-1}, \tag{9.61}$$

with

$$\begin{aligned} D_N &= \alpha_N \prod_{j=1}^{N-1} \mu_j = \beta_N \prod_{j=1}^{N-1} \lambda_j \\ &= \prod_{j=1}^{N-1} \mu_j + \sum_{i=1}^{N-2}\Big(\prod_{j=N-i}^{N-1} \lambda_j\Big)\Big(\prod_{j=i+1}^{N-1} \mu_j\Big) + \prod_{j=1}^{N-1} \lambda_j \\ &> 0. \end{aligned} \tag{9.62}$$

Therefore, (9.23) has the following form:

$$c^{(f_v)}(k,2) = \frac{\lambda_k}{D_N} B^{(f_v)}(k,2) \qquad k = 1, 2, \cdots, N-1, \tag{9.63}$$

where

$$\begin{aligned} B^{(f_v)}(k,2) &= \alpha_k \sum_{i=k}^{N-1} a_i \beta_{N-i}\Big(\prod_{j=N-i+1}^{N-1} \mu_j\Big)\Big(\prod_{j=i+1}^{N-1} \lambda_j\Big) \\ &+ \mu_{N-k}\beta_{N-k}\Big(\prod_{j=k+1}^{N-1} \lambda_j\Big)\sum_{i=1}^{k-1} \frac{a_i}{\mu_{N-i}} \alpha_i\Big(\prod_{j=N-i+1}^{N-1} \mu_j\Big)\Big(\prod_{j=i}^{k-1} \lambda_j\Big) \\ &\qquad for\ k = 1, 2, \cdots, N-1. \end{aligned} \tag{9.64}$$

Thus, from (9.60) and (9.63), we obtain

$$\frac{\partial J(v)}{\partial \lambda_k} = -\frac{p(k)}{\eta D_N}\{B^{(f_v)}(k,2) - D_N \frac{\partial f(k,\mu_{N-k},\lambda_k)}{\partial \lambda_k}\}. \tag{9.65}$$

Now, suppose $f_v(k) = f(k, \mu_{N-k}, \lambda_k)$ does not depend on λ_k, then it is easy to check from (9.64) that $B^{(f_v)}(k,2)$ does not depend on λ_k. Since $\frac{\partial f(k,\mu_{N-k},\lambda_k)}{\partial \lambda_k} = 0$ in this case, we immediately have

$$\frac{\partial J(v)}{\partial \lambda_k} = -\frac{p(k)}{\eta D_N} B^{(f_v)}(k,2),$$

where $\frac{p(k)}{\eta D_N}$ is positive and $B^{(f_v)}(k,2)$ is independent of λ_k. The lemma is then proved.

The case where $f_v(k) = f(k, \mu_{N-k}, \lambda_k)$ depends on λ_k is more complicated. Both terms in the braces on the right-hand side of (9.65) depend on λ_k. From (9.64), $B^{(f_v)}(k,2)$ depends on λ_k via a_k and a_{k+1}. Using (9.5), we can collect the two terms in $B^{(f_v)}(k,2)$ that depend on λ_k as follows:

$$\begin{aligned} &\alpha_k f(k,\mu_{N-k},\lambda_k)\{\beta_{N-k}(\prod_{j=N-k+1}^{N-1}\mu_j)(\prod_{j=k+1}^{N-1}\lambda_j) \\ &-\beta_{N-k-1}\frac{\mu_{N-k-1}}{\mu_{N-k}}(\prod_{j=N-k}^{N-1}\mu_j)(\prod_{j=k+2}^{N-1}\lambda_j)\} \\ = \;&\alpha_k f(k,\mu_{n-k},\lambda_k)(\prod_{j=N-k+1}^{N-1}\mu_j)(\prod_{j=k+1}^{N-1}\lambda_j). \end{aligned} \tag{9.66}$$

Equation (9.19) is used in the last equation. For $k = N-1$, the dependence appears in only the term involving a_{N-1} on the right-hand side of (9.64), which gives exactly the same term as (9.66) with $k = N-1$.

On the other hand, recall that $D_N = \alpha_N \prod_{j=1}^{N-1}\mu_j$. Applying (9.20) iteratively from α_N to α_k, we get

$$\begin{aligned} D_N \;=\; &\{\prod_{j=1}^{N-1}\mu_j + \sum_{i=2}^{N-k}(\prod_{j=i}^{N-1}\mu_j)(\prod_{j=N-i+1}^{N-1}\lambda_j)\} \\ + \;&\alpha_k\lambda_k(\prod_{j=N-k+1}^{N-1}\mu_j)(\prod_{j=k+1}^{N-1}\lambda_j). \end{aligned} \tag{9.67}$$

It is now straightforward from (9.66)-(9.67) that the two terms on the right-hand side of (9.65), $B^{(f_v)}(k,2)$ and $D_N\frac{\partial f(k,\mu_{N-k},\lambda_k)}{\partial \lambda_k}$, have their terms containing λ_k cancelled if and only if

$$f(k,\mu_{N-k},\lambda_k) = \phi_1(k,\mu_{N-k}) + \lambda_k\phi_2(k,\mu_{N-k}); \tag{9.68}$$

this is the same form as (9.55). Lemma 9.1 is proved for the two-server cyclic network.

M-Server Networks

To extend the two-server network results to M-server networks, we use the idea of Norton's aggregation (see Section 2.3).

When some service rates are zero, the steady-state probability of a closed Jackson network in one of the recurrent subset Φ_0 still has a product-form. In fact, one can check that the steady-state probability $p(\mathbf{n})$, $\mathbf{n} \in \Phi_0$, is

$$p(\mathbf{n}) = \frac{1}{G_M(N)} \prod_{i=1}^{M} \frac{y_i^{n_i - n_i^*}}{A_i(n_i)},$$

where $A_i(n_i^*) = 1$,

$$A_i(k) = \prod_{j=n_i^*+1}^{k} \mu_{i,j}, \qquad k \geq n_i^* + 1,$$

and

$$G_M(N) = \sum_{\mathbf{n} \in \Phi_0} \prod_{i=1}^{M} \frac{y_i^{n_i - n_i^*}}{A_i(n_i)}.$$

By the same arguments as those in Section 9.2, Norton's aggregation holds for the performance sensitivities provided that the equivalent performance function is chosen according to (9.30). For simplicity, we shall derive equations with the assumption $n_i^* = 0$ for all i.

We construct a two-server equivalent network by aggregating servers $1, 2, \cdots, M-1$. The equivalent network consists of one server with service rate $\mu_{M,k}$ and one server with service rate μ_{N-k} determined by (9.28). The cost function for the equivalent network is defined, according to (9.30), as

$$f'(n_M, \mu_{M,n_M}) := E[f(\mathbf{n}, v(\mathbf{n}))|n_M]$$
$$= \frac{1}{G_{M-1}(N - n_M)} \sum_{n_1, \ldots, n_{M-1}} f(\mathbf{n}, v(\mathbf{n})) \prod_{i=1}^{M-1} \frac{y_i^{n_i}}{A_i(n_i)}. \qquad (9.69)$$

We use a subscript "e" to denote the value for the equivalent two-server network. For example, $J_e(v)$ is its long-run customer-average performance. By the same argument as that for (9.32), we get

$$\frac{\mu_{M,k}}{\eta_e^{(I)}} \frac{\partial J_e(v)}{\partial \mu_{M,k}} = \frac{\mu_{M,k}}{\eta^{(I)}} \frac{\partial J(v)}{\partial \mu_{M,k}}.$$

That is, the elasticity of the performance measure $J_e(v)$ with respect to $\mu_{M,k}$ for the equivalent network equals that for the original network.

Applying (9.68) to f' in the two-server equivalent network, we obtain immediately that the derivative of $J(v)$ with respect to the service rate $\mu_{M,k}$ of server M is the product of a positive term and a quantity whose value is independent of $\mu_{M,k}$ if and only if

$$f'(k, \mu_{M,k}) = \phi_1 + \mu_{M,k}\phi_2 \tag{9.70}$$

for some ϕ_1 and ϕ_2 independent of $\mu_{M,k}$. From (9.69), Equation (9.70) holds if and only if $f(\mathrm{n}, v(\mathrm{n}))$ at $n_M = k$ is linear in $\mu_{M,k}$; i.e., f satisfies the form (9.55). Lemma 9.1 is thus proved for server M. By permuting the order of the servers so that server i is designated as server M, Lemma 9.1 holds for each server i.

Bibliography

[1] F. Baccelli and P. Brémaud, *Palm Probabilities and Stationary Queues*, Vol. 41 of *Lecture Notes in Statistics*, Springer-Verlag, New York, 1987.

[2] F. Baccelli and P. Brémaud, *Elements of Queueing Theory*, Springer-Verlag, New York, 1994.

[3] S. Balsamo and G. Iazeolla, "An Extension of Norton's Theorem for Queueing Networks," *IEEE Transactions on Software Engineering*, Vol. 8, 298-305, 1982.

[4] F. Baskett, K. M. Chandy, R. R. Muntz, and F. G. Palacios, "Open, Closed, and Mixed Networks of Queues with Different Classes of Customers," *Journal of ACM*, Vol. 22, 284-260, 1975.

[5] P. Billingsley, *Probability and Measure*, John Wiley and Sons, 1979.

[6] A. A. Borovkov, *Stochastic Processes in Queueing Theory*, Springer-Verlag, 1976.

[7] A. A. Borovkov, "Ergodicity and Stability Theorems for a Class of Stochastic Equations and Their Application," *Theory of Probability and its Applications*, Vol. 23, 227-247, 1978.

[8] A. A. Borovkov, "Limit Theorems for Queueing Networks, I," *Theory of Probability and its Applications*, Vol. 31, 1987.

[9] L. Breiman, *Probability*, Addison Wesly, Reading, Mass. 1968.

[10] P. Brémaud, G. Kannurpatti, and R. Mazumdar, "Event and Time Average: A Review," *Advances in Applied Probability*, Vol. 24, 377-411, 1992.

[11] S. C. Bruell and G. Balbo, *Computational Algorithms for Closed Queueing Networks*, North Holland, New York, 1980.

[12] A. E. Bryson and Y. C. Ho, *Applied Optimal Control*, Blaisedel, 1969.

[13] J. P. Buzen and P. J. Denning, "Operations Treatment of Queue Distributions and Mean-Value Analysis," *Comput. Perform.* Vol. 1, 6-14, 1980.

[14] J. P. Buzen, "Computational Algorithm for Closed Queueing Networks with Exponential Servers," *Communications of the ACM*, Vol. 16, 527-531, 1973.

[15] X. R. Cao, "Performance Sensitivity Analysis of Open Markovian Queueing Networks," *European Journal of Opertional Research*, to appear, 1993.

[16] X. R. Cao, "A New Method of Performance Sensitivity Analysis for non-Markovian Queueing Networks," *Queueing Systems: Theory and Applications*, Vol. 10, 313-350, 1992.

[17] X. R. Cao, "A Linear Algebraic Formulation of the Performance Sensitivities of Queueing Networks," *European Journal of Opertional Research*, Vol. 56, 394-406, 1992.

[18] X. R. Cao, "Event Coupling and Performance Sensitivity Analysis of Generalized Semi-Markov Processes," Manuscript, 1992.

[19] X. R. Cao, "Realization Factors and Sensitivity of Analysis of Queueing Networks with State Dependent Service Rates," *Advances in Applied Probability*, Vol. 22, 178-210, 1990.

[20] X. R. Cao, "The Convergence Property of Sample Derivatives in Closed Jackson Queuing Networks," *Stochastic Processes and their Applications*, Vol. 33, 105-122, 1989.

[21] X. R. Cao, "Estimates of Performance Sensitivity of a Stochastic System," *IEEE Transactions on Information Theory*, Vol. 35, 1058-1068, 1989.

[22] X. R. Cao, "A Compariosn of the Dynamics of Continuous and Discrete Event Systems," *IEEE Proceedings*, Vol. 77, 7-13, 1989.

[23] X. R. Cao, "Realization Probability and Throughput Sensitivity in a Closed Jackson Network," *Journal of Applied Probability*, Vol. 26, 615-624, 1989.

[24] X. R. Cao, "Calculation of Sensitivities of Throughputs and Realization Probabilities in Closed Queueing Networks with Finite Buffers," *Advances in Applied Probability,* Vol. 21, 181-206, 1989.

[25] X. R. Cao, "System Representations and Performance Sensitivity Estimates of Discrete Event Systems," *Mathematics and Computers in Simulation*, Vol. 31, 113-122, 1989.

[26] X. R. Cao, "The Static Property of a Perturbed Multiclass Closed Queuing Network and Decomposition," *IEEE Transactions on Automatic Control*, Vol. 34, 246-249, 1989.

[27] X. R. Cao, "Realization Probability in Multi-Class Closed Queuing Networks," *European Journal of Operational Research,* Vol. 36, 393-401, 1988.

[28] X. R. Cao, "A Sample Performance Function of Jackson Queueing Networks," *Operations Research,* Vol. 36, No. 1, 128-136, 1988.

[29] X. R. Cao, "Realization Probability in Closed Jackson Queueing Networks and Its Application," *Advances in Applied Probability,* Vol. 19, 708-738, 1987.

[30] X. R. Cao, "First-Order Perturbation Analysis of a Single Multi-Class Finite Source Queue," *Performance Evaluation,* Vol. 7, 31-41, 1987.

[31] X. R. Cao, "Convergence of Parameter Sensitivity Estimates in a Stochastic Experiment," *IEEE Transactions on Automatic Control,* Vol. AC-30, 834-843, 1985.

[32] X. R. Cao and Y. Dallery, "An Operational Approach to Perturbation Analysis of Closed Queueing Networks," *Mathematics and Computers in Simulation,* Vol. 28, 433-451, 1986.

[33] X. R. Cao and D. J. Ma, "New Performance Sensitivity Formulae for a Class of Product-Form Queueing Networks," *Discrete Event Dynamic Systems: Theory and Applications,* Vol. 1, 289-313, 1992.

[34] X. R. Cao and D. J. Ma, "Sensitivity Analysis of General Performance Measures of Queueing Networks with State-Dependent Service Rates," *Applied Mathematics Letters*, Vol. 4, 57-60, 1991.

[35] X. R. Cao and D. J. Ma, "Performance Sensitivity Formulas, Algorithms and Estimates for Closed Queueing Networks with Exponential Servers," Technical report, Digitial Equipment Corporation, June 1990.

[36] C. G. Cassandras, *Discrete Event Systems: Modeling and Performance Analysis,* Aksen Associates, Inc., 1993.

[37] K. M. Chandy, U. Herzog, and L. Woo, "Parametric Analysis of Queueing Networks," *IBM J. Res. Develop.* 36-42, 1975.

[38] K. M. Chandy, J. H. Howard Jr., and D. F. Towsley, "Product Form and Load Balance in Queueing Networks," *Journal of ACM,* Vol. 24, 250-263, 1977.

[39] K. M. Chandy and A. J. Martin, "A Characterization of Product-Form Queueing Networks," *Journal of ACM,* Vol. 30, 286-299, 1983.

[40] K. L. Chung, *A Course in Probability Theory,* Academic Press, New York, 1974.

[41] E. Çinlar, *Introduction to Stochastic Processes,* Prentice Hall, Inc., 1975.

[42] J. W. Cohen, *The Single Queue,* American Elsevier, New York, 1969.

[43] A. E. Conway and N. D. Georganas, *Queueing Newtorks - Exact Computational Algorithms: A Unified Theory Based on Deccomposition and Aggregation,* The MIT Press, Cambridge, Mass., 1989.

[44] R. Courant and F. John, *Introduction to Calculus and Analysis,* 1965.

[45] D. R. Cox, "A use of complex probabilities in the theory of stochastic process," *Proc. Cambridge Phil. Soc.,* Vol. 51, 313-315, 1955.

[46] R. B. Cooper, *Introduction to Queueing Theory,* 2nd edition, North Holland, New York, 1981.

[47] P. J. Courtois, *Decomposability: Queueing and Computer System Applications,* Academic Press, New York, 1977.

[48] M. A. Crane and A. J. Lemoine, *An Introduction to the Regenerative Method for Simulation Analysis,* Springer-Verlag, New York, 1977.

[49] D. J. Daley and D. Vere-Jones, *An Introduction to the Theory of Point Processes,* Springer-Verlag, New York, 1988.

[50] Y. Dallery and X. R. Cao, "Operational Analysis of Stochastic Closed Queueing Networks," *Performance Evaluation,* Vol. 14, 43-61, 1992.

[51] Y. Dallery and R. David, "Some new results on operational analysis," *Performance'84*, 119-134, E. Gelenbe (ed.), North-Holland, Amsterdam, 1984.

[52] Y. Dallery and R. David,"Operational Analysis of multiclass queueing networks," *Proceedings of the IEEE Conference on Decision and Control*, 1986.

[53] P. Denning and J. Buzen, "The Operational Analysis of Queuing Network Models," *Computing Survey*, Vol. 10, No. 3, 225-261, 1978.

[54] N. M. van Dijk, "On a Simple Proof of Uniformization for Continuous and Discrete-State Continuous-Time Markov Chains," *Advances in Applied Probability*, Vol. 22, 749-750, 1990.

[55] E. Gelenbe and G. Pujolle, *Introduction to Queueing Networks*, Wiley, New York, 1987.

[56] P. Glasserman, "The Limilting Value of Derivative Estimates Based on Perturbation Analysis," *Communications in Statistics: Stochastic Models*, Vol. 6, No. 2, 1991.

[57] P. Glasserman, *Gradient Estimation Via Perturbation Analysis*, Kluwer Academic Publisher, Boston, 1991.

[58] P. Glasserman, "Infinitesimal Perturbation Analysis of a Birth and Death Process," *Operations Research Letters*, Vol. 7, 43-49, 1988.

[59] P. Glasserman and D. D. Yao, "Monotonicity in Generalized Semi-Markov Processes," *Mathematics of Operations Research*, Vol. 17, 1-21, 1992.

[60] P. W. Glynn, "A GSMP formalism for Discrete Event Systems," *IEEE Proceedings*, Vol. 77, 14-23, 1989.

[61] W. B. Gong and Y. C. Ho, "Smoothed Perturbation Analysis for Discrete Event Dynamic Systems," IEEE Transactions on Automatic Control, Vol. 32, 858-866, 1987.

[62] W. B. Gong, S. Nananukul, and A. Yan, "Pade Approximations for Stochastic Discrete Event Systems", IEEE Transactions on Automatic Control, to appear, 1993.

[63] W. M. Gordon, "The Evaluation of Normalizing Constants in Closed Queueing Networks," *Operations Research*, Vol. 38, 863-869, 1990.

[64] W. J. Gordon and G. F. Newell, "Closed Queueing Systems with Exponential Servers," *Operation Research,* Vol. 15, 252-265, 1967.

[65] D. Gross and C. M. Harris, *Fundamentals of Queueing Theory,* Wiley, New York, 1974.

[66] P. Heidelberger, X. R. Cao, M. Zazanis, and R. Suri, "Convergence Properties of Infinitesimal Perturbation Analysis Estimates," *Management Science,* Vol. 34, No. 11, 1281-1302, 1988.

[67] Y. C. Ho (editor), *Discrete-Event Dynamic Systems: Analyzing Complexity and Performance in the Modern World,* IEEE Press, New York, 1992.

[68] Y. C. Ho and X. R. Cao, *Perturbation Analysis of Discrete-Event Dynamic Systems,* Kluwer Academic Publisher, Boston, 1991.

[69] Y. C. Ho and X. R. Cao, "Performance Sensitivity to Routing Changes in Queueing Networks and Flexible Manufacturing Systems Using Perturbation Analysis," *IEEE Journal of Robotics and Automation,* Vol. RA-1, 165-172, 1985.

[70] Y. C. Ho and X. R. Cao, "Perturbation Analysis and Optimization of Queueing Networks," *Journal of Optimization Theory and Applications,* 40, 4, 559-582, 1983.

[71] J. R. Jackson, "Networks of Waiting Lines," *Operations Research,* Vol. 5, 518 - 521, 1957.

[72] R. Jain, *The Art of Computer Systems Performance Analysis,* Wiley, New York, 1991.

[73] A. F. Karr, *Point Processes and Their Statistical Inference,* Marcel Dekker Inc., New York, 1986.

[74] F. P. Kelly, *Reversibility and Stochastic Networks,* Wiley, New York, 1979.

[75] J. Kemeny, J. L. Snell, and A. Knapp, *Denumerable Markov Chains,* Van Nostrand, Princeton, New Jersey, 1960.

[76] L. Kleinrock, *Queueing Systems, Volume I, Theory,* Wiley, New York, 1975.

[77] L. Kleinrock, *Queueing Systems, Volume II, Computer Applications,* Wiley, New York, 1976.

[78] E. Kreyszig, *Introductory Functional Analysis with Application,* Wiley, New York, 1978.

[79] S. Lavenberg and M. Reiser, "Stationary State Probabilities at Arrival Instants for Closed Queueing Networks With Multiple Types of Customers," *Journal of Applied Probability,* Vol. 17, 1048-1061, 1980.

[80] Z. Liu and P. Nain, "Sensitivity Results in Open, Closed and Mixed Product Form Queueing Networks," *Performance Evaluation,* Vol. 13, 237-251, 1991.

[81] D. J. Ma and X. R. Cao, "A Direct Approach to Decentralized Control of Service Rates in a Closed Jackson Network," *IEEE Transactions on Automatic Control,* submitted.

[82] D. J. Ma and X. R. Cao, "A Direct Approach to Decentralized Control of Service Rates in a Closed Jackson Network by Using Realization Factors," Working paper, Digital Equipment Corporation, 1992.

[83] D. J. Ma and X. R. Cao, "Performance Sensitivity Formulas and Optimal Control of Closed Queueing Networks," *Proceedings of the 29th IEEE Conference on Decision and Control,* 167-172, 1990.

[84] R. Marie, "An Approximate Analytical Method for General Queueing Networks," *IEEE Transactions on Software Engineering,* Vol. 5, 530-538, 1979.

[85] K. Matthes, "Zur Theorie der Bedienungsprozesse," *Third Prague Conference on Information Theory, Statistical Decision Functions and Random Processes,* Prage, 1962.

[86] B. Melamed, "On Markov Jump Process Imbedded at Jump Epochs and Their Queuing-theoretic Applications," *Math. of Oper. Res., Vol.7, No.1,* 111-128, 1982.

[87] I. Mitrani, *Simulation Techniques for Discrete-Event Systems,* Cambridge Press, London, 1982.

[88] M. F. Neuts, *Matrix-Geometric Solutions in Stochastic Models: An Algorithmic Approach,* The John Hopkins University Press, Baltimore, 1981.

[89] G. F. Newell, *Applications of Queueing Theory,* Chapman and Hall, London, 1972.

[90] N. U. Prabhu, *Queues and Inventories*, Wiley, New York, 1965.

[91] N. J. Pullman, *Matrix Theory and its Applications*, Dekker, New York, 1976.

[92] R. Rubinstein, *Monte Carlo Optimization, Simulation, and Sensitivity Analysis of Queueing Networks*, Wiley, New York, 1986.

[93] T. L. Saaty, *Elements of Queueing Theory with Applications*, McGraw-Hill, New York, 1961.

[94] K. C. Sevcik and M. M. Klawe, "Operations Analysis Versus Stochastic Modeling of Computer Systems," *Proc. Symp. Interface*, 1979.

[95] K. C. Sevcik and I. Mitrani, "The Distribution of Queuing Network States at Input and Output Instants," *Journal of ACM*, Vol.28, No.2, 358-371, 1981.

[96] G. Shanthikumar and D. D. Yao, "General Queueing Networks: Representation and Stochastic Monotonicity," *Proceedings of the 26th IEEE conference on Decision and Control*, 1084-1087, 1987.

[97] K. Sigman, "Notes on the Stability of Closed Queueing Networks," *Journal of Applied Probability*, Vol. 26, 678-681, 1989.

[98] K. Sigman, "The Stability of Open Queueing Networks," *Stochastic Processes and Their Applications*, Vol. 35, 11-25, 1990.

[99] M. Spivak, *Calculus*, W. A. Benjamin Inc. New York, 1967.

[100] R. Suri, "Robustness of Queueing Networks Formulas," *Journal of ACM*, Vol. 30, 564-594, 1983.

[101] R. Suri, "Infinitesimal Perturbation Analysis for General Discrete Event Systems," *Journal of ACM*, Vol. 34, 686-717, 1987.

[102] F. Szidarovszky and A. T. Bahill, *Linear Systems Theory*, CRC press, 1992.

[103] Y. C. Tay and R. Suri, "Error Bounds for Performance Prediction in Queueing Networks," *ACM Transactions on Computer Systems*, Vol. 3, 227-254, 1985.

[104] J. Walrand, *An Introduction to Queueing Networks*, Prentice-Hall, Englewood Cliffs, New Jersey, 1988.

[105] A. C. Williams and R. A. Bhandiwad, "A Generating Function Approach to Queueing Network Analysis of Multiprogrammed Computers," *Networks,* Vol. 6, 1-22, 1976.

[106] W. Whitt, "Continuity of Generalized Semi-Markov Processes," *Mathematics of Operations Research,* Vol. 5, 1980.

[107] D. D. Yao and Z. Schechner, "Decentralized Control of Service Rates in a Closed Jackson Network," *IEEE Transactions on Automatic Control,* Vol. 34, 236-240.

Selected Notations

$A,\ B$	Sets;
A^c	The complement of A;
$\mathcal{A}$	The infinitesimal generator, $[\mathcal{A}(i,j)]$;
a_l	The lth arrival instants;
$\mathcal{B}$	The Borel field;
$\mathbf{C}$	The realization matrix;
$\mathcal{C}_i$	The set of service completion times of server i, $\{t_{i,k}\}_{k=1}^{\infty}$;
$c(\mathbf{n}, V)$	The realization probability of the perturbations in set V at state $\mathbf{n}$;
$c^{(f)}(\mathbf{n}, V)$	The realization factor of the perturbations in set V at state $\mathbf{n}$ with respect to f;
$c_{\mathbf{m}}(\mathbf{n}, V)$	The realization factor with respect to $\delta_{\mathbf{m}}(\mathbf{n})$;
$\mathbf{D}$	The sensitivity matrix;
$\delta_{\mathbf{m}}(\mathbf{n})$	The basic performance function, $\delta_{\mathbf{m}}(\mathbf{m}) = 1$ and $\delta_{\mathbf{m}}(\mathbf{n}) = 0$ for $\mathbf{n} \neq \mathbf{m}$;
$\Delta_{i,k}$	The perturbation of $t_{i,k}$;
$\mathcal{E}$	The space of the events in a GSMP;
$E(X)$	The mean of a random variable X;
η	The steady-state system throughput;
$\eta^{(f)}$	The steady-state customer-average performance;
$\eta_T^{(f)}$	The steady-state time-average performance;
η_L	The sample throughput in $[0, T_L)$, $\frac{L}{T_L}$;
$\eta_L^{(f)}$	The customer-average sample performance function in $[0, T_L)$;
$\eta_{T_L}^{(f)}$	The time-average sample performance function in $[0, T_L)$;
$f(\mathbf{n})$	The performance function;
$G_M(N)$	The normalizing factor for an M-customer N-server closed Jackson network;
Γ	The set of servers, $\{1, 2, \cdots, M\}$;
$G(r, \theta)$	The perturbation generation function;
L	The number of customers served;
λ	A service rate or an arrival rate;

Λ	A subset of Σ;
$\mu(\mathbf{n})$	The state transition rate at state $\mathbf{n}$;
$\mu_{i,\mathbf{n}}$	The service rate of server i at state $\mathbf{n}$;
μ_{i,n_i}	The service rate of server i when there are n_i customers in the server;
μ_i	The service rate of server i;
$\mathbf{n}$	The vector of all n_i, $(n_1, \cdots, n_M)$;
n_i	The number of customers at server i;
$\mathcal{N}(t)$	The process of $\mathbf{n}$;
$(\Omega, \mathcal{F}, P)$	The probability space;
$p(\mathbf{n})$	The steady-state probability of $\mathbf{n}$;
Φ	The space of $\mathbf{n}$;
$\phi(t, s)$	The state transition matrix for continuous dynamic systems;
Π	$\Phi \times R^M$, the space representing the state and the sizes of perturbations;
$\pi(\mathbf{n})$	The steady-state probability of $\mathbf{n}$ in the embedded Markov chain;
Ψ	The space of the server-class pairs, $\{(i, k)\}$, with i denoting the server and k denoting the class;
Q	The routing probability matrix of a queueing network, $Q = [q_{i,j}]_{i,j=1}^M$; or the state transition matrix of a Markov process, $Q = [Q(i,j)]_{i,j=1}^M$;
$q_{i,j}$	The routing probability from server i to server j;
$Q(i, j)$	The transition probability of a Markov process;
R	The space of real numbers, $(-\infty, \infty)$;
$\mathbf{r}$	The vector of all r_i, $(r_1, \cdots, r_M)$;
$\mathcal{R}(t)$	The process of $\mathbf{r}$;
r_i	The residual service time of the customer at server i;
R_+	The space of nonnegative real numbers, $[0, \infty)$;
R_L	The average time that the system stays in a state, $\frac{T_L}{L}$;
$RI(\mathbf{n}, V, l)$	The realization index;
S_l	The intertransition time, $T_{l+1} - T_l$;
$\bar{s}_i$	The mean service time of server i;
$s_{i,k}$	The service time of the kth customet of server i;
Σ	The space of the state-event pairs; $\Sigma = \Phi \times \Gamma$ for queueing networks, and $\Sigma = \Phi \times \mathcal{E}$ for GSMP;

$\mathcal{T}$	A transformation or a mapping;
T_l	The lth transition time of a state process;
$t_{i,k}$	The service completion time of the kth customet of server i;
τ_l	The interarrival time, $a_{l+1} - a_l$;
V	A subset of Γ, representing the servers that have the perturbation;
$Var(X)$	The variance of X;
ϑ	The shift transformation;
X, Y	Random variables;
$\mathbf{x}$	The vector of all x_i, $(x_1, \cdots, x_M)$;
$\mathbf{X}$	The embedded Markov chain, $\{X_0, X_1, \cdots\}$;
x_i	The state of server i; $x_i = (n_i, r_i)$ for single-class servers with nonexponentially distributed service times; $x_i = (k_{i,1}, \cdots, k_{i,n_i})$ for multiclass servers with exponentially distributed service times, $k_{i,j}$ is the class of the jth customer;
X_l	The state at the lth transition;
$\mathcal{X}(t)$	A stochastic process;
ξ	The random vector that determines a sample path of a queueing network;
Ξ	The space of ξ, $[0,1)^\infty \times \Phi$;
$\xi_{i,k}$	The uniformly distributed random variable on $[0,1)$ that determines $s_{i,k}$;
y_i	The visit ratio to server i;

Index

Lecture Notes in Control and Information Sciences

Edited by M. Thoma

1989–1993 Published Titles:

Vol. 135: Nijmeijer, Hendrik; Schumacher, Johannes M. (Eds.)
Three Decades of Mathematical System Theory. A Collection of Surveys at the Occasion of the 50th Birthday of Jan C. Willems.
562 pp. 1989 [3-540-51605-0]

Vol. 136: Zabczyk, Jerzy W. (Ed.)
Stochastic Systems and Optimization. Proceedings of the 6th IFIP WG 7.1 Working Conference, Warsaw, Poland, September 12-16, 1988.
374 pp. 1989 [3-540-51619-0]

Vol. 137: Shah, Sirish L.; Dumont, Guy (Eds.)
Adaptive Control Strategies for Industrial Use. Proceedings of a Workshop held in Kananaskis, Canada, 1988.
360 pp. 1989 [3-540-51869-X]

Vol. 138: McFarlane, Duncan C.; Glover, Keith
Robust Controller Design Using Normalized Coprime Factor Plant Descriptions.
206 pp. 1990 [3-540-51851-7]

Vol. 139: Hayward, Vincent; Khatib, Oussama (Eds.)
Experimental Robotics I. The First International Symposium, Montreal, June 19-21, 1989.
613 pp. 1990 [3-540-52182-8]

Vol. 140: Gajic, Zoran; Petkovski, Djordjija; Shen, Xuemin (Eds.)
Singularly Perturbed and Weakly Coupled Linear Control Systems. A Recursive Approach.
202 pp. 1990 [3-540-52333-2]

Vol. 141: Gutman, Shaul
Root Clustering in Parameter Space.
153 pp. 1990 [3-540-52361-8]

Vol. 142: Gündes, A. Nazli; Desoer, Charles A.
Algebraic Theory of Linear Feedback Systems with Full and Decentralized Compensators.
176 pp. 1990 [3-540-52476-2]

Vol. 143: Sebastian, H.-J.; Tammer, K. (Eds.)
System Modelling and Optimizaton. Proceedings of the 14th IFIP Conference, Leipzig, GDR, July 3-7, 1989.
960 pp. 1990 [3-540-52659-5]

Vol. 144: Bensoussan, A.; Lions, J.L. (Eds.)
Analysis and Optimization of Systems. Proceedings of the 9th International Conference. Antibes, June 12-15, 1990.
992 pp. 1990 [3-540-52630-7]

Vol. 145: Subrahmanyam, M. Bala
Optimal Control with a Worst-Case Performance Criterion and Applications.
133 pp. 1990 [3-540-52822-9]

Vol. 146: Mustafa, Denis; Glover, Keith
Minimum Enthropy H Control.
144 pp. 1990 [3-540-52947-0]

Vol. 147: Zolesio, J.P. (Ed.)
Stabilization of Flexible Structures. Third Working Conference, Montpellier, France, January 1989.
327 pp. 1991 [3-540-53161-0]

Vol. 148: Not published

Vol. 149: Hoffmann, Karl H; Krabs, Werner (Eds.)
Optimal Control of Partial Differential Equations. Proceedings of IFIP WG 7.2 - International Conference. Irsee, April, 9-12, 1990.
245 pp. 1991 [3-540-53591-8]

Vol. 150: Habets, Luc C.
Robust Stabilization in the Gap-topology.
126 pp. 1991 [3-540-53466-0]

Vol. 151: Skowronski, J.M.; Flashner, H.; Guttalu, R.S. (Eds.)
Mechanics and Control. Proceedings of the 3rd Workshop on Control Mechanics, in Honor of the 65th Birthday of George Leitmann, January 22-24, 1990, University of Southern California.
497 pp. 1991 [3-540-53517-9]

Vol. 152: Aplevich, J. Dwight
Implicit Linear Systems.
176 pp. 1991 [3-540-53537-3]

Vol. 153: Hajek, Otomar
Control Theory in the Plane.
269 pp. 1991 [3-540-53553-5]

Vol. 154: Kurzhanski, Alexander; Laseicka, Irena (Eds.)
Modelling and Inverse Problems of Control for Distributed Parameter Systems. Proceedings of IFIP WG 7.2 - IIASA Conference, Laxenburg, Austria, July 1989.
170 pp. 1991 [3-540-53583-7]

Vol. 155: Bouvet, Michel; Bienvenu, Georges (Eds.)
High-Resolution Methods in Underwater Acoustics.
244 pp. 1991 [3-540-53716-3]

Vol. 156: Hämäläinen, Raimo P.; Ehtamo, Harri K. (Eds.)
Differential Games - Developments in Modelling and Computation. Proceedings of the Fourth International Symposium on Differential Games and Applications, August 9-10, 1990, Helsinki University of Technology, Finland.
292 pp. 1991 [3-540-53787-2]

Vol. 157: Hämäläinen, Raimo P.; Ehtamo, Harri K. (Eds.)
Dynamic Games in Economic Analysis. Proceedings of the Fourth International Symposium on Differential Games and Applications. August 9-10, 1990, Helsinki University of Technology, Finland.
311 pp. 1991 [3-540-53785-6]

Vol. 158: Warwick, Kevin; Karny, Miroslav; Halouskova, Alena (Eds.)
Advanced Methods in Adaptive Control for Industrial Applications.
331 pp. 1991 [3-540-53835-6]

Vol. 159: Li, Xunjing; Yong, Jiongmin (Eds.)
Control Theory of Distributed Parameter Systems and Applications. Proceedings of the IFIP WG 7.2 Working Conference, Shanghai, China, May 6-9, 1990.
219 pp. 1991 [3-540-53894-1]

Vol. 160: Kokotovic, Petar V. (Ed.)
Foundations of Adaptive Control.
525 pp. 1991 [3-540-54020-2]

Vol. 161: Gerencser, L.; Caines, P.E. (Eds.)
Topics in Stochastic Systems: Modelling, Estimation and Adaptive Control.
1991 [3-540-54133-0]

Vol. 162: Canudas de Wit, C. (Ed.)
Advanced Robot Control. Proceedings of the International Workshop on Nonlinear and Adaptive Control: Issues in Robotics, Grenoble, France, November 21-23, 1990.
Approx. 330 pp. 1991 [3-540-54169-1]

Vol. 163: Mehrmann, Volker L.
The Autonomous Linear Quadratic Control Problem. Theory and Numerical Solution.
177 pp. 1991 [3-540-54170-5]

Vol. 164: Lasiecka, Irena; Triggiani, Roberto
Differential and Algebraic Riccati Equations with Application to Boundary/Point Control Problems: Continuous Theory and Approximation Theory.
160 pp. 1991 [3-540-54339-2]

Vol. 165: Jacob, Gerard; Lamnabhi-Lagarrigue, F. (Eds.)
Algebraic Computing in Control. Proceedings of the First European Conference, Paris, March 13-15, 1991.
384 pp. 1991 [3-540-54408-9]

Vol. 166: Wegen, Leonardus L. van der
Local Disturbance Decoupling with Stability for Nonlinear Systems.
135 pp. 1991 [3-540-54543-3]

Vol. 167: Rao, Ming
Integrated System for Intelligent Control.
133 pp. 1992 [3-540-54913-7]

Vol. 168: Dorato, Peter; Fortuna, Luigi; Muscato, Giovanni
Robust Control for Unstructured Perturbations: An Introduction.
118 pp. 1992 [3-540-54920-X]

Vol. 169: Kuntzevich, Vsevolod M.; Lychak, Michael
Guaranteed Estimates, Adaptation and Robustness in Control Systems.
209 pp. 1992 [3-540-54925-0]

Vol. 170: Skowronski, Janislaw M.; Flashner, Henryk; Guttalu, Ramesh S. (Eds.)
Mechanics and Control. Proceedings of the 4th Workshop on Control Mechanics, January 21-23, 1991, University of Southern California, USA.
302 pp. 1992 [3-540-54954-4]

Vol. 171: Stefanidis, P.; Paplinski, A.P.; Gibbard, M.J.
Numerical Operations with Polynomial Matrices: Application to Multi-Variable Dynamic Compensator Design.
206 pp. 1992 [3-540-54992-7]

Vol. 172: Tolle, H.; Ersü, E.
Neurocontrol: Learning Control Systems Inspired by Neuronal Architectures and Human Problem Solving Strategies.
220 pp. 1992 [3-540-55057-7]

Vol. 173: Krabs, W.
On Moment Theory and Controllability of Non-Dimensional Vibrating Systems and Heating Processes.
174 pp. 1992 [3-540-55102-6]

Vol. 174: Beulens, A.J. (Ed.)
Optimization-Based Computer-Aided Modelling and Design. Proceedings of the First Working Conference of the New IFIP TC 7.6 Working Group, The Hague, The Netherlands, 1991.
268 pp. 1992 [3-540-55135-2]

Vol. 175: Rogers, E.T.A.; Owens, D.H.
Stability Analysis for Linear Repetitive Processes.
197 pp. 1992 [3-540-55264-2]

Vol. 176: Rozovskii, B.L.; Sowers, R.B. (Eds.)
Stochastic Partial Differential Equations and their Applications. Proceedings of IFIP WG 7.1 International Conference, June 6-8, 1991, University of North Carolina at Charlotte, USA.
251 pp. 1992 [3-540-55292-8]

Vol. 177: Karatzas, I.; Ocone, D. (Eds.)
Applied Stochastic Analysis. Proceedings of a US-French Workshop, Rutgers University, New Brunswick, N.J., April 29-May 2, 1991.
317 pp. 1992 [3-540-55296-0]

Vol. 178: Zolésio, J.P. (Ed.)
Boundary Control and Boundary Variation. Proceedings of IFIP WG 7.2 Conference, Sophia- Antipolis,France, October 15-17, 1990.
392 pp. 1992 [3-540-55351-7]

Vol. 179: Jiang, Z.H.; Schaufelberger, W.
Block Pulse Functions and Their Applications in Control Systems.
237 pp. 1992 [3-540-55369-X]

Vol. 180: Kall, P. (Ed.)
System Modelling and Optimization. Proceedings of the 15th IFIP Conference, Zurich, Switzerland, September 2-6, 1991.
969 pp. 1992 [3-540-55577-3]

Vol. 181: Drane, C.R.
Positioning Systems - A Unified Approach.
168 pp. 1992 [3-540-55850-0]

Vol. 182: Hagenauer, J. (Ed.)
Advanced Methods for Satellite and Deep Space Communications. Proceedings of an International Seminar Organized by Deutsche Forschungsanstalt für Luft-und Raumfahrt (DLR), Bonn, Germany, September 1992.
196 pp. 1992 [3-540-55851-9]

Vol. 183: Hosoe, S. (Ed.)
Robust Control. Proceesings of a Workshop held in Tokyo, Japan, June 23-24, 1991.
225 pp. 1992 [3-540-55961-2]

Vol. 184: Duncan, T.E.; Pasik-Duncan, B. (Eds.)
Stochastic Theory and Adaptive Control. Proceedings of a Workshop held in Lawrence, Kansas, September 26-28, 1991.
500 pages. 1992 [3-540-55962-0]

Vol. 185: Curtain, R.F. (Ed.); Bensoussan, A.; Lions, J.L.(Honorary Eds.)
Analysis and Optimization of Systems: State and Frequency Domain Approaches for Infinite-Dimensional Systems. Proceedings of the 10th International Conference, Sophia-Antipolis, France, June 9-12, 1992.
648 pp. 1993 [3-540-56155-2]

Vol. 186: Sreenath, N.
Systems Representation of Global Climate Change Models. Foundation for a Systems Science Approach.
288 pp. 1993 [3-540-19824-5]

Vol. 187: Morecki, A.; Bianchi, G.; Jaworeck, K. (Eds.)
RoManSy 9: Proceedings of the Ninth CISM-IFToMM Symposium on Theory and Practice of Robots and Manipulators.
476 pp. 1993 [3-540-19834-2]

Vol. 188: Naidu, D. Subbaram
Aeroassisted Orbital Transfer: Guidance and Control Strategies.
192 pp. 1993 [3-540-19819-9]

Vol. 189: Ilchmann, A.
Non-Identifier-Based High-Gain Adaptive Control.
220 pp. 1993 [3-540-19845-8]

Vol. 190: Chatila, R.; Hirzinger, G. (Eds.)
Experimental Robotics II: The 2nd International Symposium, Toulouse, France, June 25-27 1991.
580 pp. 1993 [3-540-19851-2]

Vol. 191: Blondel, V.
Simultaneous Stabilization of Linear Systems.
212 pp. 1993 [3-540-19862-8]

Vol. 192: Smith, R.S.; Dahleh, M. (Eds.)
The Modeling of Uncertainty in Control Systems.
412 pp. 1993 [3-540-19870-9]

Vol. 193: Zinober, A.S.I. (Ed.)
Variable Structure and Lyapunov Control
428 pp. 1993 [3-540-19869-5]